Environmental Policy and Public Health

Environmental Policy and Public Health
Emerging Health Hazards and Mitigation, Volume 2

Third Edition

Barry L. Johnson, MSc, PhD, FCR
RADM (Ret.), U.S. Public Health Service
Adjunct Professor
Rollins School of Public Health
Emory University

Maureen Y. Lichtveld, MD, MPH
Dean, Graduate School of Public Health
Professor, Environmental and Occupational Health
Jonas Salk Chair in Population Health
University of Pittsburgh

CRC Press
Taylor & Francis Group
Boca Raton London New York

CRC Press is an imprint of the
Taylor & Francis Group, an **informa** business

Third edition published 2022
by CRC Press
6000 Broken Sound Parkway NW, Suite 300, Boca Raton, FL 33487-2742

© 2022 Barry L. Johnson and Maureen Y. Lichtveld

First edition published by CRC Press 2006

CRC Press is an imprint of Taylor & Francis Group, LLC

Reasonable efforts have been made to publish reliable data and information, but the author and publisher cannot assume responsibility for the validity of all materials or the consequences of their use. The authors and publishers have attempted to trace the copyright holders of all material reproduced in this publication and apologize to copyright holders if permission to publish in this form has not been obtained. If any copyright material has not been acknowledged please write and let us know so we may rectify in any future reprint.

Except as permitted under U.S. Copyright Law, no part of this book may be reprinted, reproduced, transmitted, or utilized in any form by any electronic, mechanical, or other means, now known or hereafter invented, including photocopying, microfilming, and recording, or in any information storage or retrieval system, without written permission from the publishers.

For permission to photocopy or use material electronically from this work, access www.copyright.com or contact the Copyright Clearance Center, Inc. (CCC), 222 Rosewood Drive, Danvers, MA 01923, 978-750-8400. For works that are not available on CCC please contact mpkbookspermissions@tandf.co.uk

Trademark notice: Product or corporate names may be trademarks or registered trademarks and are used only for identification and explanation without intent to infringe.

ISBN: 978-1-032-08034-5 (hbk)
ISBN: 978-1-032-08035-2 (pbk)
ISBN: 978-1-003-21262-1 (ebk)

DOI: 10.1201/9781003212621

Typeset in Times
by codeMantra

Contents

Preface xv
Authors xvii

Chapter 1   Environment-Related Infectious Diseases 1

    1.1   Introduction 1
    1.2   Infectious Diseases – A Global Perspective 1
    1.3   The Environment and Infectious Diseases 2
        1.3.1   Cholera 2
        1.3.2   Typhus 4
        1.3.3   Scabies 5
        1.3.4   Implications for Environmental Health Policy:
                Clean Water and Sanitation 6
    1.4   Zoonotic Diseases and Human Health 6
        1.4.1   H1N1 Global Influenza Pandemic 7
        1.4.2   Coronavirus Disease 2019 (COVID-19) Pandemic 9
        1.4.3   Social Justice Issues 13
        1.4.4   West Africa Ebola Epidemic 13
        1.4.5   Measles Epidemics 15
        1.4.6   Implications for Environmental Health Policy:
                Quarantine and Public Health 17
    1.5   Emergence of Vector-Borne Diseases 18
        1.5.1   Mosquito-Borne Diseases 18
                1.5.1.1   Malaria 18
                1.5.1.2   Yellow Fever 20
                1.5.1.3   Dengue 21
                1.5.1.4   West Nile Virus 23
                1.5.1.5   Chikungunya Virus 24
                1.5.1.6   Zika Virus 24
                1.5.1.7   Eastern Equine Encephalitis (EEE)
                          Virus 26
                1.5.1.8   Implications for Environmental Health
                          Policy: Vector Control and Pesticides 27
        1.5.2   Tick-Borne Diseases 28
        1.5.3   Rodent- and Bat-Borne Diseases 30
        1.5.4   Parasitic Diseases Caused by Worms 31
                1.5.4.1   Helminth Infections 32
                1.5.4.2   Guinea Worm Disease 32
        1.5.5   The Kissing Bug 34
        1.5.6   Superbugs 34
        1.5.7   Research to Counter Superbugs 38
                1.5.7.1   New Approaches to Antibiotics 38

| | | 1.5.7.2 | Bacteriophage Therapy .... 39 |
| | | 1.5.7.3 | Basic Research on Microorganism Mechanisms .... 39 |

	1.5.8	Nosocomial Infections 40
		1.5.8.1	*Candida auris* Infection 40
		1.5.8.2	*Klebsiella pneumoniae* 41
	1.5.9	Necrotizing Fasciitis (Flesh-Eating Bacteria) 41
	1.5.10	Swimming Pool Bacteria 42
	1.5.11	Legionnaires Disease 42
	1.5.12	Naegleria Fowleri, the "Brain-Eating Ameba" 43
	1.5.13	Implications for Environmental Health Vector Control Policies 43
1.6	Tropical Infectious Diseases 44
1.7	Implications for Environmental Policy: Climate Change 44
1.8	Implications for Ecosystem Health 45
1.9	Hazard Interventions 46
1.10	Summary 46
References 47

Chapter 2 Solid and Hazardous Waste 55

2.1	Introduction 55
2.2	Contemporary Challenges to Waste Management 56
	2.2.1	Plastics Waste 57
		2.2.1.1	Impacts of Plastics Waste on Human and Ecological Health 57
		2.2.1.2	Plastics as Solid Waste 58
		2.2.1.3	Plastics in Oceans 59
		2.2.1.4	Microbeads in Waste 60
		2.2.1.5	Microbead-Free Waters Act, 2015 62
		2.2.1.6	Recycling of Plastics Waste 62
		2.2.1.7	Bans on Single-Use Plastics 64
	2.2.2	Food Waste 68
	2.2.3	Electronic Products Waste (E-Waste) 70
2.3	U.S. Solid Waste Management Policies 72
	2.3.1	Impacts of Solid Waste on Human Health 73
	2.3.2	Impacts of Solid Waste on Ecosystem Health 74
	2.3.3	Solid Waste Disposal Act, 1965 74
	2.3.4	Resource Conservation and Recovery Act, 1976 75
		2.3.4.1	History 75
		2.3.4.2	Amendments to the RCRAct 80
		2.3.4.3	Key Provisions of the RCRAct Relevant to Public Health 82
		2.3.4.4	Coal Ash Regulations under RCRA 83
	2.3.5	U.S. State's Solid Waste Act Example 87
	2.3.6	Solid Waste Policy Issues 88

Contents vii

	2.4	U.S. Hazardous Waste Policies 88
		2.4.1 Impacts of Uncontrolled Hazardous Waste Sites on Human Health 88
		2.4.2 Impacts of Uncontrolled Hazardous Waste Sites on Ecosystem Health 90
		2.4.3 Nuclear Waste Policies and Issues 90
		2.4.4 Comprehensive Environmental Response, Compensation, and Liability Act, 1980 91
		2.4.4.1 History 91
		2.4.4.2 Key Provision of the CERCLAct Relevant to Public Health 94
		2.4.5 Health Significance of NPL Site Remediation 95
		2.4.6 Trump EPA Superfund Sites & Climate Change 96
		2.4.7 Emergency Planning and Community Right-to-Know Act, 1986 96
		2.4.8 EPA's Brownfields Program 97
	2.5	Key Global Hazardous Waste Policy 99
	2.6	U.S. Oil Pollution Policies 100
		2.6.1 Oil Pollution Act, 1990 100
		2.6.2 National Oil and Hazardous Substances Pollution Contingency Plan 101
	2.7	U.S. Pollution Prevention Act, 1990 102
	2.8	U.S. Ocean Waste Pollution Policies 103
		2.8.1 Ocean Dumping Act, 1972 104
		2.8.2 Act to Prevent Pollution from Ships, 1980 105
		2.8.3 Obama's Oceans Policy, 2010 105
	2.9	International Policies on Maritime Pollution 106
		2.9.1 London Convention and Protocol, 1972 106
		2.9.2 International Maritime Organization (IMO) 107
		2.9.3 EU Maritime Pollution Policies 107
	2.10	Waste Reduction: Recycling of Waste 108
		2.10.1 Recycling Issues 109
		2.10.2 Innovative Technology for Waste Reduction 110
		2.10.3 The Circular Economy 111
	2.11	Hazard Interventions 112
	2.12	Summary 113
	References 114

Chapter 3 Drugs and Alcohol 123

       3.1 Introduction 123
       3.2 Drugs and Public Health 124
            3.2.1 Précis History of Therapeutic Drugs 124
            3.2.2 Précis History of Drug Abuse in the U.S. 125
            3.2.3 Impact of Drug Abuse on Human Health 127
                  3.2.3.1 U.S. Opioids Epidemic 128

		3.2.3.2	Effects of Opioid Epidemic on Children	129
	3.2.4		Impact of Drug Abuse on Ecosystem Health	130
	3.2.5		U.S. Policies on Drug Abuse	132
		3.2.5.1	Pure Food and Drug Act, 1906	132
		3.2.5.2	Food, Drug, Cosmetic Act, 1938	133
		3.2.5.3	Comprehensive Drug Abuse Prevention and Control Act, 1970	134
		3.2.5.4	The Controlled Substances Act, 1970	134
		3.2.5.5	U.S. Federal Agencies with Illegal Drug Authorities	135
	3.2.6		Oregon's Decriminalization of Hard Drugs	138
	3.2.7		UN Policies on Illegal Drugs	138
3.3			Alcohol and Public Health	140
	3.3.1		Précis History of Alcohol Discovery and Use	140
	3.3.2		Forms and Prevalence of Alcohol Consumption	142
	3.3.3		COVID-19 and Alcohol Consumption	144
	3.3.4		Impact of Alcohol on Human Health	144
	3.3.5		Impact of Alcohol on Ecosystem Health	146
	3.3.6		U.S. Policies on Alcohol Production	147
		3.3.6.1	18th Amendment to the U.S. Constitution	147
		3.3.6.2	Federal Alcohol Administration Act, 1935	148
		3.3.6.3	U.S. Bureau of Alcohol, Tobacco, Firearms, and Explosives	149
	3.3.7		Global Policies on Alcohol Production	150
3.4			Low-Risk Drinking Guidelines	151
3.5			Hazard Interventions	151
3.6			Summary	152
References				152

Chapter 4 Firearms Violence 155

4.1	Introduction	155
4.2	Definition	156
4.3	Précis History of Firearms	156
4.4	Forms of Firearms	157
	4.4.1 Possession and Storage of Firearms in the U.S.	158
	4.4.2 Possession of Firearms Globally	160
4.5	Human Health Impacts of Firearms	161
	4.5.1 Premature Firearms-Related Deaths in the U.S.	161
	4.5.2 Premature Firearms-Related Deaths Globally	163
	4.5.3 Children and Adolescent Premature Deaths	163
	4.5.4 Mass Casualty Shootings	164

Contents ix

			4.5.4.1	School Shootings	166
			4.5.4.2	Psychosocial Impacts of School Shootings	167
	4.6	Ecological Health Impacts of Firearms			169
	4.7	Socioeconomic Impacts of Firearms			172
	4.8	COVID-19 Impacts on Firearm Issues			172
	4.9	Government Policies to Control Firearms			173
		4.9.1	Major U.S. Federal Policies to Control Firearms		173
			4.9.1.1	Second Amendment to the U.S. Constitution	174
			4.9.1.2	The National Firearms Act, 1934	175
			4.9.1.3	Gun Control Act, 1968	176
			4.9.1.4	Omnibus Crime Control and Safe Streets Act, 1968	176
			4.9.1.5	Firearm Owners' Protection Act, 1986	177
			4.9.1.6	Undetectable Firearms Act, 1988	177
			4.9.1.7	Gun-Free School Zones Act, 1990	177
			4.9.1.8	Brady Handgun Violence Prevention Act, 1993	178
			4.9.1.9	Law Enforcement Officers Safety Act, 2004	179
			4.9.1.10	Protection of Lawful Commerce in Arms Act, 2005	179
			4.9.1.11	Bureau of Alcohol, Tobacco, and Firearms	180
			4.9.1.12	Trump and Biden Firearms Executive Actions	180
		4.9.2	U.S. State and Local Firearm Control Policies		181
			4.9.2.1	"Red Flag" Gun Seizure Statutes	182
			4.9.2.2	California's Ammunition Control Statute	182
			4.9.2.3	Retail Sales of Assault Weapons	182
		4.9.3	International Government Policies to Control Firearms		183
		4.9.4	U.S. Federal Policies on Firearms Research		184
	4.10	Firearms Hazard Interventions			185
	4.11	Summary			186
	References				187
Chapter 5	Noise and Light Pollution				191
	5.1	Introduction			191
	5.2	Noise Pollution			191
		5.2.1	Definition		193
		5.2.2	Sources of Noise Pollution		193

		5.2.3	Effects of Noise Pollution on Human Health 194
		5.2.4	Effects of Noise Pollution on Ecological Health 196
		5.2.5	Economic Impacts of Noise Pollution 198

5.3 Policies to Control Noise Pollution 199
- 5.3.1 U.S. Federal Policies to Control Noise Pollution 200
  - 5.3.1.1 Aircraft Noise Abatement Act, 1968 200
  - 5.3.1.2 Noise Control Act, 1972 200
  - 5.3.1.3 Quiet Communities Act, 1978 202
  - 5.3.1.4 Quiet Communities Act, 2015, 2016, 2017 203
  - 5.3.1.5 What Sources of Noise are Subject to Federal Regulation? 203
- 5.3.2 International Government Policies to Control Noise Pollution 205

5.4 Noise Pollution Hazard Interventions 205
5.5 Light Pollution 206
- 5.5.1 Definition 207
- 5.5.2 Sources of Light Pollution 207
- 5.5.3 Effects of Light Pollution on Human Health 208
- 5.5.4 Effects of Light Pollution on Ecological Health 210
- 5.5.5 Effects of Light Pollution on the Night Sky 211

5.6 Policies to Control Light Pollution 212
- 5.6.1 U.S. Federal Policies 212
- 5.6.2 U.S. State and Local Policies 213

5.7 Light Pollution Hazard Interventions 213
5.8 Summary 214
References 215

Chapter 6 The Built Environment 219

6.1 Introduction 219
6.2 Terms and Concepts 220
- 6.2.1 Land-Use Policy Tools 220
- 6.2.2 Land-Use Policy Actors 221

6.3 History of U.S. Planning Practices 222
- 6.3.1 Industrialization and the Birth of a Movement 223
- 6.3.2 U.S. Federal Government's Growth into Planning 225
- 6.3.3 New Federalism and the Reemergence of Public Health 227

6.4 Impacts of the Built Environment on Human Health 228
6.5 Health Benefits of Urban Nature 230
6.6 Impacts of the Built Environment on Ecosystem Health 232
6.7 Impacts of the Built Environment on the Social Environment 232
6.8 Social Justice Issue 233

	6.9	U.S. Policy Overview 234
	6.10	Current Practices and Issues 235
		6.10.1 Sprawl 235
		6.10.2 Air Quality 235
		6.10.3 Housing and Health Issues 236
		6.10.4 Brownfields 236
		6.10.5 U.S. Transportation Priorities 237
		6.10.6 Sustainability 237
		6.10.7 Climate Change 238
	6.11	Policies 239
		6.11.1 Health in All Policies 239
		6.11.2 Complete Streets 240
		6.11.3 LEED 241
		6.11.4 Green Roofs Policies 241
		6.11.5 Urban Agriculture 243
		6.11.6 Other Built Environment Policies 243
	6.12	Global Perspective 246
	6.13	Seminal Issue of Selected Global Built Policies 246
	6.14	Hazard Interventions 247
	6.15	Summary 248
	References 248
Chapter 7	Transportation 253
	7.1	Introduction 253
	7.2	Précis History of Motor-Powered Vehicles 254
	7.3	Impacts of Transportation on Human Health 257
	7.4	Impacts of Transportation on Ecosystem Health 260
	7.5	U.S. Federal Transportation Policies 260
		7.5.1 Federal-Aid Highway Act, 1956 260
		7.5.2 Urban Mass Transportation Act, 1964 262
		7.5.3 Department of Transportation Act, 1966 263
		7.5.4 National Traffic and Motor Vehicle Safety Act, 1966 263
		7.5.5 Motor Vehicle Information and Cost Savings Act. 1972 264
		7.5.6 National Minimum Drinking Age Act, 1984 264
		7.5.7 Intermodal Surface Transportation Efficiency Act, 1991 265
		7.5.8 Motor Carrier Safety Improvement Act, 1999 266
		7.5.9 Fixing America's Surface Transportation Act, 2015 267
	7.6	U.S. States' Transportation Policies with Health Implications 268
		7.6.1 Motor Vehicle Speed Limits 268
		7.6.2 Driving under the Influence of Drugs or Alcohol 269

		7.6.3	Use of Cell Phones while Driving a Vehicle 271
7.7	Safety Technology and Health Implications 273
	7.7.1	Seat Belts 273
	7.7.2	Child Safety Seat or Booster Seat 275
	7.7.3	Airbags 275
	7.7.4	Vehicle Designs 276
		7.7.4.1	Vehicle Crash Design 277
		7.7.4.2	Crash Avoidance Technology 278
		7.7.4.3	Autonomous Vehicles 278
	7.7.5	Road Design 280
7.8	U.S. Federal Agencies with Transportation Responsibilities 281
	7.8.1	U.S. Department of Transportation 281
		7.8.1.1	Federal Highway Administration 281
		7.8.1.2	Federal Motor Carrier Safety Administration 282
		7.8.1.3	Federal Railroad Administration 283
		7.8.1.4	Federal Aviation Administration 283
		7.8.1.5	Federal Transit Administration 284
		7.8.1.6	Maritime Administration 284
		7.8.1.7	National Highway Traffic Safety Administration 285
		7.8.1.8	National Transportation Safety Board 286
	7.8.2	Other Federal Agencies with Transportation Policies 286
7.9	International Government Organizations with Transportation Policies 287
	7.9.1	International Civil Aviation Organization 287
	7.9.2	International Maritime Organization 288
	7.9.3	World Health Organization 288
7.10	NGOs Concerned with Traffic and Vehicle Safety 289
	7.10.1	Global Road Safety Facility 289
	7.10.2	Insurance Institute for Highway Safety 289
	7.10.3	Air Transport Action Group 290
	7.10.4	The International Freight Forwarders Association 290
	7.10.5	The International Air Transport Association 290
	7.10.6	International Chamber of Shipping and International Shipping Federation 290
	7.10.7	The International Road Transport Union 291
	7.10.8	The International Union of Railways 291
	7.10.9	The SMDG 291
7.11	Motor Vehicles and Climate Change 291
	7.11.1	Motor Vehicle Emissions and Climate Change 292
		7.11.1.1	Corporate Average Fuel Economy Standards, 1975 292

		7.11.1.2	The Safer Affordable Fuel-Efficient Vehicles Rule, 2018 293
	7.11.2	The Future of Transportation is Electric Vehicles 294
		7.11.2.1 Emerging Government Policies 294
		7.11.2.2 Electric Trucks 295
		7.11.2.3 Electric Trains 296
	7.11.3	Aircraft Emissions and Climate Change 296
	7.11.4	Urban Traffic Control and Air Pollution 297
7.12	Global Implications of Transportation 298
7.13	Hazard Interventions 298
7.14	Summary 299
References 299

Lessons Learned and Authors' Reflections 305

Environmental Policy and Public Health: Emerging Health Hazards and Mitigation, Volume 2 Workbook 307

Index 321

Preface

Environmental Policy and Public Health is a two-volume textbook/reference book addressing key physical hazards in the environment that impact public health. Volume 1: *Principal Health Hazards and Mitigation* is complemented by Volume 2, *Emerging Health Hazards and Mitigation*. In addition to the principal environmental health hazards described in Volume 1, several other environmental health hazards have emerged and are discussed in Volume 2. *Environmental Policy and Public Health: Emerging Health Hazards and Mitigation* describes emerging health hazards of concern. These include environment-related infectious diseases such as COVID-19 and migratory tick-borne infections.

In addition, firearms violence has become a public health concern, given the large number of lives lost prematurely. Challenges to the management of solid and hazardous waste have emerged as the improper disposal of plastics into the planet's oceans has increased. In addition, as human populations increase and technology adds more devices to daily use that generate noise and light, adverse human and ecological health effects have become recognizable and are described as issues of policy and mitigation. The volume concludes with a description of how built environments, for example, green space, can affect human health and social well-being. This volume is accompanied by practice questions to facilitate interactive classroom and group learning.

As with Volume 1, we have also endeavored in this third edition to objectively describe the major environmental health policy redirections that have occurred in the U.S. as a consequence of sociopolitical changes in the U.S. federal government's environmental policies. In particular, we record here the major differences in environmental policies associated with the Obama, Trump, and Biden presidential administrations. And because we consider climate change as the most threatening global environmental hazard to human and ecological health, we have labored to include climate change impacts wherever possible throughout several chapters. Additionally, the year 2020 was host to two significant social and public health challenges that merited special attention and description in this third edition: social justice and the COVID-19 pandemic. Both subjects are described in those chapters whose contents were influenced by social justice and/or COVID-19.

ACKNOWLEDGMENTS:

The authors gratefully acknowledge the contributions of the following individuals in the preparation of this book: Irma Shagla Britton, Senior Editor, CRC Press/Taylor & Francis Group, Vijay Shanker P, Sr. Project Manager, CodeMantra, and Jacob Romanowski, Graduate School of Public Health, University of Pittsburgh.

Authors

Barry L. Johnson, MSc, PhD, FCR, is adjunct professor at the Rollins School of Public Health, Emory University. Dr. Johnson conducted environmental research at the USPHS, USEPA, and CDC. An elected Fellow, Collegium Ramazani, he retired from the USPHS with the rank of Rear Admiral. Dr. Johnson is author of *Legacy of Hope* (2016), *Environmental Policy and Public Health* (2007, 2018), *Impact of Hazardous Waste on Human Health* (1999), and senior editor of *Hazardous Waste: Impacts on Human and Ecological Health* (1997); *Hazardous Waste and Public Health* (1996); *Advances in Neurobehavioral Toxicology* (1990); and *Prevention of Neurotoxic Illness in Working Populations* (1987).

Maureen Y. Lichtveld, MD, MPH, is Dean, Professor, Environmental and Occupational Health and Jonas Salk Chair in Population Health, University of Pittsburgh, Graduate School of Public Health. Previous roles: Professor, Freeport McMoRan Chair of Environmental Policy, Tulane University's School of Public Health and Tropical Medicine. She is a member of the National Academy of Medicine with decades of environmental public health experience. Research: environmentally induced disease, health disparities, environmental health policy, disaster preparedness, climate and health, and community resilience. Dr. Lichtveld is a board member of the Consortium of Universities for Global Health. Honors: Johns Hopkins' Society of Scholars; CDC's Environmental Health Scientist of the Year.

1 Environment-Related Infectious Diseases

1.1 INTRODUCTION

Prior chapters in Volume 1 have addressed chemical hazards in the environment that have led to policymaking to control the effects on human and ecological health. This chapter describes a different kind of environmental-related outcome, infectious diseases. These diseases are caused by organisms or pathogens (such as viruses, bacteria, fungi, parasites, protozoa) that can be transmitted through physical contact with or exposure to bodily fluids of an infected person. These causal agents are collectively called vectors, which are defined as living organisms that can transmit infectious diseases between humans or from animals to humans. A particularly hazardous disease agent is the mosquito. As will be subsequently discussed, mosquitoes are a major environmental vector of human disease. Infectious diseases are constantly circulating within a population. However, when an infectious disease causes an increased, or unexpected, number of cases, it can be classified as an outbreak. Health departments at federal, U.S. state, territorial, and local levels monitor cases of infectious diseases and implement actions to control emerging and re-emerging pathogens within their populations. Described in this chapter are various infectious agents in the environment and policies and public health actions that are purposed to mitigate the adverse human and ecological health effects.

1.2 INFECTIOUS DISEASES – A GLOBAL PERSPECTIVE

In high-income countries, medical and public health advances and sound environmental policies created a shift toward noncommunicable, chronic conditions. Over the last 10 years, lower and middle-income countries also saw a significant increased mortality and morbidity associated with noncommunicable diseases such as cardiovascular diseases while these countries continue to face threats of endemic infectious diseases. However, unlike developed countries, most of the health infrastructure needed to combat both types of diseases is often underdeveloped and fragile.

The U.S. recognizes that infectious diseases remain a problem worldwide. The U.S. commitment to combat the spread of infectious diseases is visible through global programs and funds. In the U.S., the Centers for Disease Control and Prevention (CDC) is the leading national public health agency in charge of

> Health departments at federal, state, and local levels monitor cases of infectious diseases and implement actions to control emerging and re-emerging pathogens within their populations.

early detection, control and prevention, and preparedness. CDC's National Notifiable Diseases Surveillance System (NNDSS) collects, analyzes, and shares health data on notifiable diseases among local, state, territorial, federal, and international public health departments.

Globally, the World Health Organization (WHO) is the UN's public health agency and the leading international health organization, supported by 193 member states. The WHO has established a disease surveillance network comprised of national and international medical laboratories in its member states. CDC, the UK Public Health Laboratory Service, the French Pasteur Institute, and schools of public health globally report to WHO on a series of infectious diseases [1]. Under WHO's International Health Regulations (IHR), member states are legally required to report infectious diseases of international importance. The WHO also provides operational support to response efforts during an epidemic and strengthens national core capacities to prevent, prepare, and recover for and from emergencies [2]. Furthermore, WHO's environmental guidelines and policies on drinking water, sanitation and hygiene, and the use of chemicals such as insecticides directly and indirectly contribute to the prevention of infectious diseases.

1.3 THE ENVIRONMENT AND INFECTIOUS DISEASES

The primary determinant of infectious diseases is the pathogen (the infectious agent). A pathogen is any organism, usually a microbe, that can cause disease in a host. The host, the second determinant of infectious disease, is exposed to the pathogen and either harbors the disease,

> The primary determinant of infectious diseases is the pathogen (the infectious agent). A pathogen is any organism, usually a microbe, that can cause disease in a host.

or becomes ill [3]. Public health is concerned mostly with human hosts; however, animal hosts can have an impact on human health. The environment is necessary for the exposure to and the spread of the infectious agent and is therefore the third determinant of infectious disease. As depicted in Figure 1.1, Triangle of Disease, the three vertices of the triangle connect the pathogen, host, and environment to allow the transmission of infectious diseases. This chapter describes the role of the environment in transmission and control of infectious diseases [4].

1.3.1 Cholera

Cholera is caused by the bacterium *Vibrio cholerae*. During the 1800s, cholera outbreaks occurred frequently in the U.S. Cholera is transmitted by ingestion of water or food that has been contaminated with human feces. Upon infection, the cholera bacterium releases a toxin (classified as O1 or O139) that can cause severe diarrhea and dehydration [5,6]. If left untreated, it can be fatal. Currently, cholera is mostly travel-associated [7,8]. The CDC's Cholera and Other Vibrio Illness Surveillance System was created in 1988 in partnership with the FDA and the Gulf Coast States to obtain information on any *Vibrionaceae* associated-illness and provide information about risk groups and exposure risk [9]. Other systems that conduct surveillance on

Environment-Related Infectious Diseases

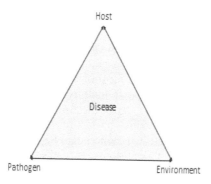

FIGURE 1.1 Epidemiological triangle of vector-borne disease. (CDC, 2006. *Principles of Epidemiology in Public Health Practice*, 3rd Edition. Office of the Associate Director for Communication, Centers for Disease Control and Prevention, Atlanta, Georgia. https://www.cdc.gov/about/organization/oadc.html)

Vibrio associated-illness include CDC's National Notifiable Diseases Surveillance System, the National Antimicrobial Resistance (AMR) Monitoring System, and the National Outbreak Reporting System. All cholera cases also have to be reported to the WHO in compliance with international health regulations.

Cholera is endemic in more than 50 countries worldwide. The last endemic cholera outbreak in the U.S. occurred in 1911. Most cases occur in African and Asian countries where populations do not have access to improved drinking water sources. The WHO maintains and publishes weekly reports of aggregate national cholera data. The 1969 WHO IHRs and the 2005 ratification require WHO notification of all cholera cases. Cholera, however, is estimated to be heavily underreported, with official estimates representing 5%–10% of all annual cases. It is estimated that an annual 1–4 million cholera infections occur worldwide, resulting in an estimated 21,000–143,000 annual deaths [10].

> Cholera is transmitted by ingestion of water or food that has been contaminated with human feces. If left untreated, it can be fatal.

Cholera disease is particularly fatal to children. A simple scientific advancement, known as Oral Rehydration Therapy, replaces essential salts and body fluids through a sodium glucose mixture known as Oral Rehydration Salts (ORS). ORS are cheap, globally distributed, and credited for reducing cholera as well as other diarrheal disease mortality, in children by more than 54% from 1.3 million in 2000 to fewer than 600,000 in 2013 [11]. However, even with an inexpensive and effective solution to decrease cholera morbidity, cholera prevention is based entirely on providing clean water supplies for human consumption [12]. Many people in developing countries do not have access to safe drinking water on a daily basis [13] (Volume 1, Chapter 39). Recent epidemics have demonstrated poor infrastructure accelerates the spread of water-borne diseases during a natural disaster [10]. An example of an inadequate public health infrastructure follows.

On January 12, 2010, a 7.0 magnitude earthquake struck Haiti. The country was already the poorest country in the Western hemisphere. The Haitian population depended largely on the agricultural sector, residents lived in poverty in overcrowded slums, and Haitians had limited access to sanitation and clean water. Despite a severely limited public health infrastructure, there had never been a case of cholera reported in Haiti [14]. The earthquake and its aftermath killed an estimated 230,000 people and displaced most of the nation's population. The international community was quick to respond and large-scale search and rescue missions began within days. On October 17, 2010, the first case of cholera was reported to international agencies [15].

Investigations by WHO found that cholera had been introduced into Haiti by the UN's Nepalese peacekeepers. The Nepalese UN base was located upstream of the Artibonite River, which supplied most of the country with fresh drinking water. Improper sanitation practices at the base, including open dump pits and leaky latrine pipes, allowed human feces infected with *Vibrio cholerae* O1, serotype Ogawa, a strain found in cholera cases in Nepal to enter the river. By March 2011, more than 500,000 cholera infections were reported with approximately 5,000 deaths. Since its introduction, several subsequent cholera outbreaks have been reported in Haiti [16].

Demonstrating that cholera epidemics do not respect geography or history, a major cholera outbreak occurred in Yemen in 2017. The WHO reports a total number of suspected cholera cases in Yemen in 2017 had risen to 500,000, and nearly 2,000 people had died since the outbreak began to spread rapidly at the end of April 2017 [17]. Yemen's cholera epidemic was the largest in the world in 2017. The epidemic spread rapidly due to deteriorating hygiene and sanitation conditions and disruptions to the water supply across the country. Millions of people were cut off from clean water, and waste collection had ceased in major cities as a consequence of civil war in Yemen. Later, a restored water treatment plant dramatically reduced a local cholera epidemic. Al Barzakh is one of around 10 water treatment centers in Yemen, and it serves four different districts in Aden, in southern Yemen, as well as the Lahij and the Abyan governorates. As a contribution to public health, UNICEF undertook the plant's restoration, while also analyzing the infrastructure needs of the region. The plant became fully operational in September 2017, and by January 2018, cholera reports had decreased to 164 cholera cases and zero deaths, which was a 92% decrease in new cholera cases [18].

> A total number of suspected cholera cases in Yemen in 2017 had risen to 500,000, and nearly 2,000 people had died.

1.3.2 TYPHUS

Mites, lice, and fleas are as ubiquitous as ticks. Mites can live freely in the environment or as parasites feeding on mammalian blood and keratin (nails and hair). Due to their parasitic nature, mites can cause common diseases such as scabies (*Sarcoptes scabiei*) or transmit bacterial diseases such as scrub typhus [19]. Scrub typhus bacteria (*Orientia tsutsugamushi*) is transmitted by mites throughout Asia and Australia. This wide ecological distribution is due to the geographic spread of the vector and a high population density. Scrub typhus presents as a febrile illness with headaches and rash. Even though it is not endemic in the U.S., the global burden of disease is high with scrub typhus infections accounting for more than 20% of febrile illness hospitalizations in some Asian countries [20,21].

Fleas are parasites that live on birds or mammals and exclusively feed on host blood. Rat and cat fleas (*Xenopsylla cheopis* and *Ctenocephalides felis*) can transmit *Rickettsia typhi*, a bacterium that can cause human illness when infected flea feces are rubbed into scraps or bites in the skin. In the U.S., cases of murine typhus have been reported in Hawaii, California, and Texas. Murine typhus infection includes common symptoms such as fever, rash, and headache, but in some cases, when left untreated can lead to organ damage [21]. Lastly, body lice can transmit *Rockettsia prowazekii*, the typhus fever causative agent. Typhus fever bacteria is spread by lice through contaminated feces similar to murine typhus. When left untreated, typhus fever can cause common symptoms and a distinct upper trunk rash. Typhus fever is not endemic in the U.S. but is present in colder regions of South America, Africa, and Asia. When left untreated, the case-fatality rate is 40% [22].

The wide range and variety of ticks, mites, fleas, and lice, the number of specific diseases, and the different hosts from which they can feed make these arthropod-borne diseases impossible to eradicate. Control and prevention strategies target personal protection to avoid coming in contact and using repellents containing 20%–30% DEET (N, N-diethyl-m-toluamide) when going into wooded areas [23]. Personal hygiene and sanitation are also essential in combating mites, fleas, and lice. Currently, some of the mite and lice-borne diseases discussed are not endemic in the U.S., however, globalization and population movement, combined with an increase in population density, increase the likelihood of these arthropods to disseminate across a larger geographic range.

> When left untreated, the typhus case-fatality rate is 40%.

1.3.3 SCABIES

Scabies is a skin condition caused by mites. It commonly leads to intense itching and a pimple like skin rash that may affect various areas of the body. Scabies is contagious and can spread quickly in areas where people are in close physical contact. Through secondary bacterial skin infection (impetigo), scabies can lead to serious complications such as septicemia, renal disease, and rheumatic heart disease. An Australian research team reviewed scabies prevalence data worldwide, with data available for all regions except North America [24]. The prevalence of scabies ranged from 0.2% to 71.4%. All regions except for Europe and the Middle East included populations with a prevalence greater than 10%. Overall, scabies prevalence was highest in the Pacific and Latin American regions and was substantially higher in children than in adolescents and adults. In 2017, the WHO added scabies to its list of neglected tropical diseases. Of note, a treatment against scabies was discovered in 2019 by an Australian and

> Scabies is a skin condition caused by mites. It can lead to serious complications such as septicemia, renal disease, and rheumatic heart disease.

Fijian medical team [25]. The team found that the drug ivermectin, which kills many types of parasitic worms, was also effective in killing the mites that cause scabies.

1.3.4 Implications for Environmental Health Policy: Clean Water and Sanitation

The U.S. federal government's involvement in preventing and controlling the spread of infectious diseases is a recent occurrence in the nation's history. Toward the end of the 19th century, industrialization of the disparate, rural, agrarian towns that made up most of the U.S. forced the migration and concentration of people into cities. This made infectious disease outbreaks larger in magnitude, more difficult to control and more costly to society. U.S. state and local governments were in charge of combating disease and illness, but this task was still viewed as a personal responsibility. However, as national outbreaks of cholera occurred, the U.S. federal government was forced to establish clean water and waste disposal systems in the U.S.

The Public Health Service Act of 1912 implemented the first federal policies to dispose of human waste. The modern water and sewage treatment systems that emerged in the U.S. and other Western countries during the 20th century made it possible to adequately dispose and treat sanitary waste (Chapter 2). The separation of water and sewage eliminated many waterborne diseases, including cholera in the U.S. [26]. However, waterborne diseases, such as cholera, remain a problem in the world due mostly to limited access to potable drinking water. In response, WHO has issued guidelines for drinking water quality [27]. As previously noted, the introduction of cholera during natural disasters, such as the 2010 Haiti Earthquake, is an example of the importance that strong water and sewage infrastructure has on the elimination of infectious diseases.

> The separation of water and sewage eliminated many waterborne diseases, including cholera in the U.S.

1.4 ZOONOTIC DISEASES AND HUMAN HEALTH

The CDC and the WHO define zoonosis (or zoonotic disease) as any disease or infection that is naturally transmissible from vertebrate animals to humans. Zoonoses can be caused by bacteria, parasites, fungi, or viruses. The vertebrate animal is the pathogen's "reservoir host" or "intermediate host" in which it can multiply and develop when there is no active transmission to humans. The CDC has adopted the One Health approach presented during the 2012 Global Risk Forum One Health Summit [28]. One Health recognizes the interdependence of human health and that of animals and the environment [29].

Even though the concept of multisectoral and multistakeholder public health cooperation to manage public health threats, food safety, and food security are not new, three global causes have advanced the One Health strategy: (1) over the past 50 years, human populations have grown exponentially and expanded into new habitats; (2) land use practices, such

> Zoonotic diseases are caused by a pathogen that is transmitted from vertebrate animals to humans.

Environment-Related Infectious Diseases

as deforestation and farming, combined with climate change, have disrupted habitat conditions allowing for new human–animal interactions; and (3) the increase of international travel due to globalization allows infectious diseases to quickly spread across the globe [30]. Therefore, public health surveillance, public education on how to handle animals and pets, and treatment and vaccination research are essential to prevent the transmission and control of zoonotic diseases.

Zoonotic diseases can be transmitted from animals to humans through the consumption of contaminated water (as described before in the case of cholera) and through direct contact with infected animals. Animal husbandry practices in the U.S. are closely monitored and regulated to prevent the transmission of disease between animals and humans. The U.S. Department of Agriculture (USDA) collaborates in the One Health initiative with CDC, FDA, and EPA, among others, to ensure the safety of national and imported animal commodities [31]. The USDA monitors animals and animal products in the U.S. through surveillance and research to help control and eliminate conditions that can lead to the spread of disease. Healthy livestock and sound livestock practices can prevent economic loss as well as the emergence of new microbial diseases. Influenza, or the "flu," illustrates the case of a naturally occurring virus that circulates among livestock and has the potential of causing severe human disease.

> The One Health approach recognizes the interdependence of human health and that of animals and the environment.

1.4.1 H1N1 Global Influenza Pandemic

Of special public health concern, influenza, or the seasonal flu, is a contagious respiratory illness that spreads in the U.S. through human contact, particularly in the winter months. Influenza viruses are of the genus Orthomyxoviridae, consisting of three types: influenza A, B, and C. Influenza A is the most common type. Different influenza viruses express different hemagglutinin ("H") and neuraminidase ("N") proteins on the outer envelope of the virus and these are used to classify the virus. There are 18 and 11 known "H" and "N" variants, respectively [32].

During the 20th century, four flu pandemics were recorded. As commented by the CDC, the 1918 influenza pandemic known as the "Spanish Flu" was the most severe pandemic in recent history [33]. It was caused by an H1N1 virus with genes of avian origin. Although there is no universal consensus regarding where the virus originated, it spread worldwide during 1918–1919. In the U.S., it was first identified in military personnel in spring 1918. It is estimated that about 500 million people or one-third of the world's population became infected with this virus. The number of deaths was estimated to be at least 50 million worldwide with about 675,000 occurring in the U.S. Mortality was high in people younger than 5 years old, 20–40 years old, and 65 years and older. The high mortality in healthy people, including those in the 20–40 year age group, was a unique feature of this pandemic. Illustrated in Figure 1.2 are military nurses

> The 1918 flu epidemic caused 50 million deaths worldwide with about 675,000 occurring in the U.S.

FIGURE 1.2 Military nurses training to assist flu victims in 1918. (Blackburn, C. C., G. W. Parker, M. Wendelbo. 2018. *The Conversation*, March 1.)

undergoing training to provide assistance to flu victims. Noteworthy is the presence of masks for respiratory protection [34].

In 2009, the H1N1 influenza virus was detected in the U.S. population. By June 2009, WHO had declared a H1N1 pandemic [35]. H1N1 is also known as the "swine flu" due to genetic characteristics previously isolated in swine. H1N1 had circulated in swine in the U.S. for years, causing sporadic human infections since 2005. Phylogenetic analysis determined that the epidemic causing H1N1 was a combination of swine, avian, and human viral lineages. Originally thought to have begun in Mexico, Mexican government officials closed private and public facilities to prevent the spread of H1N1. The international community responded by scanning people arriving from Mexico and canceling many aircraft flights, heavily affecting Mexican tourism. Even though the virus was less lethal compared to the 1918 epidemic, millions fell ill due to ease of transmission and short supply of the H1N1 vaccine [36]. Furthermore, the term "swine flu" was heavily used in mass media leading countries to ban livestock or livestock-related products from North America, halt importation of all swine, or in the case of Egypt, slaughter all the swine in the country [37].

The H1N1 influenza threat led to an aggressive vaccination campaign in the U.S. It is estimated that the campaign prevented 1 million illnesses and 300 deaths. The H1N1 pandemic was declared over in August 2010 [38]. The CDC recommends all adults get the annual flu vaccine, which is developed through surveillance of the most frequently circulated influenza strains, and in collaboration with the five WHO Collaborating Centers for Reference and Research on Influenza [39]. The CDC maintains FluView, a weekly influenza surveillance report, to inform health departments and practitioners across the U.S. [40]. USDA's Animal and Plant Health Inspection Service monitors avian and swine influenza, providing updated information on reportable animal diseases for public safety [41]. The 2009 H1N1 demonstrated how a novel flu variant can affect even the most prepared countries. The interaction between poultry, livestock, and humans allowed genetic recombination of flu virus lineages that are able to mutate and transfer from animals to humans. Such recombination is ongoing prompting public health departments to be constantly vigilant of any influenza case in the U.S. and abroad.

More recently, the 2017–2018 flu season was the worst in the U.S. in 40 years. The CDC estimated that overall there were 959,000 hospitalizations and 79,400 deaths during the 2017–2018 season. More than 48,000 hospitalizations occurred in children (aged <18 years); however, 70% of hospitalizations occurred in older adults aged ≥65 years. Older adults also accounted for 90% of deaths, highlighting that older adults are particularly vulnerable to severe disease with influenza virus infection. An estimated 10,300 deaths occurred among working age adults (aged 18–64 years), an age group that often has low influenza vaccination. The ineffectiveness of the flu vaccine was a major element in the public health flu jolt of 2017–2018 [42].

> The 2017–2018 flu season was the worst in the U.S. in 40 years. The CDC estimated that overall there were 959,000 hospitalizations and 79,400 deaths.

1.4.2 Coronavirus Disease 2019 (COVID-19) Pandemic

In late December of 2019, an outbreak with a novel virus named *Severe acute respiratory syndrome coronavirus 2 (SARS-CoV-2)* occurred. Coronaviruses are named for the crown-like spikes on their surface. There are four main subgroupings of coronaviruses, known as alpha, beta, gamma, and delta. Human coronaviruses were first identified in the mid-1960s. Seven coronaviruses can infect people. Some only cause mild respiratory symptoms. Three coronavirus infections of public health significance are [43]:

- Severe acute respiratory syndrome (SARS) is a viral respiratory illness caused by a coronavirus called SARS-associated coronavirus (SARS-CoV). SARS was first reported in Asia in February 2003. The illness spread to more than two dozen countries in North America, South America, Europe, and Asia before the SARS global outbreak of 2003 was contained. Since 2004, there have not been any known cases of SARS reported anywhere in the world.
- Middle East Respiratory Syndrome (MERS) is a viral respiratory illness that is new to humans. It was first reported in Saudi Arabia in 2012 and has since spread to several other countries, including the United States. Most people infected with MERS-CoV developed severe respiratory illness, including fever, cough, and shortness of breath. Many of them have died.
- An outbreak caused by a novel (new) coronavirus was first identified in 2020 in Wuhan, Hubei Province, China, a city of more than 11 million residents.

Chinese authorities identified the new coronavirus, which has resulted in thousands of confirmed cases in China, including cases outside Wuhan, with additional cases being identified in a growing number of countries internationally. Persons who required hospitalization showed symptoms of fever, severe cough, shortness of breath, and pneumonia. The outbreak was linked epidemiologically to the Hua Nan seafood and wet animal wholesale market in Wuhan. The first case in the U.S. was announced on January 21, 2020 [43,44]. The local government ordered public health actions that were intended to interdict the person-to-person spread of the virus. These actions included cancelation of public meetings and travel and stay-in-home orders to Wuhan residents.

The new virus, which first emerged at the end of December 2019, initially killed at least 17 people and sickened more than 540 through January 2020, including persons in Taiwan, Japan, Thailand, South Korea, and the U.S. [44]. The virus is a coronavirus whose scientific designation is 2019-nCoV [43]. In an effort to contain the viral infection, Chinese authorities in late January 2020 canceled planes and trains leaving the city, and suspending buses, subways, and ferries within it. On January 25, 2020, China's health authorities greatly expanded a travel lockdown in central China to include 12 cities near the center of the outbreak, effectively quarantining 35 million residents, the largest quarantine ever in the history of public health [45]. Illustrated in Figure 1.3 are medical personnel attending a patient in a Wuhan

> Middle East Respiratory Syndrome (MERS) is viral respiratory illness that is new to humans. It was first reported in Saudi Arabia in 2012.

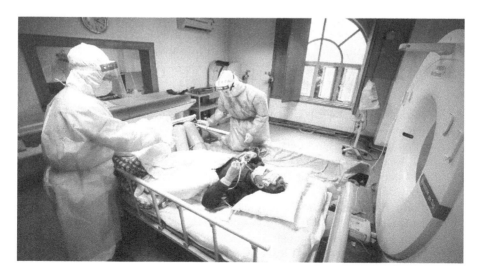

FIGURE 1.3 Wuhan, China, medical team attending a COVID-19 hospital patient, 2020. (Belluz, J., 2020. Stringer/AFP via Getty Images. *Vox*, May 3.)

Environment-Related Infectious Diseases

hospital isolation ward. Noteworthy is the degree of protective garb being work by the hospital staff.

On January 30, 2020, the WHO declared the novel coronavirus outbreak a global health emergency [46]. The agency declared a "public health emergency of international concern" (PHEIC), which is the WHO's highest level of alarm – a step it reserves for events that pose a risk to multiple countries and that require a coordinated international response. The WHO's action was prompted by four other places outside mainland China – Japan, Taiwan, Germany, and the U.S. – having reported person-to-person transmission, as the size and reach of the outbreak grew. Illustrated in Figure 1.4 is a person grocery shopping while wearing a face mask to protect against inhaling the virus. The WHO announced in February 2020 that 44,000 cases had been confirmed in China. In the same month, WHO named the disease COVID-19, with the virus designated as SARS-Cov-2 by the International Committee on Taxonomy of Viruses [47]. In late January 2020, the WHO estimated the death rate of the new virus to be about 2%, a rate less than that for MERS and SARS, both of which are also caused by strains of coronaviruses [48].

> On January 30, 2020, the WHO declared the novel coronavirus outbreak a global health emergency. The agency declared it a "public health emergency of international concern."

During the pandemic, the WHO published daily records of numbers of persons with confirmed cases of COVID-19, numbers of persons who succumbed to the infection, and geographic sources of the infection data [49]. Globally, as of January 3, 2021, there have been 83,322,499 confirmed cases of COVID-19, including 1,831,412 deaths, reported to the WHO. Further, the data indicate a mortality rate of about 2.20%. In the U.S. on January 3, 2021, the totals were 20,346,372 COVID-19 cases and 349,246 deaths since the beginning of the pandemic.

FIGURE 1.4 Store shopper wearing face mask, 2020. (CDC, Requirement for Face Masks on Public Transportation Conveyances and at Transportation Hubs. 6/10/2002, https://www.cdc.gov/coronavirus/2019-ncov/travelers/face-masks-public-transportation.html.)

In December 2020, the FDA gave emergency use approval to two vaccines that were shown in testing to be effective in preventing COVID-19 infections. Globally, both China and Russia developed COVID-19 vaccines and commenced large-scale inoculations of their populations. The first COVID-19 vaccinations were given to more than 1.1 million people in four countries, according to a source that tracks vaccination rates. This was the start of the biggest vaccination campaign in global history and one of the largest logistical challenges ever undertaken. Vaccinations in the U.S. began in mid-December with healthcare workers, and 24 states reported the first 49,567 doses administered. Those numbers surged through the early months of 2021. Regrettably, a public opinion poll found that about 40% of U.S. residents were averse to receiving the vaccine.

Perspective: The U.S. federal government did not develop a national public health plan to prevent the spread of COVID-19, relying on U.S. states to implement their own plans. Among the policies executed by multiple U.S. states were stay-at-home orders, closing schools and "non-essential" businesses, and requiring face masks and physical distancing in public. Overall, 45 states implemented stay-at-home orders, 44 states closed schools and 37 states have mandated cloth face coverings in some capacity. A recent study compared the impact on the COVID-19 cases between two states, Iowa which did not issue a state-wide stay-at-home order, and Illinois which did so. A lack of the state-wide stay-at-home order in Iowa led to a faster increase in cases compared to Illinois that saw progression slowed after issuing the order [50].

The COVID-19 pandemic resulted in a disruption of national economies, excess deaths, and challenging social policies of wearing face masks and respecting social distancing. Another outcome was the speed with which the pharmaceutical industry developed, tested, and commenced COVID-19 vaccinations of high-risk populations, e.g., healthcare workers. Notably, CDC data revealed that deaths from COVID-19 were the third leading cause of deaths in the U.S. in 2020, following heart disease and cancer. COVID-19 was reported as either the underlying cause of death or a contributing cause of death for some 11.3% of U.S. fatalities, and replaced suicide as one of the top 10 leading causes of death. Further, the CDC data showed the age-adjusted death rate in the U.S. had risen by 15.9% in 2020, its first increase in three years. Similarly, COVID-19 death rates were highest among individuals ages 85 and older, with the age-adjusted death rate higher among males than females. The COVID-19 death rate was highest among Hispanic and American Indian/Alaska Native people [50a].

Additionally, the interplay between public health agencies and programs with elected political officials will provide an interesting and important forum for health historians. History will reveal that the U.S. was ill-prepared to face the presence of a major pandemic. While the Barack Obama (D-IL) administration had prepared a plan and set aside expert advice resources on how to cope with a pandemic, this material was discarded by the incoming Donald Trump (R-NY) administration. Without going into detail in regards to the lack of federal government leadership in dealing with the COVID-19 pandemic, it is asserted here that failure by the federal

> CDC: Deaths from COVID-19 were the third leading cause of deaths in the U.S. in 2020, following heart disease and cancer.

government to accept the reality of the pandemic's invasion of the U.S. was directly responsible for the needless loss of lives to the virus. Most responsible for this failure was U.S. President Trump, who initially denied the outbreak of viral disease, then make recommendations for responding to the pandemic based on arguments not supported by science. His strong-handed interdiction of federal public health agencies such as the CDC resulted in confusion and uncertainty on ways to manage the pandemic.

1.4.3 Social Justice Issues

As the COVID-19 pandemic spread largely unabated throughout the U.S., data for persons infected with the virus began to indicate disparities in infected populations. In particular, African-Americans, Hispanic/Latinx, American Indians, and other minority groups suffered disproportionately from COVID-19 mortality. This social disparity was evidenced by CDC data released in June 2020, "Of the 1,320,488 reported cases analyzed between January 22 and May 30 that included information on race and ethnicity, 33% of the patients were Hispanic, 22% were Black, and 1.3% were American Indian and Alaska Native (AI/AN). These findings suggest that persons in these groups, who account for 18%, 13%, and 0.7% of the U.S. population, respectively, are disproportionately affected by the COVID-19 pandemic" [51].

Further, the COVID-19 pandemic, which claimed more than 336,000 lives in the U.S. in 2020, significantly affected life expectancy, researchers reported [51a]. The researchers projected life expectancy at birth for Americans will decrease by 1.13 years to 77.48 years. That is the largest single-year decline in life expectancy in at least 40 years and is the lowest life expectancy estimated since 2003. The declines in life expectancy are likely even starker among minority populations. For African-Americans, the researchers project their life expectancy would shorten by 2.10 years to 72.78 years, and for Latinos, by 3.05 years to 78.77 years.

1.4.4 West Africa Ebola Epidemic

Ebola virus disease (EVD) was first documented in 1976 when two simultaneous outbreaks erupted in what is now South Sudan and the Democratic Republic of Congo, near the Ebola River. The Ebola virus (family: Filoviridae, genus: *Ebolavirus*) can cause hemorrhagic fever in humans and is often fatal. The virus is highly contagious and spreads from human to human. Treatment consists of supportive care with rehydration and no vaccine exists. Five outbreaks have occurred in Africa since 1976. Laboratory infections of EVD have taken place in Reston, Virginia; England; and Russia [52].

> The Ebola virus can cause hemorrhagic fever in humans and is often fatal. The virus is highly contagious and spreads from human to human.

The EVD natural reservoir host has not been identified. However, recent studies point to a small insect-eating bat as the EVD reservoir, yet EVD has never been isolated from a bat or any other mammal [53]. It is believed that human contact with this bat began the worst EVD outbreak in 2013. EVD outbreaks had never been reported in West Africa, but the first case appeared in Guinea and spread quickly to neighboring countries Sierra Leone and Liberia. CDC collaborated with WHO, Ministries of Health, and other international partners to establish rapid control strategies in respond to mounting cases of EVD. CDC activated the EOC to coordinate technical assistance deploying more than 900 CDC personnel to provide logistics, communication, management, and support functions to the response activities. At the end of the outbreak, more than 27,000 infections and 11,000 deaths were reported. In addition, the already struggling healthcare infrastructure in these countries was left depleted as numerous health practitioners were among the dead [54].

Questions remain about how the Ebola virus was introduced into West Africa, far from its origins in Central Africa [55]. It is possible that the Ebola virus was circulating in Guinea in mammal reservoirs with which humans have limited contact. Such reservoirs are suspected to be either a fruit or insect-eating bat. In 2015, Guinea was ranked 182 out of 188 countries on the Human Development Index [56]. Civil conflict and failed economic development decimated its public health infrastructure, forcing many into unexplored habitats for survival. This encroachment could have exposed humans to the Ebola virus's reservoir host. The socioeconomic conditions and lack of public health infrastructure and surveillance contributed to the EVD rate of transmission [57]. The EVD was also introduced in Nigeria and Senegal, two African countries with robust public health systems. In Nigeria, health authorities traced all contacts of the EVD index case, and more than 800 people were identified, interviewed, and/or tested. A total of 20 EVD cases were confirmed as well as eight deaths [58].

In September 2014, the U.S. diagnosed the first confirmed EVD infection in a traveler from Liberia. The incubation period (from infection to initial symptoms) is 2–21 days during which the infected person is contagious. Therefore, complete isolation of the infected individual is necessary to prevent the spread of EVD [59]. The index patient was isolated, treated, but died from the infection. A healthcare worker caring for the index case also became infected with EVD, developing symptoms on October 10 and a second healthcare worker tested positive for the Ebola virus 5 days later. Both fully recovered from the disease. A third medical aid worker who had returned from Guinea tested positive for EVD. The patient was isolated and fully recovered.

> Governors in Illinois, New York, and New Jersey issued home quarantine for 21 days for all EVD aid workers returning from Guinea, Sierra Leone, or Liberia.

These cases prompted a U.S. national debate on how to prevent the spread of EVD by returning aid workers. The debate led governors in Illinois, New York, and New Jersey to issue home quarantine for 21 days for all EVD aid workers returning from Guinea, Sierra Leone, or Liberia [60]. CDC and other health organizations opposed such recommendations because they would prevent health practitioners from traveling to these countries where much effort was needed to curtail the EVD epidemic. The U.S. federal, state, and local public health departments and the vast network of medical centers throughout the country made the spread of EVD among local populations unfeasible. CDC issued an *Interim U.S. Guidance for Monitoring and Movement of Persons with Potential Ebola Virus Exposure* to address public concerns related to the EVD epidemic in West Africa [61]. Two years after the index case appeared in Guinea, WHO declared the last affected country, Guinea, free of Ebola virus transmission [62]. The CDC *Guidance* was retired the same day.

Such a humanitarian catastrophe is unlikely in the U.S. or other Western countries. The 2014 West Africa EVD epidemic illustrated how lack of economic development can affect human health in a two-fold fashion: (1) people venture into unknown habitats hunting for food exposing them to unknown pathogens and (2) the lack of economic development implies a public health system not capable of serving the population's health needs and unable to respond to a humanitarian health crisis.

These socioeconomic factors contributed to the reoccurrence of an Ebola epidemic in West Africa in 2019. As noted by the CDC, as of March 24, 2019, public health officials have documented that the Ebola outbreak in the eastern Democratic Republic of Congo (DRC) has surpassed 1,000 cases; the current total number of confirmed and probable cases is 1009, including 625 deaths and 318 survivors [63]. The outbreak is the largest in the DRC's history and the second largest outbreak recorded of Ebola ever (after the 2014–2016 outbreak in West Africa). The outbreak in DRC is occurring in a region where there is armed conflict, outbreaks of violence, and other problems that complicate public health response activities and increase the risk of disease spread both locally within DRC and to neighboring countries.

The outbreak of such a highly infectious disease such as Ebola is a grievous human event. The containment and subsequent prevention of Ebola outbreaks is an epic challenge to public health agencies such as a country's ministry of health, regional health agencies, the WHO, the CDC, and others. Medical teams must be mobilized, transported to areas of Ebola infection, provided protection from environmental hazards, and provided personal and team support. Often there is little good news to share, given the vigor of the infections, and socioeconomic limitations. But a ray of light can sometimes emerge, as in the 2019 Ebola outbreak in the DRC.

An experimental Ebola vaccine being used to try to contain the outbreak in the DRC was protective 97.5% of the time, according to data released by the WHO on April 12, 2019 [64]. Of more than 90,000 people who were vaccinated, only 71 went on to develop Ebola. In November 2019, the European Commission gave its approval to market an Ebola vaccine, less than a month after a European medicines panel backed the first-ever vaccine against the deadly virus. The vaccine, Ervebo, is approved for individuals aged 18 years and older and has already been used under emergency guidelines to try to protect against the spread of a deadly Ebola outbreak in the DRC [65]. Of note, FDA announced on December 19, 2019, their approval of Ervebo, an action in support of WHO's effort to contain the disease in Africa and for prevention of the disease via international travelers.

1.4.5 MEASLES EPIDEMICS

Among zoonotic diseases, measles stands out as a public health challenge due to its widespread history of global infection and ease of viral transmission. Measles is caused by a virus in the paramyxovirus family and it is normally passed through direct contact and through the air. The virus infects the respiratory tract, then spreads throughout the body. Measles is a human disease and is not known to occur in animals [66]. A measles vaccine developed in 1963 has proved highly effective in preventing the disease. Of note, the WHO has provided the following public health data pertaining to the global presence of measles.

> A measles vaccine developed in 1963 has proved highly effective in preventing the disease.

- "Even though a safe and cost-effective vaccine is available, in 2018, there were more than 140,000 measles deaths globally, mostly among children under the age of five.
- Measles vaccination resulted in a 73% drop in measles deaths between 2000 and 2018 worldwide.
- In 2018, about 86% of the world's children received one dose of measles vaccine by their first birthday through routine health services – up from 72% in 2000.
- During 2000–2018, measles vaccination prevented an estimated 23.2 million deaths making measles vaccine one of the best buys in public health" [66].

Most measles-related deaths are caused by complications associated with the disease. Serious complications are more common in children under the age of 5 or adults over the age of 30. The most serious complications include blindness, encephalitis (an infection that causes brain swelling), severe diarrhea and related dehydration, ear infections, or severe respiratory infections such as pneumonia. Severe measles is more likely among poorly nourished young children, especially those with insufficient vitamin A, or whose immune systems have been weakened by HIV/AIDS or other diseases [66].

In the decade before 1963 when a vaccine became available, nearly all children contracted measles by the time they were 15 years of age. It is estimated that 3 to 4 million people in the U.S. are infected each year. Also each year, among reported cases, an estimated 400–500 people die, 48,000 are hospitalized, and 1,000 suffer encephalitis (swelling of the brain) from measles [66]. Unvaccinated young children are at the highest risk of measles and its complications, including death. Unvaccinated pregnant women are also at risk. As a consequence of public health programs of measles vaccination of young children, in the year 2000, the WHO declared the U.S. as free of measles. Regrettably, this achievement did not last. In 2019, the CDC recorded 1,282 individual confirmed cases of measles in 31 states [67]. This is the greatest number of cases reported in the U.S. since 1992. Of this number, 128 of the people who contracted measles were hospitalized, and 61 reported having complications, including pneumonia and encephalitis. More than 73% of the cases were linked to recent outbreaks in New York. Measles is more likely to spread and cause outbreaks in U.S. communities where groups of people are unvaccinated.

Perspective: The return of measles as a public health challenge occurred in 2019 in the U.S. and some other countries where misinformation regarding measles vaccination programs occurred. Social media advocates of not vaccinating children against measles and other childhood diseases became widespread, resulting in increased numbers of children not receiving their childhood suite of vaccines.

Environment-Related Infectious Diseases

The "anti-vaccers" used social media outlets to argue that autism in children was linked to vaccinations. Their thesis was based on a discredited medical study published in 1998 in the UK that had used fabricated data to allegedly link autism to childhood immunizations. The principal author had been funded by lawyers who had been engaged by parents in lawsuits against vaccine-producing companies. In response, some U.S. states enacted legislation that permitted parents to forego their children's immunizations based on religious or medical concerns.

This kind of legislation resulted in increased numbers of children exempted from childhood vaccinations, including measles. As episodes of measles epidemics occurred in some of these states, state legislators generally revised exempting legislation to include only medical reasons. As of January 2020, 45 U.S. states and Washington, DC, grant religious exemptions to vaccines, according to the National Conference on State Legislatures, which tracks state laws [68]. In 2019, 14 states proposed eliminating religious exemptions for vaccines – a marked increase from years past, according to the American Academy of Pediatrics. New York, California, and Washington states all voted to end some vaccine exemptions for children who attend public schools, and Maine enacted similar legislation, which took effect in September 2021.

1.4.6 IMPLICATIONS FOR ENVIRONMENTAL HEALTH POLICY: QUARANTINE AND PUBLIC HEALTH

At the end of the 1700s, mariners were disproportionately afflicted by illness and disability. At sea, they were confined to small places aboard ship and unsanitary conditions, which were detrimental to

> There are 20 quarantine stations across the U.S. to stop the introduction and spread of infectious diseases.

their health and also spread disease to the locals in port cities. The need to establish federal regulation to govern the retention of vessels with cases of or coming from ports with infectious diseases led to the adoption of The National Quarantine Act (NQA) of 1878. The act gave the U.S. federal government the authority to establish rules and regulations of incoming vessels. The following year Congress enacted the Act to Prevent the Introduction of Infectious or Contagious Disease into the U.S. and to establish the National Board of Health (NBH). The NBH was charged with collecting information on all public health matters, advising state governments on public health preservation and improvement, and creating a national public health organization. Currently, there are 20 quarantine stations across the U.S. to stop the introduction and spread of infectious diseases. The DHHS and CDC have statutory authority to regulate quarantine practices in the U.S. The final rule for Control of Communicable Diseases: Interstate and Foreign became effective on March 21, 2017. The Ebola virus outbreak of 2014 prompted the update of this final rule in order to increase the transparency of such practices in the U.S. [69].

Perspective: It is important to note that when the need demands it, such as during the COVID-19 pandemic, expanded and nontraditional quarantine locations can be deployed, such as homes and university dormitories. In addition, measure quarantine can be required upon visiting a country with high case counts.

1.5 EMERGENCE OF VECTOR-BORNE DISEASES

As depicted in Figure 1.1, there are three determinants that are necessary for disease transmission – the pathogen, the host, and the environment. Vector-borne diseases are zoonotic infectious diseases that have an additional determinant, the

> Worldwide, vector-borne diseases account for 17% of all infectious diseases and more than 1 billion infections and 1 million deaths.

vector, necessary for the transmission of disease. The vector transmits the pathogen to the host by bite or through the exposure of a bodily fluid, such a saliva or urine. As noted in the figure, the vector is connected to the three triangle vertices. The environment plays a crucial role in the transmission of vector-borne diseases. The vector needs a supportive environment in which it can thrive and interact with both the pathogen and the host. Vectors include different arthropod species such as mosquitoes, ticks, mites, sandflies, and triatome bugs; mammal species such as livestock, rodents, cats and dogs, and other species such as turtles and birds. An increase in the U.S. of three vectors, mosquitoes, ticks, and fleas has steadily occurred according to CDC surveillance data [70].

In the U.S., 14 vector-borne diseases of national public health concern are reportable to CDC's National Notifiable Diseases Surveillance System [71]. Within CDC, the National Center for Emerging and Zoonotic Infectious Diseases [72] is tasked with the prevention and reduction of illness and death caused by these diseases as well as to limit the spread of vector-borne diseases. Worldwide, vector-borne diseases account for 17% of all infectious diseases and more than 1 billion infections and 1 million deaths [73]. The following is a short summary of selected vector-borne diseases that highlights the importance of the environment in the transmission of infectious diseases.

1.5.1 Mosquito-Borne Diseases

Similar to the Ebola virus, mosquito-borne viruses have jumped from feral animals to humans due to activities that increase the interactions between them. These activities include deforestation, human encroachment on animal habitats, and bush meat hunting. Deforestation and bush meat hunting are a direct consequence of expanding human populations, and logging and mining activities can indirectly lead to the emergence of new infectious diseases. This section describes the principal infectious diseases caused by the transmission of viruses and parasites via the most dangerous vector to human health, mosquitoes.

1.5.1.1 Malaria

Malaria is an ancient disease with records of a malaria-like disease described in ancient Chinese medical textbooks dating back to 2700 BCE [67,68]. The causal association between malaria and mosqui-

> Malaria is a parasitic disease transmitted to humans by infected female *Anopheles* mosquitoes.

toes was discovered by Ronald Ross, an Anglo-Indian physician, who on August 20, 1897, in Secunderabad, India, made a landmark public health discovery. While

Environment-Related Infectious Diseases

dissecting the stomach tissue of an anopheline mosquito fed 4 days previously by biting a malarious patient, he found the malaria parasite. He pursued his discovery and went on to prove the role of female *Anopheles* mosquitoes in the transmission of malaria parasites to humans [74].

Malaria is transmitted mostly by mosquito species of the *Anopheles* genus. All *Plasmodium* species have a similar life cycle involving the mosquito vector and the human host [75,76]. Malaria transmission also depends on the *Anopheles* mosquito. The intensity of malaria transmission is dependent on the numbers and types of *Anopheles* mosquitoes present. Furthermore, some *Anopheles* species prefer human blood meals (anthropophilic) and others preferentially feed on animals (zoophilic). Most *Anopheles* species are active between dusk and dawn but some feed indoors (endophagic) while others prefer to feed outdoors (exophagic). In contrast to *Ae. aegypti*, *Anopheles* lays its eggs in muddied or dirty waters and is not reliant on clean standing water. *Anopheles* can therefore lay its eggs in any makeshift water container from ponds and lakes to deep tire-tracks on dirt roads. Hence, unlike *Ae. aegypti*, the *Anopheles* mosquito does not concentrate in urban areas but can be found in all types of habitats. The *Anopheles* species feeding preferences have implications for vector control strategies [77].

Unlike other mosquito-borne diseases discussed in this chapter, malaria is a parasitic disease transmitted to humans by infected female *Anopheles* mosquitoes. Malaria infection is characterized by high fever, headache, sweating, and or moderate to severe shaking chills. These symptoms can last a few weeks or the parasite can remain dormant in the body for years. Complications can arise from malaria infections including cerebral malaria, which can lead to a coma, anemia, low blood sugar, organ failure, and even death. The persons most at risk of malaria complications are infants and young children, travelers to endemic regions, and pregnant women and the fetus. In addition to the *Anopheles* mosquito, malaria can be transmitted through blood transfusions and from mother to child during pregnancy [78].

Because malaria was disrupting the construction of the Panama Canal, a vector-control campaign at the time was undertaken to reduce both yellow fever and malaria infections. This campaign was the first in many efforts by the U.S. to eradicate malaria from its population. The discovery of the insecticide properties of DDT in the 1940s accelerated the efforts to decrease malaria transmission after World War II. In the U.S., DDT was used to prevent the re-introduction of malaria by soldiers returning from training in endemic areas. These efforts, paired with the U.S. Public Health Service's malaria water level control and insecticide spraying campaigns, eliminated endemic malaria from the continental U.S. by 1947 [79]. Currently, about 1,500 cases of malaria are diagnosed in the U.S. annually, mostly in travelers and immigrants returning from malaria-endemic countries.

Historically, malaria was endemic in the tropical and subtropical region of the world but currently is mostly a problem in sub-Saharan African countries, South Asia, and Papua New Guinea, Solomon Islands, and Haiti. These areas accounted

> About 1,500 cases of malaria are diagnosed in the U.S. annually, mostly in travelers and immigrants returning from malaria-endemic countries.

for an estimated 600,000 deaths worldwide in 2012. Even though mortality and morbidity are still extremely high, successful campaigns have significantly reduced mortality by 45% and saved more than 3 million lives since 1990. In 1998, WHO, UN Children's Fund, the World Bank, and the UN Development Program founded the Roll Back Malaria partnership. Roll Back Malaria focused on strengthening endemic and nonendemic malaria countries' health systems for sustainable health improvement by partnering with governments and civil society (NGOs, academia, etc.).

Countries that committed to the Roll Back Malaria Initiative took the lead in identifying strategies to intensify malaria control efforts, understanding how their population viewed, prevented, and treated malaria, and considered the context in which private and public care was offered in the country. The WHO provided support for Roll Back Malaria partner countries by endorsing the technical malaria strategy approach, advising on technical and financial assistance, and encouraging partners to stay engaged in the partnership [80].

In 2019, the WHO's director-general observed that "malaria has been eliminated progressively in many parts of the world. By the 1950s, malaria had been eliminated from much of the temperate zone, and from that point until now, WHO has certified 38 countries and overseas territories as malaria-free. The global malaria death rate declined by more than 60% between 2000 and 2015. However, he cautioned that the currently available tools and approaches will not be sufficient to achieve malaria eradication. As with smallpox and polio, a commitment to eradicate must spur research and development to deliver new tools" [81].

For policy reminder purposes, it needs to be noted that in 1980, the World Health Assembly declared smallpox eradicated, and no cases of naturally occurring smallpox have occurred since. The eradication occurred through coordinated public health actions led by the WHO, CDC, and national health ministries.

> WHO has certified 38 countries and overseas territories as malaria-free. The global malaria death rate declined by more than 60% between 2000 and 2015.

Similar good news was expressed on August 25, 2020, when the Africa Regional Certification Commission certified the WHO African Region as wild polio-free after 4 years without a case. With this historic milestone, five of the six WHO regions – representing more than 90% of the world's population – are now free of the wild poliovirus, moving the world closer to achieving global polio eradication. As of 2020, only two countries worldwide continue to see wild poliovirus transmission: Pakistan and Afghanistan [81a].

1.5.1.2 Yellow Fever

The yellow fever virus was first introduced into the Americas during the 1600s. The first documented yellow fever outbreak occurred in 1648 in the Yucatan peninsula. Within years, outbreaks were reported in Boston, Charleston, and New York City. Over the course of the years, yellow fever outbreaks were relegated to southern U.S. port cities, where the subtropical climate sustained a mosquito population year round [82]. At the turn of the 20th century, U.S. efforts to build the Panama Canal were being halted due to the vast number of workers falling ill with yellow fever. In

Environment-Related Infectious Diseases

response to this problem, Major Walter Reed, a U.S. Army physician, was sent to Cuba to investigate an outbreak of yellow fever there. Following investigations that involved infecting team members with yellow fever, Reed in 1900 and his team confirmed that yellow fever is transmitted by mosquitoes, rather than by direct contact with persons suffering from yellow fever [74].

A large yellow fever outbreak in 1905 reduced the Panama Canal work force, threatening significant construction delays. With the knowledge that mosquitoes were the cause of yellow fever, President Theodore Roosevelt (R-NY) authorized the U.S. Army's chief sanitation officer, Dr. William Gorgas, to implement the largest fumigation effort in history. The mosquito populations significantly declined and, as a result, yellow fever cases were reduced by half within 1 year [83,84].

> At the turn of the 20th century, U.S. efforts to build the Panama Canal were being halted due to the vast number of workers falling ill with yellow fever.

Efforts to find a vaccine against yellow fever began as early as the *Ae. aegytpi* fumigation campaigns. In 1937, Max Theiler, at the Rockefeller Foundation, developed a more efficacious strain of a previous vaccine called 17D. The first mass vaccination campaign with 17D occurred in Brazil in 1939 when more than 1 million people were vaccinated. An effective vaccine and aggressive vector-control campaigns reduced yellow fever infections in certain tropical regions in the South American continent [85]. Currently, the 17D strain is still part of the WHO's List of Essential Medicines and in the U.S. it is given to travelers to countries with endemic yellow fever [86,87]. However, in 2015, a yellow fever outbreak in Angola and the Democratic Republic of Congo exhausted the world's emergency stock pile [88,89]. A large yellow fever outbreak occurred in the urban center in Minas Gerais, Brazil, in 2017, which resulted in 130 cases and 53 deaths. The Ministry of Health responded with the vaccination of 23.8 million Brazilians against yellow fever [90]. These two outbreaks illustrate the importance of mosquito control and environmental manipulation to prevent the spread of vector-borne diseases even when an effective vaccine against the disease exists.

1.5.1.3 Dengue

In the 1950s, a new virus came into the Americas. The dengue virus was first isolated from an infected patient in Trinidad in 1953 [91]. Dengue or severe dengue is caused by the dengue virus (DENV) of which there are four types (DENV1–4). Like the yellow fever virus, DENV has a sylvatic[1] cycle in Africa but is maintained in the Americas through the domestic cycle by *Ae. aegypti*. Upon a second infection with a different DENV type, the infected person can develop much more severe illness, including hemorrhagic presentations, hypovolemia, shock, and even death. Dengue, or severe dengue, has a mortality rate of 20% if untreated. Treatment is primarily supportive therapy, which reduces mortality of severe dengue to less than 1% [92]. Research is also under way to develop a vaccine to

---

[1] **Sylvatic**: relating to or denoting certain diseases when contracted by wild animals, and the pathogens causing them.

prevent dengue [93,94]. By the 1980s, all four DENV types had been introduced and were in circulation in the Americas [91].

Most cases of dengue in the U.S. are not endemic. However, a small outbreak in Brownsville, Texas, in 2005 demonstrated the ease with which the disease can become endemic when the vector is abundant in the region [95]. Similarly, endemic dengue cases surfaced in south Florida in 2009. Florida had not documented an indigenous dengue case since 1934 and investigations led to the confirmation of a dengue outbreak in Monroe and Miami-Dade counties. The outbreak ended in 2011 [96]. Despite continuous surveillance and vector-control efforts in Florida, locally acquired cases of dengue were reported in 2014 and 2016.

> Dengue causes the greatest human disease burden of any arbovirus, with an estimated 10,000 deaths and 100 million symptomatic infections per year.

The global outlook on the spread of dengue was estimated in a study by a multinational academic research team in 2019 [97]. Dengue causes the greatest human disease burden of any arbovirus,[2] with an estimated 10,000 deaths and 100 million symptomatic infections per year in more than 125 countries. Observing that dengue now affects more than half the world's population, the researchers opined that dengue infections would further expand due to ongoing global phenomena including climate change and urbanization. The team made use of climate, population, and socioeconomic projections for the years 2020, 2050, and 2080 to project future changes in virus suitability and human population at risk.

The researchers predicted that 3.83 (3.45–4.09) billion people (about 53% of the global population) live in areas that are suitable for dengue transmission, with the vast majority residing in Asia, followed by Africa and the Americas. Much of the southeastern U.S. was predicted to become suitable by 2050, with many of the larger cities in coastal eastern China and Japan are also likely to become suitable for dengue infection by 2050. The continent that is likely to see the biggest change in dengue risk is Africa.

Globally, the researchers predicted 2.25 (1.27–2.80) billion more people will be at risk of dengue in 2080 compared to 2015, bringing the total population at risk to more than 6.1 (4.7–6.9) billion, or 60% of the world's population. This growth would be largely driven by population growth in already endemic areas as opposed to the spread of the dengue virus to new populations, emphasizing the increasing public health burden many dengue-endemic countries are likely to face [97].

> Researchers predicted the total population at risk of dengue infection in 2080 was 6.1 billion, or 60% of the world's population.

A vaccine has been developed for dengue prevention, but the WHO has advised caution in its application, stating, "The live attenuated dengue vaccine CYD-TDV has been shown in clinical trials to be efficacious and safe in persons who have had a previous dengue virus infection (seropositive individuals), but carries an increased risk of severe dengue in those who experience their first natural dengue infection

---

[2] **Arbovirus**: any of a group of viruses that are transmitted by mosquitoes, ticks, or other arthropods.

Environment-Related Infectious Diseases

after vaccination (seronegative individuals). For countries considering vaccination as part of their dengue control program, pre-vaccination screening is the recommended strategy. With this strategy, only persons with evidence of a past dengue infection would be vaccinated (based on an antibody test, or on a documented laboratory confirmed dengue infection in the past)" [98]. The FDA has approved the use of the vaccine, but with major restrictions on its use, specifically, it can only be used in individuals aged 9–16 living in parts of the U.S. where the dengue virus is endemic. Furthermore, the vaccine can only be given to children and teens who have had one previous laboratory-confirmed case of dengue [99].

In August 2019, the government of the Philippines declared a national dengue epidemic, reporting that more than 146,000 dengue cases had occurred in the first 7 months, with 622 premature deaths [100]. Two years prior, the Philippines had immunized more than a million children against dengue, but the vaccine was withdrawn by its manufacturer over potential side effects.

1.5.1.4 West Nile Virus

The West Nile Virus (WNV), a *flavivirus*, was introduced to the U.S. in 1999. It was first isolated in 1937 in the West Nile District of Uganda. For years, WNV circulated in Africa, Asia, and the Middle East, arriving in Europe in the 1990s. Like many other viral illnesses, WNV infection can remain asymptomatic or infected people develop mild disease including fever, headache, body aches, and a rash. However, 1 in 150 infected people develops severe symptoms characteristic of neuro-invasive diseases such as encephalitis or meningitis [101]. In 1999, the WNV was isolated from an infected patient in New York. It remains uncertain how this virus came to the U.S., either through an infected person, a WNV vector or intermediate host.

By 2002, WNV had established itself in 44 U.S. states and spread to five Canadian provinces [102]. The accelerated spread across North America was possible due to two reasons: the ubiquitous distribution of the intermediate host and a variety of potential vectors. WNV is maintained in nature in avian hosts and ornithophilic (avian feeding) mosquitoes. A large variety of birds act as a reservoir host for WNV and both ornithophilic and anthropophilic (human feeding) mosquitoes transmit the virus to humans. The main vector of transmission to humans are mosquitoes of the *Culex* genus but WNV has also been detected in species of mosquitoes in the *Aedes, Anopheles,* and *Culiseta* genus among others [103,104].

> WNV has established itself in 44 U.S. states and spread to five Canadian provinces.

Preventing the spread of WNV will require a combination of traditional public health measures such as mosquito control as well as the application of basic research findings. Regarding the latter, Washington State University researchers demonstrated in 2019 that mammalian insulin activated an antiviral immunity pathway in mosquitoes, increasing the insects' ability to suppress the viruses, in effect, dampening the ability of host mosquitoes to transport viruses that cause human disease [105]. The application of this finding in a program of WNV prevention would be another research endeavor.

1.5.1.5 Chikungunya Virus

The failure to eliminate and eradicate the *Ae. aegypti* mosquito has also led to the recent introduction of new zoonotic viruses to the Americas. Unlike DENV and yellow fever viruses, the chikungunya virus (CHIKV) belongs to the family *Togoviridae*, but similar to the other two mosquito viruses, it has a sylvatic, semi-domestic, and domestic cycle. It was first identified in 1953 in West Africa and is now found throughout Africa, south and south-east Asia, and Pacific islands. Chikungunya means "the bend walker" due to the arthralgia (joint pain) and flu-like symptoms experienced upon infection [106]. People infected with CHIKV can remain asymptomatic or experience mild disease, but severe infections can cause arthralgia and sequelae that can last for months or years. Currently, there is no available antiviral therapy or vaccine against CHIKV.

The most common vectors of CHIKV are *A. aegypti* and *Aedes albopictus*, both of which are distributed along the tropics [107]. For years, low levels of human CHIKV infections had been reported in African and Asian countries, where the virus is endemic. Chikungunya outbreaks from Indian Ocean islands and countries were reported between 2005 and 2007. Subsequently, a small outbreak in southern Italy was confirmed in 2007 where the CHIKV was being transmitted and became endemic due to *Ae. albopictus*. By 2013, the CHIKV reached the Caribbean when in December two autochthonous cases were confirmed in the French island of St. Martin. Within months, the CHIKV had spread to the Caribbean countries, all countries in Central America, South America (except for Chile and Uruguay), Mexico, and the U.S. By mid-2016, more than 1.7 million suspected cases were reported by the PAHO in 45 countries throughout the Americas [105,106].

> By mid-2016, more than 1.7 million suspected cases of Chikungunya were reported by the Pan-American Health Organization (PAHO) in 45 countries throughout the Americas.

1.5.1.6 Zika Virus

Similar to other mosquito-borne viruses, the Zika virus was first discovered in West Africa in 1947 in rhesus monkeys and the first reported human cases occurred in Uganda in 1952 [108]. Through the 1980s, the Zika virus spread throughout equatorial Asia, causing sporadic cases of mild disease. Until 2007, no outbreaks of Zika virus disease had ever been documented. In 2007, a Zika virus outbreak was identified in Yap Island. Most of the cases showed mild disease with symptoms including rash, fever, and arthralgia [109]. The next reported outbreak identifying ZIKV as the causative agent was in French Polynesia in 2013 with 5,895 suspected cases and 294 confirmed cases [110]. Two years later, 7,000 cases of an unknown illness are reported between February and April 2015 in northwest Brazil. Retrospectively, this outbreak was linked to the introduction of the ZIKV to Brazil from the South Pacific

> By May 2017, 48 countries and territories in the Americas had confirmed autochthonous, vector-borne transmission of Zika virus disease (ZIKD).

during a major sporting event in either 2013 or 2014 [106,111]. By May 2017, 48 countries and territories in the Americas had confirmed autochthonous, vector-borne transmission of Zika virus disease and 26 had reported confirmed cases of congenital syndrome associated with the infection.

Zika, like yellow fever and dengue, is caused by a *flavivirus* transmitted by *Ae. aegypti* and *Ae. albopictus*. The continental distribution of *Ae. aegypti* is one of the primary reasons for the rapid spread of ZIKV in the Americas along with an immunologically naïve population and high levels of travel [112]. In a majority of cases, ZIKV infection causes mild symptoms associated with other viral diseases. This includes fever, rash, headache, joint and muscle pain, and conjunctivitis. More severe symptoms including Guillain-Barré syndrome (GBS), a rare immune disorder that affects the peripheral nervous system, have been linked to ZIKV infections [113,114].

The most concerning consequence of ZIKV infection is the association with adverse fetal outcomes, including fetal loss for maternal ZIKV infections between 6 and 32 weeks of gestation and microcephaly for maternal ZIKV infections between 7 and 13 weeks of gestation [106,107]. A study by CDC of 1,450 babies examined neurological or developmental problems, such as seizures, hearing loss, impaired vision or difficulty crawling, in children at least 1-year old who were born in Puerto Rico and other U.S. territories and exposed to Zika *in utero*. Overall, 14% of children exposed to Zika in the womb – about 1 in 7 – were harmed in some way by the virus. The babies were either born with a birth defect such as microcephaly – a condition in which a baby's head is significantly smaller than it should be – or developed neurological symptoms that may be related to Zika, or both [115].

> The most concerning consequence of ZIKV infection is the association with adverse fetal outcomes.

Furthermore, substantial evidence exists that the ZIKV can be transmitted from mother to fetus, and, unlike any other mosquito-borne arbovirus, through sexual intercourse [116]. The unprecedented speed of ZIKV transmission across the Americas, combined with the modes of transmission and severe consequences of infection (such as Guillain-Barré syndrome [GBS] and congenital defects), prompted the WHO in 2017 to declare a Public Health Emergency of International Concern [117].

Following the initial clusters of GBS and microcephaly after ZIKV outbreaks in Brazil, experts expected an increase in the incidence of GBS and microcephaly during the next seasonal cycle. In 2016, CDC researchers published their findings of a causal relationship between ZIKV and birth defects. The researchers stated, "To determine whether Zika virus infection during pregnancy causes these adverse outcomes, we evaluated available data using criteria that have been proposed for the assessment of potential teratogens. On the basis of this review, we conclude that a causal relationship exists between prenatal Zika virus infection and microcephaly and other serious brain anomalies" [118].

As of 2017, clinical vaccine trials were under development for dengue [119], chikungunya [120], and Zika [121]. Regarding the Zika vaccine, notable progress has been reported, as reflected with the publication of three reports of phase 1 clinical trials in the fourth quarter of 2017 [122]. Clinical trials involving candidate DNA and purified inactivated virus vaccines showed all were safe and well-tolerated in the small number of volunteers and all induced neutralizing antibodies, although these varied by vaccine candidate and dosing regimen. Phase 2 clinical trials were underway in 2019. However,

> Brazil has been especially hard hit by the Zika epidemic, but the number of reported ZIKV decreased from 205,578 cases in 2016 to 13,353 in 2017.

as demonstrated by the yellow fever vaccine, even an effective vaccine cannot substitute for sound vector control and environmental practices in the prevention of disease. Control of *Ae. aegypti* and *Ae. albopictus* populations in the Americas is crucial to prevent future arbovirus outbreaks and stop the geographic spread of these mosquitoes.

Brazil has been especially hard hit by the Zika epidemic, but a national public health response has contributed to the number of reported ZIKV cases in Brazil decreasing from 205,578 cases in 2016 to 13,353 in 2017 (up to epidemiological week 25), with population immunity thought to be the main cause of the decline. The decline in new Zika cases might have been partly due to the use of two novel tactics that targeted the reduction of mosquitoes. One method exploits the egg-laying behavior of mosquitoes to have them disseminate tiny particles of a potent larvicide, pyriproxyfen (PPF), from resting to breeding sites. Juvenile mosquito mortality in sentinel breeding sites (SBSs) (about 4% at baseline) increased by more than one order of magnitude during PPF dissemination (about 75%). This led to a >10-fold decrease of adult mosquito emergence from SBSs [123]. A second novel method involved releasing sterile, transgenic mosquitoes and mosquitoes bearing Wolbachia to evaluate their impact on local vector populations and arbovirus transmission. The results showed a reduction of field populations of *Ae. aegypti* by 81%–95% [124]. It is unclear if these two novel methods were applied on larger geographic scales.

1.5.1.7 Eastern Equine Encephalitis (EEE) Virus

EEE virus is a rare cause of brain infections (encephalitis). Only a few cases are reported in the U.S. each year. Most occur in eastern or Gulf Coast states. Approximately 33% of people with EEE die and many survivors have ongoing neurologic problems. According to the CDC, Eastern equine encephalitis virus (EEEV) is maintained in a cycle between Culiseta melanura mosquitoes and avian hosts in freshwater hardwood swamps. Cs. melanura is not considered to be an important vector of EEEV to humans because it feeds almost exclusively on birds. Transmission to humans requires mosquito species capable of creating a "bridge" between infected birds and uninfected mammals such as some Aedes, Coquillettidia, and Culex species [125]. In 2019, there were three fatal cases of EEEV each in Michigan and Massachusetts and two deaths in Connecticut. The Connecticut Agricultural Experiment Station said that while EEE was detected in mosquitoes in 15 towns, six other locations have seen either human or horse cases of the virus. Local authorities in all three states have used pesticides for mosquito control [126].

Environment-Related Infectious Diseases

1.5.1.8 Implications for Environmental Health Policy: Vector Control and Pesticides

One of the most characteristic features of *Ae. aegypti* is its adaptation to live in urban areas, where it breeds in clean-water containers and feeds during the day. In 1947, an intensive *Ae. aegypti* Pan-American eradication campaign heavily reliant on the use of DDT succeeded in reducing the mosquito to undetectable levels and 17 countries in South and Central America were declared *Ae. aegypti* free by 1961. The U.S. did not participate in eradication efforts [127]. Over the years, the eradication campaigns lost political support and funding, and over the next 10 years, only three more countries achieved eradication of this mosquito. The decrease in surveillance and mosquito resistance to DDT and other insecticides also contributed to the continental infestation of *Ae. aegypti*.

Malaria, yellow fever, dengue, chikungunya, and Zika viral disease reinforce the importance of vector-control strategies. However, DDT is banned and misuse and overuse of other chemicals have led to larvicide- and insecticide-resistant mosquitoes. A reduction in mosquito breeding sites, insecticide and larvacides, and education on personal protection has contributed to the reduction of yellow fever and dengue. However, endemic seasonal outbreaks still exist. To halt transmission of yellow fever, dengue, chikungunya, and Zika, environmental control of the breeding sites, and mosquito surveillance campaigns need to occur year round. To date, no disease spread by a vector (or intermediate host) has ever been eradicated, although one disease, guinea worm disease (dracunculiasis), is nearing global eradication, as discussed in a subsequent section of this chapter.

> Malaria, yellow fever, dengue, chikungunya, and Zika viral disease reinforce the importance of vector-control strategies.

Other insecticides and larvacides have been developed and used to control mosquito breeding sites. The Roll Back Malaria Initiative demonstrates the importance of a comprehensive approach to malaria elimination that requires a political and financial commitment. Much of this campaign success is due to large financial support and Roll Back Malaria Partner efforts that focused mostly on four large-scale interventions: (1) large-scale, country-wide insecticide-treated nets to people in malaria-endemic areas, (2) User-friendly drug packages to increase medication compliance and slow down *Plasmodium* drug resistance, (3) Access to early treatment to reduce childhood mortality, and (4) Residual house spraying of insecticide and environmental management [128]. However, as successful as this global campaign has been, many challenges remain. Both *Plasmodium* and *Anopheles* have developed resistance to treatment and insecticides, respectively. Similar to the *Ae. aegypti* eradication campaigns, dwindling financial support, political instability, war, and climate change threaten to reverse some of the progress made in the combat of malaria.

The application of genetic engineering of mosquitoes has been proposed or control of mosquito populations. The mosquito, named OX5034, has been altered to produce female offspring that die in the larval stage, well before hatching and growing large enough to bite and spread disease. This approach to mosquito control avoids the use of chemical pesticides that environmental groups find objectionable.

1.5.2 Tick-Borne Diseases

Ticks are the second most common vector of human disease in the U.S. Ticks are small hematophagous (blood-feeding) arthropods that live in foliage globally but prefer warm, moist climates. Over the past 20 years, tick-borne diseases have become a serious problem in the U.S. due mostly to human migration into wilderness areas [129]. Ticks feed on the blood of different mammals, including humans, and can transmit a variety of diseases caused by bacteria, parasites, or viruses. In North America, the most common diseases transmitted by tick bites are Lyme diseases (spirochete *Borrelia burgdorferi*), ehrlichiosis (bacteria *Ehrlichia chaffeensis*, *Ehrlichia ewingii*, and *Ehrlichia muris*-like), babesiosis (different protozoan of the *Babesia* species), Rocky Mountain Spotted Fever (bacteria *Rickettsia rickettsia*), Tularemia (bacteria *Francisella tularensis*), and Q fever (bacteria *Coxiella burnetii*). Powassan (POW) virus is transmitted to humans by infected ticks. Approximately 75 cases of POW virus disease were reported to CDC in the U.S. over the past 10 years. Most cases have occurred in the Northeast and Great Lakes region, but climate change could enlarge the area of infected ticks. There is no specific treatment, but people with severe POW virus illnesses often need to be hospitalized to receive respiratory support, intravenous fluids, or medications to reduce swelling in the brain [130].

> Ticks are the second most common vector of human disease in the U.S.

Different human-biting tick species are distributed around the U.S., each transmitting a specific tick-borne disease. The known ticks in North America are American dog tick (*Dermacentor variabilis*), which transmits tularemia and Rocky Mountain spotted fever east of the Rocky Mountains and the Pacific Coast; Blacklegged Tick (*Ixodes scapularis*) transmits anaplasmosis, babesiosis, Lyme disease, and Powassan disease in the Northeast and upper Midwest; Brown Dog Tick (*Rhipicephalus sanguineus*), which transmits Rocky Mountain spotted fever with a country-wide distribution; Gulf Coast tick (*Amblyomma maculatum*), which transmits a form of spotted fever known as *Rickettsia parkeri* rickettsiosis along the Atlantic and Gulf of Mexico coastal areas; Lone star tick (*Ambylyomma americanum*), which transmits erhlichiosis in the Southeast and East of the U.S.; Rocky Mountain wood tick (*Dermacntor andersoni*), which transmits Rocky Mountain spotted fever, Colorado tick fever and tularemia in Rocky Mountain states; and Western Blacklegged tick (*Ixodes pacificus*), which transmits Anaplasmosis and Lyme disease along the U.S. Pacific Coast [131].

> Different human-biting tick species are distributed around the U.S., each transmitting a specific tick-borne disease.

Illustrated in Figure 1.5 is a long-horned tick, a species of tick native to Asia, but spreading across the U.S. The CDC notes the following about the invading tick: Not normally found in the Western Hemisphere, these ticks were reported for the first time in the U.S. in 2017. Asian long-horned ticks have been found on pets, livestock,

Environment-Related Infectious Diseases

FIGURE 1.5 What you need to know about Asian longhorned ticks – a new tick in the United States (CDC, 2021, https://www.cdc.gov/ticks/longhorned-tick/index.html.)

wildlife, and people. In other countries, bites from these ticks can make people and animals seriously ill. As of June 24, 2019, long-horned ticks have been found in Arkansas, Connecticut, Kentucky, Maryland, North Carolina, New Jersey, New York, Pennsylvania, Tennessee, Virginia, and West Virginia [132].

Regarding tick bites, Lyme disease merits further comment, given that it is the most common vector-borne illness in the U.S. [133]. Lyme disease is caused by the Borrelia burgdorferi bacterium, which is spread by infected ticks. Sufferers can have a fever, headache, chills, fatigue, joint and muscle ache, and swollen lymph nodes. In many cases, patients experience a skin rash that can grow up to 12 inches around the area of the tick bite. Around 329,000 cases are thought to occur each year in the U.S., although the true number of infections is thought to be higher. In most cases, antibiotics are prescribed and the symptoms pass. However, for others, the symptoms can linger in what is known as post-treatment Lyme disease (PTLD) or post-treatment Lyme disease syndrome. The condition is characterized by cognitive dysfunction, incapacitating fatigue, and chronic pain.

To estimate the prevalence of PTLD, researchers drew on several pools of data to create a model to estimate the number of PTLD cases that occurred between 2016 and 2020. A conservative estimate made by the researchers suggested that 69,011 cases of PTLD occurred in 2016, which would rise to 81,509 cases in 2020. In a different model, the same researchers accounted for cases rising from the 1980s, combined with treatment failing in 20% of cases. The result was 1,523,869 cases in 2016 and 1,944,189 forecast for 2020. The researchers acknowledged the considerable uncertainty in their findings but cautioned that further research on the treatment of PTLD was needed [133].

> Lyme disease is the most common vector-borne illness in the U.S. Around 329,000 cases are thought to occur each year in the U.S.

Most tick species feed on feral animals including birds, rodents, deer, and lizards, but ticks are also found on domesticated animals such as dogs. The variety of tick species and tick-borne pathogens in the U.S. combined with human encroachment in wilderness areas and close proximity to domesticated animals have made tick-borne illnesses a serious, difficult to control problem. The CDC's Division of Vector-Borne Diseases (DVBD) leads a federal effort to prevent and control tick-borne diseases. Three of the four DVBD branches, Arboviral Diseases Branch, Bacterial Diseases Branch, and Rickettsial Zoonoses Branch focus on research that targets the control of tick-borne diseases at multiple levels, from animal health and prevention to vaccination. In addition, NIH's National Institute of Allergy and Infectious Diseases has a tick-borne disease-specific research program that focuses on basic biological research, vaccine development, and diagnosis, treatment, and prevention strategies.

The wide range and variety of ticks, the number of tick-specific diseases, and the different hosts from which ticks feed make tick-borne diseases impossible to eradicate. Control and prevention strategies target personal protection to avoid coming in contact with ticks and using repellents containing 20%–30% DEET (N,N-diethyl-m-toluamide) when going into wooded areas [23,134].

1.5.3 Rodent- and Bat-Borne Diseases

Worldwide, rats and mice spread more than 35 diseases. These diseases can be spread to humans directly, through handling of rodents, through contact with rodent feces, urine, or saliva, or through rodent bites. Diseases carried by rodents can also be spread to humans indirectly, through ticks, mites or fleas that have fed on an infected rodent [135]. According to the CDC, the following diseases can be directly transmitted by rats to humans: Hantavirus Pulmonary Syndrome, Hemorrhagic Fever with Renal Syndrome, Lassa Fever, Leptospirosis, Lymphocytic Chorio-meningitis (LCM), Omsk Hemorrhagic Fever, Plague, Rat-Bite Fever, Salmonellosis,

South American Arenaviruses (Argentine hemorrhagic fever, Bolivian hemorrhagic fever, Sabiá-associated hemorrhagic fever, Venezuelan hemorrhagic fever), and Tularemia. Of this list, two diseases are transmitted by rodents as detailed in the following.

Hanta Pulmonary Syndrome and Bubonic Plague: Although Hantavirus Pulmonary Syndrome (HPS) is caused by a virus (Hantavirus Family *Bunyaviridae*) and the Bubonic plague by a bacterium (*Yersinia pestis*), both infectious diseases can be transmitted by rodents. Concerning the former, rodents in the U.S. that can carry the HPS virus include the cotton rat, rice rat, white-footed mouse, and deer mouse. As of January 2017, 728 cases of HPS have been reported to the CDC since surveillance in the U.S. began in 1993. These are all laboratory-confirmed cases and include hantavirus pulmonary syndrome (HPS) and nonpulmonary hantavirus infection [136]. Four to 10 days after the initial phase of illness, the late symptoms of HPS appear. These include coughing and shortness of breath, with the sensation of, as one survivor put it, a "…tight band around my chest and a pillow over my face" as the lungs fill with fluid. HPS can be fatal. It has a mortality rate of 38%.

> Worldwide, rats and mice spread more than 35 diseases. These diseases can be spread to humans directly, through handling of rodents, contact with rodent feces, urine, or saliva, or through rodent bites.

The Bubonic Plague, also known as The Plague or Black Death, arrived in Europe from Asia to Europe during the 14th century by ships infested with infected rats. *Yersinia pestis* can be transmitted to humans by infected flea or rat bite, or pneumonically from person to person. Due to its high case-fatality rate of 30%–60%, The Plague killed approximately 50 million people in Europe within 4 years of becoming endemic, reducing the population by 60%.

Large epidemics of The Plague occurred throughout the centuries, but the impact of the disease was greatly mitigated in the 19th century due to modern sanitation, public health practices, and medical advances and antibiotics have reduced mortality to 11% [137]. Nowadays, small outbreaks or single cases of infections of the Bubonic Plague are still reported worldwide. In the U.S., 1,006 human Plague cases occurred between 1900 and 2012. Most of the Bubonic Plague cases occur in rural or semirural areas in the southwestern states of Arizona, Colorado, and New Mexico. Transmission occurs mostly through flea or rodent (including squirrels, chipmunks, or rats) bites and is rarely due to person-to-person contact [138]. All cases of Bubonic Plague must be reported to the CDC and the WHO.

Turning from rodents to another species, bats can also be vectors of disease. A 2013 study found that when it comes to carrying viruses that can jump to other species – so-called "zoonotic" viruses – bats may be in a class of their own [139]. **Bats are reservoirs for more than 60 viruses that can infect humans, and host more viruses per species than even rodents.** The flying mammals are reservoirs for more than 60 viruses that can infect humans, and host more viruses per species than even rodents do, according to the study. In recent years, bats have received a lot of attention for their virus-hosting abilities. They have been shown to carry a number of harmful infections, including rabies and viruses related to SARS (severe acute respiratory syndrome). Moreover, research suggests bats might be the original hosts of nasty viruses such as Ebola and Nipah, which cause deadly brain fevers in people.

An outbreak of Nipah infection occurred in India in 2018 [140]. Nipah was discovered in 1998, when it sickened nearly 300 people and killed 100 in Malaysia (its name was taken from one of the villages where it first struck). Many of the victims had been farmers who contracted the virus through close contact with their pigs, which led to the euthanization of millions of pigs. But it turns out the pigs actually got the virus from another animal: It is now known that fruit bats belonging to the genus Pteropus (otherwise called flying foxes) are the native carriers of Nipah. As of June 1, 2018, an outbreak of the Nipah virus infected at least 18 people and killed 17 in Kerala, India, the WHO reported. The outbreak, which is the first to strike South India, raised fears of the disease becoming more far-reaching.

1.5.4 PARASITIC DISEASES CAUSED BY WORMS

Humans and other animals have long been afflicted with various kinds of worms. While to some the thought of warms growing inside one's body and living there as parasites might seem grotesque, it is nonetheless a fact that worm infections are a major public health problem domestically and globally. To be described are some of the more common worm infections, together with a story of a pending remarkable global success, the eradication of a human parasitic disease that used to afflict millions of people.

1.5.4.1 Helminth Infections

According to the WHO, soil-transmitted helminth[3] infections are among the most common infections worldwide and affect the poorest and most deprived communities [141]. They are transmitted by eggs present in human feces that in turn contaminate soil in areas where sanitation is poor. The main species that infect people are the roundworm (*Ascaris lumbricoides*), the whipworm (*Trichuris trichiura*), and hookworms (*Necator americanus* and *Ancylostoma duodenale*). Adult worms live in the intestine where they produce thousands of eggs each day. In areas that lack adequate sanitation, these eggs contaminate the soil.

The WHO further notes that more than 1.5 billion people, or 24% of the world's population, are infected with soil-transmitted helminth infections worldwide. Infections are widely distributed in tropical and subtropical areas, with the greatest numbers occurring in sub-Saharan Africa, the Americas, China, and East Asia. More than 267 million preschool-age children and more than 568 million school-age children live in areas where these parasites are intensively transmitted, and are in need of treatment and preventive interventions.

Remarkably, a study conducted by the National School of Tropical Medicine at Baylor College of Medicine in conjunction with the Alabama Center for Rural Enterprise found that more than one in three people sampled in a poor area of Alabama tested positive for traces of hookworm, a gastrointestinal parasite that was thought to have been eradicated from the U.S. decades ago [142]. Some 73% of residents included in the Baylor survey reported that they had been exposed to raw sewage washing back into their homes as a result of faulty septic tanks or waste pipes becoming overwhelmed in torrential rains.

> More than 24% of the world's population are infected with soil-transmitted helminth (worm) infections.

1.5.4.2 Guinea Worm Disease

This parasitic disease has been a human scourge for ages, but its eradication from humankind experience is imminent. As a public health success, there is cause for celebration of the achievement. Eradication of this disease will likely become the second human disease eradicated by public health efforts, with the elimination of smallpox being the first success.

Dracunculiasis, also known as Guinea worm disease (GWD), is an infection caused by the parasite *Dracunculus medinensis*. GWD affects poor communities in remote parts of Africa that do not have safe water to drink. People become infected with Guinea worm by drinking water from ponds and other stagnant water containing tiny "water fleas" that carry the Guinea worm larvae. The larvae are eaten by the water fleas that live in these water sources. Once drunk, the larvae are released from copepods in the stomach and penetrate the digestive track, passing into the body cavity. During the next 10–14 months, the female larvae grow into full-size adults. These adults are 60–100 centimeters (2–3 feet) long and as wide as a cooked spaghetti noodle [143].

---

[3] Helminth is a general term meaning worm.

Environment-Related Infectious Diseases

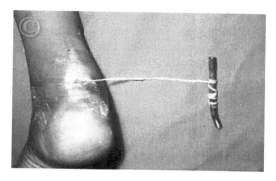

FIGURE 1.6 Guinea worm being removed by winding it around a stick. (Carter Center [143].)

When the adult female worm is ready to exit a victim's body, it creates a blister on the skin anywhere on the body, but usually on the legs and feet. This blister causes a very painful burning feeling and it bursts within 24–72 hours. Immersing

> Eradication of Guinea worm disease will likely become the second human disease eradicated by public health efforts.

the affected body part into water helps relieve the pain. It also causes the Guinea worm to come out of the wound and release a milky white liquid into the water that contains millions of immature larvae. This contaminates the water supply and starts the cycle over again. For several days, the female worm can release more larvae whenever it comes in contact with water [143]. Illustrated in Figure 1.6 is the removal, using a stick, of a female guinea worm from the leg of a victim.

The global eradication of Guinea worm disease has been led by the Carter Center, based in Atlanta, Georgia. When The Carter Center began leading the international campaign to eradicate Guinea worm disease in 1986, there were an estimated 3.5 million cases in at least 21 countries in Africa and Asia. Today, that number has been reduced by more than 99.99%. In 2018, 28 human cases of Guinea worm disease were reported worldwide. During 2018, 17 human cases of Guinea worm disease were reported in Chad, 10 in South Sudan, and one in Angola. Guinea worm disease also afflicts domestic and feral animals. The affliction of domestic and feral animals with GWD led the WHO in 2019 to push back the target date for stamping out the disease from 2020 to 2030 [144]. The agency explained that a series of puzzling discoveries had made it impossible to meet the 2020 target. The most urgent issue was the soaring, unexplained, rate of infections among dogs in Chad – which has helped to keep Guinea worm circulating in the environment. Then there is the emergence of the first known cases among people in Angola, perplexing infections in baboons in Ethiopia, and conflicts that have hampered eradication efforts in parts of Mali, Sudan and South Sudan. Notwithstanding these setbacks, WHO reaffirmed its commitment to GWD eradication.

There is no drug to treat Guinea worm disease and no vaccine to prevent infection. Near eradication of the disease has

> In 1986, there were an estimated 3.5 million cases of GWD in at least 21 countries in Africa and Asia. In 2018, 28 human cases were reported worldwide.

been achieved by education of villagers at risk of contact with water containing the Guinea worm larvae. The disease can be prevented by avoiding drinking unsafe water. Villagers are advised to drink only water from protected sources (such as from boreholes or hand-dug wells) that are free from contamination, treat unsafe drinking water sources with an approved larvicide, and to use a cloth filter or a pipe filter, to remove the tiny "water fleas" that carry the Guinea worm larvae [143].

1.5.5 THE KISSING BUG

Triatomine bugs (also called "kissing bugs," cone-nosed bugs, and bloodsuckers) can live indoors, in cracks and holes of substandard housing, or in a variety of outdoor settings. They are typically found in the southern U.S., Mexico, Central America, and South America (as far south as southern Argentina). According to the CDC, eleven different species of triatomine bugs have been found in the southern U.S. [145]. Triatomines are mostly active at night and feed on the blood of mammals (including humans), birds, and reptiles. They seem to have a predilection to bite a human's face, hence the name "kissing bug." Triatomine bugs live in a wide range of environmental settings, generally within close proximity to an animal the bug can feed on, called a blood host. The bugs nest in cracks and holes of substandard housing.

WHO estimates that about 6 million to 7 million people worldwide, mostly in Latin America, are estimated to be infected with the parasite that causes Chagas disease.

Triatomines can transmit Chagas disease (American trypanosomiasis). The insect vector is a triatomine bug that carries *Trypanosoma cruzi*, the parasite that causes Chagas disease. The WHO estimates that about 6–7 million people worldwide, mostly in Latin America, are estimated to be infected with the parasite [146]. *Trypanosoma cruzi* infection is curable if treatment is initiated soon after infection, but up to 30% of chronically infected people develop cardiac alterations and up to 10% develop digestive, neurological, or mixed alterations that may require specific treatment.

The CDC estimates that more than 300,000 persons with *Trypanosoma cruzi* infection reside in the U.S. [145]. Most people with Chagas disease in the U.S. were infected in the parts of Latin America where Chagas disease is endemic Although there are triatomine bugs in the U.S., only a few cases of Chagas disease from contact with the bugs have been documented in the U.S., but warmer global temperatures due to climate change have resulted in more triatomine bugs migrating northward into North America, thereby increasing the potential for increased incidence of Chagas disease in the U.S.

1.5.6 SUPERBUGS

Superbug is a term coined by the news media to describe bacteria that cannot be killed using multiple antibiotics. [It is interesting that news media sometimes refer to a large event or outcome by coining the prefix "super." As a consequence, Superbowl and Superfund (Chapter 2) were coined and added to the American-English lexicon.] Rather than using the term superbug, medical doctors often use phrases like "multidrug-resistant bacteria." That is because a superbug is not necessarily resistant to all antibiotics. It refers to bacteria that cannot be treated using two or more antibiotics [147]. But by whatever nomenclature used, the CDC and the WHO state that antibiotic resistance is one of the biggest public health challenges of our time. Each year in the U.S., at least 2 million people get an antibiotic-resistant infection, and at least 23,000 people die. Fighting this threat is a public health priority [148].

In 1928, Alexander Fleming discovered in a moldy petri dish a substance (penicillin) that killed the bacteria he was examining. Soon, penicillin became a wonder drug of its time, which could cure patients with bacterial infections. Antibiotics work by destroying the cell walls of disease-causing bacteria, disrupting their repair mechanisms, or preventing cell multiplication. The era of medical treatment using antibiotics had begun, saving millions of human lives. But bacteria and other pathogens have always evolved so that they can resist the new drugs that medicine has used to combat them. Resistance has increasingly become a problem in recent years because the pace at which we are discovering novel antibiotics has slowed drastically, while antibiotic use is rising.

And it is not just a problem confined to bacteria, but all microbes that have the potential to mutate and render our drugs ineffective. The damaging effects of AMR are already manifesting themselves across the world [148]. For instance, ETH Zurich researchers have shown that antimicrobial-resistant infections are rapidly increasing in animals in low- and middle-income countries. Antibiotics are increasingly used for purpose to increase animal production in order to meet greater food demand in these countries. According to this study, the regions associated with high rates of AMR in animals are northeast China, northeast India, southern Brazil, Iran, and Turkey [149].

> The CDC and the WHO state that antibiotic resistance is one of the biggest public health challenges of our time.

Two government organizations, the Organization for Economic Cooperation and Development (OECD) and a UN group have made separate estimates of human deaths attributable to AMR by the year 2050. While the two estimates differ in number, both estimates reinforce the need for much more careful management of the use of antibiotics in medical and veterinary practices. One report, released by the U.N. Interagency Coordination Group on AMR in 2019 states that the threat of AMR is already deadly, with 700,000 people dying each year as a result of drug-resistant diseases. Fast forward to 2050, the report adds, and AMR could cause as many as 10 million deaths each year under a worst-case scenario [150]. Further, the report asserts that if no action is taken, drug-resistant diseases could cause damage to the economy as catastrophic as the 2008–2009 global financial crisis. A second report, prepared by the OECD, said 2.4 million people could die from superbugs by 2050 annually and estimated that the cost of treating such infections would swell to an average of US$3.5 billion a year in each country included in its analysis [151].

In 2019, CDC updated its 2013 *AMR Threats Report*, which sounded the alarm to the danger of antibiotic resistance [152]. According to the 2019 report, more than 2.8 million antibiotic-resistant infections occur in the U.S. each year, and more than 35,000 people die as a result. In addition, 223,900 cases of *Clostridioides difficile* occurred in 2017 and at least 12,800 people died. The report lists 18 antibiotic-resistant bacteria and fungi into three categories based on level of concern to human health – urgent, serious, and concerning. The five bacteria or fungus comprising CDC's urgent category are listed in Table 1.1. The other 13 bacteria or fungus of concern are available in the source reference [152].

The global seriousness of AMR has been described by the WHO as a "global threat" and in 2019 launched a global campaign urging governments to adopt a tool to reduce the spread of AMR, adverse events, and costs [153]. The AWaRe tool was developed by the WHO Essential Medicines List to contain rising resistance and make antibiotic use safer and more effective. It classifies antibiotics into three groups – Access, Watch, and Reserve – and specifies which antibiotics to use for the most common and serious infections, which ones should be available at all times in the healthcare system, and those that must be used sparingly or preserved and used only as a last resort. The new campaign aims to increase the proportion of global consumption of antibiotics in the Access group to at least 60% and to reduce the use of the antibiotics most at risk of resistance from the Watch and Reserve groups.

According to CDC, more than 2.8 million antibiotic-resistant infections occur in the U.S. each year, and more than 35,000 people die as a result. WHO estimated that some 700,000 people die annually from the same cause.

In 2020 the WHO announced that some 700,000 people die each year because medicines that once cured their conditions are no longer effective [154]. The agency further commented that without government intervention, resistant infections could kill 10 million people annually by 2050 and prompt an economic slowdown to rival the global financial crisis of 2008. Government intervention was cited as a need to support the pharmaceutical industry, which had found developing new antibiotics to be a nonprofitable endeavor.

Perspective: The global loss of 700,000 people who die prematurely because of ineffective antibiotics represents a loss to humankind and a challenge to public health. While the evolution of 'superbugs' presents a grim picture for disease prevention and treatment, there is emerging reason for optimism. At the core of this reason is the historical record of how humankind has responded to similar health and medical challenges. The record includes the discovery of penicillin, the intuitive application of what became known as vaccinations, and the observation that some chemical pollutants in air, water, and food can be deleterious to health. In each of these examples, a health problem prompted the search for a solution.

As previously described, a current problem is the evolution of bacteria that can resist the assault of antibiotics. Solutions to this problem are actively being sought. One promising area is the **basic research** being used to determine how (emphasis added) bacteria resist assaults to their existence. An example of this kind of basic research, conducted at McMaster University, concerns the discovery of a new group of antibiotics with a unique approach to attacking bacteria [155]. The newly found corbomycin and the lesser-known complestatin have a unique way to kill bacteria, which is achieved by blocking the function of the bacterial cell wall. Antibiotics like penicillin kill bacteria by preventing building of the wall, but these new antibiotics

TABLE 1.1
Urgent Threat Bacteria and Fungi Listed in CDC's 2019 AR Threats Report [152]

Threat	Type	About	Hospitalization/Clinical Cases	Deaths
Carbapenem-resistant *Acinetobacter*	Bacteria	Can cause pneumonia and wound, bloodstream, and urinary tract infections	8,500 in 2017	700
Candida auris	Fungus	Can cause severe infections and spreads easily between hospitalized patients and nursing home residents	323 Clinical cases in 2018	Unknown
Clostridioides difficile	Bacteria	Can cause life-threatening diarrhea and colitis, mostly in people who have had both recent medical care and antibiotics	223,900 Infections per year	12,800
Carbapenem-resistant Enterobacteriaceae	Bacteria	Can cause infections in almost any body part, including bloodstream infections, ventilator-associated pneumonia, and intra-abdominal abscesses	13,100 Estimated cases in hospitalized patients in 2017	1,100 estimated deaths in 2017
Drug-resistant *Neisseria gonorrhoeae*	Bacteria	Can cause gonorrhea, which can result in life-threatening ectopic pregnancy and infertility, and can increase the risk of getting and giving HIV	550,000 Estimated drug-resistant infections per year	Unknown

work by doing the opposite – they prevent the wall from being breached by the cell, which is essential for cells to divide. Cells that cannot divide eventually die.

As a second example of research targeting bacterial drug resistance, MIT researchers used a machine-learning algorithm to identify in 2020 a drug they named halicin that kills many strains of bacteria [156]. In laboratory tests, the drug killed many of the world's most problematic disease-causing bacteria, including some strains that are resistant to all known antibiotics. It also cleared infections in two different mouse models. Clinical applications of halicin are planned.

1.5.7 Research to Counter Superbugs

It is not unreasonable to assert that solutions to the public health challenge of superbugs are vital for human well-being globally. While antibiotics have quite literally saved millions of human lives, commencing with the use of penicillin to treat wounded combatants in World War II, the success of antibiotics has led to their ineffectiveness in some cases. Quite simply, antibiotics have too often been applied in medical situations that did not require them. As a consequence, the targeted bacteria and fungus evolved mechanisms to ward off the antibiotics. In response, research is underway in many academic and government laboratories to find solutions. Some new research approaches are illustrated herein.

1.5.7.1 New Approaches to Antibiotics

In one new approach to overcoming AMR, a new molecule was developed by a biotech company that can kill deadly strains of common bacteria, such as *Escherichia coli* and *Klebsiella pneumonia*, that are resistant to most existing antibiotics [157]. The newly developed drug uses a different tactic. It inhibits a key enzyme in the cell membrane that helps the bacteria secrete proteins. That means that strategies that bacteria use to evade existing antibiotics will not work here, giving the molecule an edge. When the enzyme is blocked, proteins build up in the cell membrane until the membrane bursts, ultimately killing the cell. The new drug is being tested with mice and awaits clinical trials with human patients experiencing AMR.

A second example of innovative research focused on overcoming AMR was reported by University of Lincoln researchers [158]. Researchers there successfully created a simplified, synthesized form of teixobactin – a natural antibiotic discovered by U.S. scientists in soil samples in 2015 – which has been used to treat a bacterial infection in mice, demonstrating the first proof that such simplified versions could be used to treat real bacterial infection as the basis of a new drug. The team at Lincoln developed a library of synthetic versions of teixobactin by replacing key amino acids at specific points in the antibiotic's structure to make it easier to recreate. After these simplified synthetic versions were shown to be highly potent against superbug-causing bacteria in test tube experiments, researchers from the Singapore Eye Research Institute then used one of the synthetic versions to successfully treat a bacterial infection in mice. As well as clearing the infection, the synthesized teixobactin also minimized the infection's severity. Further clinical evaluations are scheduled.

Environment-Related Infectious Diseases

1.5.7.2 Bacteriophage Therapy

A different approach to overcoming drug-resistant superbugs is the practice of phage therapy, which uses bacterial viruses (phages) to treat bacterial infections, an approach that has been around for almost a century [159]. As noted in a review of phage therapy, the authors commented, "The universal decline in the effectiveness of antibiotics has generated renewed interest in revisiting this practice. Conventionally, phage therapy relies on the use of naturally occurring phages to infect and lyse bacteria at the site of infection. Biotechnological advances have further expanded the repertoire of potential phage therapeutics to include novel strategies using bioengineered phages and purified phage lytic proteins. Current research on the use of phages and their lytic proteins, specifically against multidrug-resistant bacterial infections, suggests phage therapy has the potential to be used as either an alternative or a supplement to antibiotic treatments. Antibacterial therapies, whether phage- or antibiotic-based, each has relative advantages and disadvantages; accordingly, many considerations must be taken into account when designing novel therapeutic approaches for preventing and treating bacterial infections. Although much is still unknown about the interactions between phage, bacteria, and human host, the time to take phage therapy seriously seems to be rapidly approaching" [159].

> Phage therapy uses bacterial viruses (phages) to treat bacterial infections.

1.5.7.3 Basic Research on Microorganism Mechanisms

The evolution of superbugs has generated a global program of research on measures to prevent their development to drug and other medical treatment regimens. Advances in bacteriology, virology, and imaging methods have made basic research findings of relevance to mitigating the development of superbugs. One example of a basic research finding follows.

E. coli naturally lives in our colons, and most strains do us no harm. But there are several strains that can cause cramps, diarrhea, vomiting, even kidney failure, and death. Researchers at the University of Virginia School of Medicine scientists revealed how *E. coli* seeks out the most oxygen-free crevices of a person's colon to cause the worst infection possible. The discovery could one day let medical providers prevent the infection by allowing *E. coli* to pass harmlessly through the body. Research revealed that *E. coli*'s vital asset is a small form of RNA that activates particular genes when oxygen levels are low enough. It is at this point that the infection really gets established. Thanks to this natural sensing process, the bacteria are able to establish infection and begin to manufacture harmful Shiga toxins. "If scientists can figure how to block oxygen sensing, we may be able to prevent *E. coli* from making proteins that allow it to stick to our guts," stated the researchers [160].

As an example of basic research holding promise to overcome AMR, a Princeton university team developed what some have termed a "poisoned arrow" to defeat antibiotic-resistant bacteria [160a]. The researchers reported in

> A Princeton university team developed what some have termed a "poisoned arrow" to defeat antibiotic-resistant bacteria.

June 2020 that they had found a compound, SCH-79797, that can simultaneously puncture bacterial walls and destroy folate within their cells – while being immune to antibiotic resistance. The researchers even got a sample of the most resistant strain of N. gonorrhoeae from the vaults of the World Health Organization – a strain that is resistant to every known antibiotic – and found that their compound killed the bacteria. Further tests of the new compound are underway.

Perspective: The public health seriousness of AMR cannot be overstated. Bacterial resistance to antibiotics will result in millions of people prematurely dying from various infectious diseases and wounds. But there is hope. Innovative research on how to overcome bacteria's ability to evolve ways to resist antibiotics is occurring in research laboratories in Europe, the U.S., China, and elsewhere. Understanding the mechanisms in how bacteria develop resistance will be vital to the development of antibiotics that can overcome these mechanisms. Further, other research is seeking solutions to the application of bacteriophages, viruses that target specific bacteria, in medical practice.

1.5.8 Nosocomial Infections

A healthcare resource defines a nosocomial infection as one contracted because of an infection or toxin that exists in a certain location, such as a hospital [161]. The term "nosocomial infections" is often used interchangeably with the terms healthcare-associated infections (HAIs) and hospital-acquired infections. For a HAI, the infection must not be present before someone has been under medical care. About 1 in 10 of the people admitted to a hospital will contract a HAI. This section will describe two nosocomial infections that present challenges in terms of drug-resistance medication.

1.5.8.1 *Candida auris* Infection

Candida auris is an emerging fungus that presents a serious global health threat. It has caused severe illness in hospitalized patients in several countries, including the U.S. *C. auris*, and has caused bloodstream infections, wound infections, and ear infections. It also has been isolated from respiratory and urine specimens, but it is unclear if it causes infections in the lung or bladder. Patients can remain colonized with *C. auris* for a long time and *C. auris* can persist on surfaces in healthcare environments. This can result in spread of *C. auris* between patients in healthcare facilities. The CDC is concerned about *C. auris* for three main reasons [162]:

> A nosocomial infection as one contracted because of an infection or toxin that exists in a certain location, such as a hospital.

- It is often multidrug-resistant, meaning that it is resistant to multiple antifungal drugs commonly used to treat Candida infections. As with some bacteria, this fungus has evolved a resistance to antifungal drugs.
- It is difficult to identify with standard laboratory methods, and it can be misidentified in labs without specific technology. Misidentification may lead to inappropriate management.

- It has caused outbreaks in healthcare settings, primarily in hospitals and nursing homes. For this reason, it is important to quickly identify *C. auris* in a hospitalized patient so that healthcare facilities can take special precautions to stop its spread.

C. auris was first identified in 2009 in Japan. A retrospective review of Candida strain collections found that the earliest known strain of *C. auris* dates to 1996 in South Korea. The CDC considers *C. auris* an emerging pathogen because increasing numbers of infections have been identified in multiple countries since it was recognized [162]. *C. auris* infections have been reported from more than 30 countries, including the U.S., where in 2019 it was reported in 14 U.S. states [163]. The incidence of infections is uncertain due to the difficult laboratory procedures required for the identification of *C. auris*.

1.5.8.2 *Klebsiella pneumoniae*

Klebsiella pneumoniae is a bacterium that can live completely naturally in the intestines without causing problems for healthy people. However, when the body is unwell, it can infect the lungs to cause pneumonia, and infect the blood, cuts in the skin, and the lining of the brain to cause meningitis [164]. Antibiotic resistance has led to a six-fold increase in deaths. Deaths from carbapenem-resistant *K. pneumoniae* have increased from 341 in Europe in 2007 to 2,094 by 2015. A study of 244 hospitals in Europe led to findings that imply hospitals are the key facilitator of transmission and suggest that the bacteria are spreading from person-to-person primarily within hospitals. *K. pneumoniae* is listed by the WHO as a critical priority antibiotic-resistant bacterial pathogen for which new antibiotics are urgently needed.

> *Candida auris* is an emerging drug-resistant fungus that presents a serious global health threat.

1.5.9 NECROTIZING FASCIITIS (FLESH-EATING BACTERIA)

A bacterial infection that always catches the attention of news media, with ensuring resonance with departments of public health, is necrotizing fasciitis, a rare infection that is often described in media reports as a condition involving "flesh-eating bacteria." It can be fatal if not treated promptly. Necrotizing fasciitis spreads quickly and aggressively in an infected person. It causes tissue death at the infection site and beyond. Every year, between 600 and 700 cases are diagnosed in the U.S. About 25%–30% of those cases result in death. It rarely occurs in children. Necrotizing fasciitis is commonly caused by group A Streptococcus bacteria. That is the same type of bacteria that causes strep throat. However, several types of bacteria, such as *Staphylococcus* and others, have also been associated with the disease [165].

The bacteria that cause necrotizing fasciitis can enter the body following surgery or injury. They can also enter the body through minor cuts, insect bites,

> Every year, between 600 and 700 cases of necrotizing fasciitis are diagnosed in the U.S. About 25%–30% of those cases result in death.

abrasions. In some cases, it is unknown how the infection began. Once it takes hold, the infection rapidly destroys muscle, skin, and fat tissue. Treatment includes intravenous antibiotic therapy, surgery to remove damaged or dead tissue in order to stop the spread of infection, and medications to raise blood pressure. Amputations of affected limbs occur in some cases [165].

1.5.10 SWIMMING POOL BACTERIA

As ambient air temperatures rise due to climate change and other factors, many persons, especially young children, seek relief in public and private swimming pools. While this summertime activity can be a source of enjoyment and a respite of comfort, public health officials have voiced their concern about the presence of pool bacteria that infect some pools. In 2019 the CDC reported an increase in the incidence of swimmers' diarrhea linked to pool bacteria [166]. CDC notes that diarrheal illnesses are caused by germs such as Crypto (short for Cryptosporidium), Giardia, Shigella, norovirus, and *E. coli* O157:H7. These germs can live from minutes to days in pools, even if the pool is well-maintained. Some germs are very tolerant to chlorine and were not known to cause human disease until recently. At public swimming facilities, continuous filtration and disinfection of water should reduce the risk of spreading germs. However, swimmers might still be exposed to germs during the time it takes for chlorine to kill germs (certain germs take longer to kill than others) or for water to be recycled through filters.

Cryptosporidium is a bacteria that causes diarrhea for weeks, and outbreaks increased at an average rate of 13% a year from 2009 to 2017, according to the CDC. *Cryptosporidium* is spread through the feces of infected humans or animals. People can get sick after they swallow the parasite in contaminated water or food or after contact with infected people or animals. From 2009 through 2017, the most recent available data, there were 444 Crypto outbreaks in the U.S., with almost 7,500 people reported sick, 287 people hospitalized and one who died, according to the CDC. Of note, 35% of the outbreaks were linked to treated swimming water in places like pools and water playgrounds because the parasite can survive for days in chlorinated water in pools and water playgrounds or on surfaces disinfected with chlorine bleach, Medications, such as nitazoxanide (Alinia), can help alleviate diarrhea by attacking the metabolic processes of the cryptosporidium organisms.

1.5.11 LEGIONNAIRES DISEASE

Another water-borne disease is Legionnaires disease, a severe, often lethal, form of pneumonia. The disease is caused by the bacterium *Legionella pneumophila* found in both potable and nonpotable water systems. Scientists named the bacterium after an outbreak in Philadelphia in 1976. During that outbreak, many people who went to an American Legion convention contracted pneumonia. Each year, an estimated 10,000–18,000 people are infected with the Legionella bacteria in the U.S. [167]. About one in 10 will not survive. People can get Legionnaires disease or Pontiac fever when they breathe in small droplets of water in the air that contain Legionella. In general, people do not spread Legionnaires' disease to other people.

Environment-Related Infectious Diseases

Legionella occurs naturally in freshwater environments, like lakes and streams. It can become a health concern when it grows and spreads in human-made building water systems such as the air conditioning system in the Philadelphia hotel where the disease was first identified.

> Each year, an estimated 10,000–18,000 people are infected with the Legionella bacteria in the U.S. About one in 10 people will not survive.

Keeping Legionella out of water systems in buildings is key to preventing infection.

1.5.12 NAEGLERIA FOWLERI, THE "BRAIN-EATING AMEBA"

Another environment-related waterborne infection is caused by *Naegleria fowleri*. *N. fowleri* dwells in warm bodies of freshwater where it dines on bacteria in the sediment. As such, most infections with this amoeba in the U.S. have occurred in southern states, especially Texas and Florida, during the summer. When the sediment of a lake is disrupted, amoeba gets stirred into the water. Swimmers can then inhale the parasite through their nose. The *Naegleria fowleri* ameba then travels up the nose to the brain where it destroys brain tissue due to the absence of its normal diet of bacteria [168].

The CDC records, "*Naegleria fowleri* is a free-living ameba that causes primary amebic meningoencephalitis (PAM), a disease of the central nervous system. PAM is a rare disease that is almost always fatal. In the U.S there have been 145 PAM infections from 1962 through 2018 with only four survivors. These infections have primarily occurred in 15 southern-tier states, with more than half of all infections occurring in Texas and Florida. PAM also disproportionately affects males and children. The reason for this distribution pattern is unclear but might reflect the types of water activities (such as diving or watersports) that might be more common among young boys" [169].

1.5.13 IMPLICATIONS FOR ENVIRONMENTAL HEALTH VECTOR CONTROL POLICIES

The first line of defense against vector-borne diseases is personal protection. In the developing world, human encroachment into uninhabited areas and deforestation for economic reasons have exposed people to new diseases as previously discussed in this chapter. However, in the U.S., encroachment into vector-inhabited areas happens mostly for recreational purposes. Prevention efforts have included area-wide pesticide spraying and targeting the mammalian vector reservoir through vaccination and inoculation [170].

Multiple factors have accelerated the spread of these diseases. These include potential ecological changes due to climate change, which could expand the geographical distribution of the reservoir and the vector, closer proximity between wildlife and human populations, and human behavior for social or recreational purposes. The complex dynamic between the reservoir, the vector, and the host requires an integrated pest management approach to preventing tick-borne diseases [171]. This approach includes monitoring and controlling vector and reservoir populations and public education on personal protection and prevention.

1.6 TROPICAL INFECTIOUS DISEASES

Tropical diseases encompass all diseases that occur solely, or principally, in the tropics. In practice, the term is often taken to refer to infectious diseases that thrive in hot, humid conditions, such as malaria, leishmaniasis, schistosomiasis, onchocerciasis, lymphatic filariasis, Chagas disease, African trypanosomiasis, and dengue. Several of these infectious diseases have been already been discussed in this chapter. It is beyond the scope of this chapter to discuss other tropical infectious diseases, but additional details can be found in WHO publications. However, it merits mentioning that tropical disease outbreaks are a growing threat in Europe as temperatures rise due to climate change dynamics. The summer of 2018 saw a sharp spike in West Nile virus infections in Europe. Countries affected include Italy, Greece, Hungary, Serbia, and Romania. Elevated temperatures, followed by wet weather, resulted in favorable conditions for mosquito breeding and subsequent human contact via mosquito bites [172].

1.7 IMPLICATIONS FOR ENVIRONMENTAL POLICY: CLIMATE CHANGE

Vector-borne diseases are the best-studied diseases associated with climate change. As described in Volume 1, Chapter 1, the amounts of CO_2 released into the atmosphere have exponentially increased since the 1800s Industrial Revolution. Since the 1800s, global temperature has increased by 1.7°F, artic ice and land ice have decreased at a rate of 13.3% per decade and 281.0 Giga tons per year, respectively, and sea level is increasing 3.4 millimeters per year [173]. These changes in the planet's environment have led to increased evaporation and precipitation overall, increasing sea levels, and shifting climate patterns, resulting in a more extreme and less predictable climate.

> Tropical disease outbreaks are a growing threat in Europe as temperatures rise due to climate change dynamics. The summer of 2018 saw a sharp spike in West Nile virus infections in Europe.

Warmer temperatures: An increase in global temperatures can drastically affect the geographic distribution of known vectors of disease. Historically, the geographic range of a species is determined by climatologic conditions. Mosquitoes and arthropods thrive in warm and humid weather. As annual average temperatures increase, so has the altitude of freezing points and glacier melting in the tropics. The tropical range is also expanding longitudinally. Thus, mosquitoes can now thrive at higher altitudes over a larger geographic range, exposing non-immune populations to vector-borne diseases. The introduction of mosquito and arthropod-borne viruses into nonendemic regions can trigger large outbreaks and are difficult to control due to the lack of population immunity.

> An increase in global temperatures can drastically affect the geographic distribution of known vectors of disease.

Furthermore, extreme climate change conditions, such as El Nino, which is associated with warmer average temperatures in the tropics, can positively affect the biting rate and reduce the mortality rate of mosquitoes, leading to an increase in the severity of vector-borne disease epidemics once a new virus is introduced. This effect was modeled by Caminade et al., highlighting the amplified effect the 2015 climatological conditions had on the Zika virus outbreak in South America [174].

Changes in precipitation/seasonal weather patterns: Warmer temperatures increase the rate of evaporation and precipitation, triggering more rainfall in tropical regions. Vectors such as *Aedes* and *Anopheles* need standing bodies of water in which to lay eggs. More rainfall, and consequential flooding, creates more mosquito breeding sites and presents a challenge to vector control strategies. Furthermore, the unpredictability of rainfall and longer wet seasons increase the duration with which mosquitoes can thrive within a population. Seasonal variations make it difficult to predict potential outbreaks or periods of active disease transmission, limiting the accuracy of current vector-borne disease prediction models.

Extreme weather events: Similar to changes in weather patterns, extreme weather events, such as droughts, hurricanes, and large floods make predictions and disease transmission models less accurate. As extreme weather events increase in severity and frequency, nonendemic vectors or pathogens are more likely to become endemic.

> As extreme weather events increase in severity and frequency, nonendemic vectors or pathogens are more likely to become endemic.

An important study of the impact of climate change on the spread of mosquito-borne viruses was published in 2019 by a multiacademic team. The team observed that forecasting the impacts of climate change on Aedes-borne viruses – especially dengue, chikungunya, and Zika – is a key component of public health preparedness. The researchers used an empirically parameterized model of viral transmission by the vectors *Aedes aegypti* and *Ae. albopictus* as a function of temperature as the method to predict cumulative monthly global transmission risk in current climates. The risk was then compared with projected risk in 2050 and 2080 based on general circulation models [175]. The results showed that if mosquito range shifts track optimal temperature ranges for transmission (21.3°C–34.0°C for *Ae. aegypti*; 19.9°C–29.4°C for *Ae. albopictus*), poleward shifts in Aedes-borne virus distributions can be expected.

Further, the differing thermal niches of the two vectors produce different patterns of shifts under climate change. More severe climate change scenarios produce larger population exposures to transmission by *Ae. aegypti*, but not by *Ae. albopictus* in the most extreme cases. Climate-driven risk of transmission from both mosquitoes will increase substantially, even in the short term, for most of Europe. In contrast, significant reductions in climate suitability are expected for *Ae. albopictus*, most noticeably in southeast Asia and West Africa. The researchers concluded that in the next century nearly a billion people will be threatened with new exposure to virus transmission by both *Aedes* spp. in the worst-case climate change scenario [175].

1.8 IMPLICATIONS FOR ECOSYSTEM HEALTH

While this chapter has focused on the impact of specific environmental hazards and associated infectious diseases, these same hazards also portent adverse effects on ecosystem health. Although the ecological effects are less well known and researched, the vectors listed in this chapter can adversely affect the health of organisms that reside in ecosystems. For example, some mosquito-borne and tick-borne diseases can be transmitted to domesticated animals and livestock with adverse health and economic consequences. Public health programs of use of pesticides for mosquito and tick control can become contaminants in runoff water that reaches local water supplies and any resident marine life. Rodent infestation of a geographic area can adversely affect infrastructures such as municipal sewers that in turn indirectly supply water to support ecosystems. Additionally, instances of pandemics of infectious disease can reduce the human workforce required to service and maintain local ecosystems, for example, water quality monitoring.

1.9 HAZARD INTERVENTIONS

Interventions to mitigate the environmental hazards that cause infectious diseases were discussed throughout the chapter and can be capsulized here. As implied by the content of Figure 1.1, infectious diseases can be prevented by elimination of causative vectors. This, of course, is more easily stated than accomplished, given the complexities of mitigation of associated environmental vectors, for example, programs of rodent control. What can be stated with reliance is the fundamental essential implementation of traditional public health programs of surveillance, laboratory science, behavioral science, and epidemic investigations, as was illustrated in Figure 1.1. Further, public support of local programs of public health is vital for controlling environment-related infectious diseases.

1.10 SUMMARY

Humans occupy planet Earth with a host of other living organisms and creations of nature. As described in this chapter, some of our partners in life can be a source of infectious disease. Discussed herein are several kinds of infectious diseases and their causes that are linked to environmentally relevant sources. Infectious diseases such as cholera and typhus are serious diseases that contribute to mortality and morbidity in millions of people in tropical environments and in countries with limited public health resources. As summarized in this chapter, vector-borne diseases have a strong link to environmental conditions. Mosquitoes, ticks, rodents, and other pests are significant vectors of infectious diseases that include malaria, yellow fever, dengue, West Nile fever and encephalitis, Ebola outbreak, Zika infection, and others.

Changes in environmental conditions, for example, global and regional temperature increases, can increase the range of domain of vectors such as mosquitoes. Policies on mitigating climate change are therefore highly relevant for protection against the spread of infectious disease vectors. Public health interventions include education of populations at potential health risk, surveillance programs that monitors outbreaks of infectious diseases, personnel trained in vector eradication, available vaccines that immunize persons at risk of infection, and researchers who can develop a body of science about causal factors of an infectious disease outbreak and use the knowledge for treatment regimens. In the U.S. the Public Health Service

Act contains federal policies and authorizations for public health agencies to the purpose of preventing and responding to infectious diseases. On a global scale, WHO administers resources that are directed to infectious disease prevention and control. Policies that support WHO's efforts are vital if highly infectious diseases such as malaria, Ebola, and Zika are to be prevented or contained.

REFERENCES

1. WHO (World Health Organization). 2017. Who we are, what we do. http://www.who.int/about/en/.
2. WHO (World Health Organization). 2017. Strengthening health security by implementing the International Health Regulations (2005). http://www.who.int/ihr/about/en/.
3. NIH (National Institutes of Health). 2016. Emerging and re-emerging infectious diseases. https://science.education.nih.gov/supplements/nih1/diseases/guide/understanding1.html.
4. CDC (Centers for Disease Control and Prevention). 2012. *Principles of epidemiology in public health practice, 3rd edition—An introduction to applied epidemiology and biostatistics.* https://www.cdc.gov/ophss/csels/dsepd/ss1978/lesson1/section8.html.
5. CDC (Centers for Disease Control and Prevention). 2014. Cholera, Vibrio cholerae infection. https://www.cdc.gov/cholera/non-01-0139-infections.html.
6. Saha, A., M. I. Chowdhury, F. Khanam, et al. 2011. Safety and immunogenicity study of a killed bivalent (O1 and O139) whole cell oral cholera vaccine Shanchol in Bangladeshi adults and children as young as 1 year of age. *Vaccine* 29(46):8285–92.
7. CDC (Centers for Disease Control and Prevention). 2015. Information for travelers. https://www.cdc.gov/cholera/travelers.html.
8. Loharikar, A. L., A. E. Newton, S. Stroika, et al. 2015. Cholera in the United States, 2001–2011: A reflection of patterns of global epidemiology and travel. *Epidemiol. Infect.* 143(4):695–703.
9. CDC (Centers for Disease Control and Prevention). 2016. Cholera and other vibrio illness surveillance (COVIS). https://www.cdc.gov/vibrio/surveillance.html.
10. WHO (World Health Organization). 2016. Cholera fact sheet. http://www.who.int/mediacentre/factsheets/fs107/en/.
11. UNICEF (United Nations Children's Fund). 2014. Committing to child survival: A promise renewed. Progress report 2014. http://files.unicef.org/publications/files/APR_2014_web_15Sept14.pdf.
12. WHO (World Health Organization). 2013. Progress on improved water sanitation. http://apps.who.int/iris/bitstream/10665/81245/1/9789241505390_eng.pdf?ua=10.
13. UN (United Nations). 2014. International decade for action "water for life" 2005–2015. http://www.unwater.org/campaigns/water-for-life-decade/en/.
14. Gelting, R., K. Bliss, M. Patrick, et al. 2013. Water, sanitation and hygiene in Haiti: Past, present and future. *AJTMH* 89(4):665–70.
15. Tappero, J. and R. Tauxe. 2011. Lessons learned during public health response to cholera epidemics in Haiti and the Dominican Republic. *Emerg. Infect. Dis.* 17(11):2087.
16. Cravioto, A., C. F. Lanata, D. S. Lantagne, G. B. Nair. 2010. Final report of the independent panel of experts on the cholera outbreak in Haiti. http://www.un.org/News/dh/infocus/haiti/UN-cholera-report-final.pdf.
17. WHO (World Health Organization). 2017. Cholera count reaches 500 000 in Yemen. https://www.who.int/news-room/detail/14-08-2017-cholera-count-reaches-500-000-in-yemen.
18. Akram, S. 2019. How a Yemen water plant helped cut cholera by 92 percent. *OZY*, January 2.

19. CDC (Centers for Disease Control and Prevention). 2017. Scrub typhus. https://www.cdc.gov/typhus/scrub/index.html.
20. Chikeka, I. and J. S. Dumler. 2015. Neglected bacterial zoonoses. *Clin. Microbiol. Infect.* 21(5):404–15.
21. CDC (Centers for Disease Control and Prevention). 2017. Murine typhus. https://www.cdc.gov/typhus/murine/index.html.
22. WHO (World Health Organization). 2017. Typhus fever (Epidemic louse-borne typhus). http://www.who.int/ith/diseases/typhusfever/en/.
23. CDC (Centers for Disease Control and Prevention). 2015. Avoiding ticks. https://www.cdc.gov/ticks/avoid/index.html.
24. Romani, L., A. C. Steer, M. J. Whitfeld, et al. 2015. Prevalence of scabies and impetigo worldwide: A systematic review. *The Lancet* 15(8):960–7.
25. McNeil Jr., D. G. 2019. Scabies means misery. This pill can end it. *The New York Times*, June 26.
26. US DHHS (U.S. Department of Health and Human Services). 2016. America's health responders. https://www.usphs.gov/aboutus/history.aspx.
27. WHO (World Health Organization). 2011. *Guidelines for drinking water quality*, 4th edition. http://apps.who.int/iris/bitstream/10665/44584/1/9789241548151_eng.pdf.
28. GRF (Global Risk Forum). 2012. GRF one health summit 2012. One health – One planet – One future risks and opportunities. http://www.onehealthinitiative.com/publications/OneHealth_3rdannouncement.pdf.
29. One Health Initiative. 2016. About the one health initiative. http://www.onehealthinitiative.com/about.php.
30. CDC (Centers for Disease Control and Prevention). 2016. One health basics. https://www.cdc.gov/onehealth/basics/index.html.
31. USDA (US Department of Agriculture). 2016. One Earth. https://www.usda.gov/topics/animals/one-health.
32. CDC (Centers for Disease Control and Prevention). 2016. Types of influenza viruses. https://www.cdc.gov/flu/about/viruses/types.htm.
33. CDC (Centers for Disease Control and Prevention). 2018. History of 1918 flu pandemic. https://www.cdc.gov/flu/pandemic-resources/1918-commemoration/1918-pandemic-history.htm.
34. Blackburn, C. C., G. W. Parker, and M. Wendelbo. 2018. *How the devastating 1918 flu pandemic helped advance US women's rights*. Washington, DC: United States Foundation for the Commemoration of the World Wars.
35. WHO (World Health Organization). 2010. Pandemic H1N1. http://www.who.int/csr/disease/swineflu/en/.
36. Fineberg, H. 2014. Pandemic preparedness and response – Lessons from the H1N1 influenza of 2009. *NEJM* 370:1335–42.
37. Youssef, M. 2009. Egypt orders slaughter of all pigs over swine flu. https://web.archive.org/web/20090502022247/http://www.independent.co.uk/news/world/middle-east/egypt-orders-slaughter-of-all-pigs-over-swine-flu-1676090.html.
38. CIDRAP (Center for Infectious Disease Research and Policy). 2013. CDC: Pandemic vaccine prevented 1 million cases, 300 deaths. http://www.cidrap.umn.edu/news-perspective/2013/02/cdc-pandemic-vaccine-prevented-1-million-cases-300-deaths.
39. VIDRL (Victorian Infectious Diseases Reference Laboratory). 2017. What we do. https://www.influenzacentre.org.
40. CDC (Centers for Disease Control and Prevention). 2016. Fluview. https://www.cdc.gov/flu/weekly/.
41. USDA (U.S. Department of Agriculture). 2016. Animal and plant health inspection service (APHIS). About APHIS. https://www.aphis.usda.gov/aphis/home/://wwwnc.cdc.gov/eid/article/17/11/11-0822_article.

42. CDC (Centers for Disease Control and Prevention). 2018. 2017–2018 flu season burden estimates. https://www.cdc.gov/flu/about/burden/2017-2018.htm.
43. CDC (Centers for Disease Control and Prevention). 2020. Coronavirus: Human coronavirus types. https://www.cdc.gov/coronavirus/types.html.
44. Qin, A. and V. Wang. 2020. Wuhan, center of coronavirus outbreak, is being cut off by Chinese authorities. *The New York Times*, January 22.
45. Buckley, C., J. Hernández, V. Wang, et al. 2020. Coronavirus live updates: China's travel limits cover 35 million people. *The New York Times*, January 25.
46. Ramzy, A. and D. G. McNeil Jr. 2020. W.H.O. declares global emergency as Wuhan coronavirus spreads. *The New York Times*, January 30.
47. Staff. 2020. The illness now has a name, COVID-19. *The New York Times*, February 11.
48. Staff. 2020. 2% death rate from coronavirus, World Health Organization says. J. Emer. Med. Services, January 29.
49. WHO (World Health Organization). 2020. WHO coronavirus disease (COVID-19) dashboard. September 5. https://Covid19.who.int.
50. Lyu, W. and G. L. Wehby. 2020. Comparison of estimated rates of Coronavirus Disease 2019 (COVID-19) in border counties in Iowa without a stay-at-home order and border counties in Illinois with a stay-at-home order. *JAMA Netw.* 3(5):e2011102.
50a. Treisman, R. 2021. CDC: COVID-19 was 3rd-leading cause of death in 2020, people of color hit hardest. *NPR*, March 31.
51. Perez, M. 2020. 55% of U.S. coronavirus cases in CDC analysis are black and Hispanic. *Forbes*, June 15.
51a. University of Southern California. 2021. COVID-19 reduced U.S. life expectancy, especially among Black and Latino populations. *Science Daily*, January 14.
52. CDC (Centers for Disease Control and Prevention). 2016. Outbreak chronology. https://www.cdc.gov/vhf/ebola/outbreaks/history/chronology.html.
53. Saéz, A. M., S. Weiss, K. Nowak, et al. 2015. Investigating the zoonotic origins of the West African Ebola epidemic. *EMBO Mol. Med.* 7(1): 17–23.
54. CDC (Centers for Disease Control and Prevention). 2016. 2014–2016 Ebola outbreak in West Africa. https://www.cdc.gov/vhf/ebola/outbreaks/2014-west-africa/.
55. WHO (World Health Organization). 2015. Origins of the 2014 Ebola epidemic. http://who.int/csr/disease/ebola/one-year-report/virus-origin/en/.
56. UNDP (United Nations Development Programme). 2015. Human development report Guinea. http://hdr.undp.org/sites/all/themes/hdr_theme/country-notes/GIN.pdf.
57. Bausch, D. G. and L. Schwarz. 2014. Outbreak of Ebola virus disease in Guinea: Where ecology meets economy. *PLoS One* 8(7):e3056.
58. Shauib, F., R. Gunnala, E. O. Musa, et al. 2014. Ebola virus disease outbreak – Nigeria, July-September 2014. *MMWR* 63(39):867–72.
59. WHO (World Health Organization). 2016. Ebola factsheet. http://www.who.int/mediacentre/factsheets/fs103/en/.
60. CDC (Centers for Disease Control and Prevention). 2014. Cases of Ebola in the United States. https://www.cdc.gov/vhf/ebola/outbreaks/2014-west-africa/united-states-imported-case.html.
61. CDC (Centers for Disease Control and Prevention). 2014. Announcement: Interim US guidance for monitoring and movement of persons with potential Ebola virus exposure. *MMWR* 63(43): 984.
62. WHO (World Health Organization). 2016. End of the most recent Ebola virus disease outbreak in Liberia. *Media Center*. http://who.int/mediacentre/news/releases/2016/ebola-liberia/en/.
63. CDC (Centers for Disease Control and Prevention). 2019. Ebola outbreak in eastern Democratic Republic of Congo tops 1,000. *CDC Newsroom*, Atlanta, GA, March 24.

64. Branswell, H. 2019. The data are clear: Ebola vaccine shows 'very impressive' performance in outbreak. *Stat*, April 12.
65. Branswell, H. 2019. Ebola vaccine approved in Europe in landmark moment in fight against a deadly disease. *Stat*, November 11.
66. WHO (World Health Organization). 2019. Measles. *News Room*, Geneva, December 5.
67. CDC (Centers for Disease Control and Prevention). 2020. Measles cases in 2019. https://www.cdc.gov/measles/cases-outbreaks.html.
68. Andrew, S. 2020. Lawmakers around the nation are proposing bills for – And against – Vaccinations. *CNN News*, January 18.
69. CDC (Centers for Disease Control and Prevention). 2017. Control of communicable diseases. https://www.federalregister.gov/documents/2017/01/19/2017-00615/control-of-communicable-diseases.
70. CDC (Centers for Disease Control and Prevention). 2018. Illnesses from mosquitos, tick, and flea bites increasing in the US. https://www.cdc.gov/media/releases/2018/p0501-vs-vector-borne.html.
71. NNDSS (National Notifiable Diseases Surveillance System. 2015. https://wwwn.cdc.gov/nndss/.
72. NCEZID (National Center for Emerging and Zoonotic Infectious Diseases). 2017. https://www.cdc.gov/ncezid/.
73. WHO (World Health Organization). 2016. Vector-borne diseases. http://www.who.int/mediacentre/factsheets/fs387/en/.
74. Johnson, B. 2019. *Legacies of hope*. Seattle, WA: Amazon Kindle Publisher.
75. Wiser, M. F. 2013. Plasmodium life cycle. http://www.tulane.edu/~wiser/malaria/mal_lc.PDF.
76. CDC (Centers for Disease Control and Prevention). 2016. Laboratory identification of parasitic diseases of public health concern–DPDX. https://www.cdc.gov/dpdx/malaria/.
77. CDC (Centers for Disease Control and Prevention). 2015. Anopheles mosquito. https://www.cdc.gov/malaria/about/biology/mosquitoes/.
78. WHO (World Health Organization). 2016. Malaria. http://www.who.int/mediacentre/factsheets/fs094/en/.
79. CDC (Centers for Disease Control and Prevention). 2016. History of malaria. https://www.cdc.gov/malaria/about/history/.
80. CDC (Centers for Disease Control and Prevention). 2015. Ross and the discovery that mosquitoes transmit malaria parasites. https://www.cdc.gov/malaria/about/history/ross.html.
81. Ghebreytsus, T. A. 2019. The malaria eradication challenge. *The Lancet* 394(10203): 990–1.
81a. WHO (World Health Organization). 2020. *Global polio eradication initiative applauds WHO African region for wild polio-free certification*. Geneva: Office of Director General.
82. CDC (Centers for Disease Control and Prevention). 2010. Yellow fever: History timeline transcript. https://www.cdc.gov/travel-training/local/HistoryEpidemiologyandVaccination/HistoryTimelineTranscript.pdf.
83. HUL (Harvard University Library). 2016. Tropical diseases and the construction of the Panama Canal, 1904–1914. http://ocp.hul.harvard.edu/contagion/panamacanal.html.
84. PAHO (Pan American Health Organization). 2014. *Control of yellow fever in the Panama Canal zone*. Washington, DC: Office of Director.
85. WHO (World Health Organization). 2016. Yellow fever fact sheet. http://www.who.int/mediacentre/factsheets/fs100/en/.
86. Norrby, E. 2007. Yellow fever and Max Theiler: The only Nobel Prize for a virus vaccine. *J. Exp. Med.* 204(12):2779–84.

87. Cetron, M. S., A. A. Marfin, K. G. Julian, et al. 2002. Yellow fever vaccine: Recommendations of the advisory committee on immunization practices (ACIP). *MMWR 51.* https://www.cdc.gov/mmwr/pdf/rr/rr5117.pdf.
88. WHO (World Health Organization). 2017. Yellow fever outbreak. http://www.who.int/emergencies/yellow-fever/en/.
89. Wu, J. T., C. M. Peak, G. M. Leung, and M. Lipsitch. 2016. Fractional dosing of yellow fever vaccine to extend supply: A modelling study. *Lancet* 388(10062):2904–11.
90. Paules, C. I. and A. S. Fauci. 2017. Yellow fever: Once again on the radar in the Americas. *NEJM.* doi: 10.1056/NEJMp1702172.
91. Dick, O. B., J. L. San Martin, R. Montoya, et al. 2012. The history of dengue outbreaks in the Americas. *Am. J. Trop. Med. Hyg.* 87(4):584–93.
92. WHO (World Health Organization). 2016. Dengue. http://www.who.int/mediacentre/factsheets/fs117/en/.
93. Halstead, S. B. 2016. Critique of World Health Organization recommendation of a dengue vaccine. *J. Infect. Dis.* 214(12):1793–5.
94. Aguilar, M., N. Stollenwerk, and S. B. Halstead. 2016. The risks behind dengvaxia recommendation. *Lancet Infect. Dis.* 16(8):882–3.
95. CDC (Centers for Disease Control and Prevention). 2007. Dengue hemorrhagic fever – US-Mexico border, 2005. *MMWR* 56(31):785–9.
96. Florida Health. 2014. Dengue fever. http://www.floridahealth.gov/diseases-and-conditions/dengue/.
97. Messina, J. P., O. J. Brady, and N. Golding. 2019. The current and future global distribution and population at risk of dengue. *Nat. Microbiol.* 4:1508–15.
98. WHO (World Health Organization). 2019. Dengue and severe dengue. https://www.who.int/news-room/fact-sheets/detail/dengue-and-severe-dengue.
99. Branswell, H. 2019. FDA approves the first vaccine for dengue fever, but with major restrictions. *Stat,* May 1.
100. Gutierrez, J. 2019. Philippines declares a national dengue epidemic. *The New York Times,* August 6.
101. WHO (World Health Organization). 2017. West Nile. http://www.who.int/mediacentre/factsheets/fs354/en/.
102. Sejvar, J. 2003. West Nile virus: A historic overview. *Ochsner J.* 5(3):6–10.
103. Murray, K. O., E. Mertens, and P. Despres. 2010. West Nile virus and its emergence in the United States of America. *Vet. Res.* 41(6):67.
104. CDC (Centers for Disease Control and Prevention). 2015. West Nile virus 2015. https://www.cdc.gov/westnile/vectorcontrol/index.html.
105. Washington State University. 2019. Insulin can increase mosquitoes' immunity to West Nile virus. *Science Daily,* November 13.
106. Pialoux, G., B. A. Gauzere, S. Jaurequiberry, M. Strobel. 2007. Chikungunya, an epidemic arbovirosis. *Lancet Infect. Dis.* 7(5):319–27.
107. Vega-Rua, A., K. Zouache, R. Girod, et al. 2014. High level of vector competence of Aedes aegypti and Aedes albopictus from ten American countries as a crucial factor in the spread of chikungunya virus. *J. Virol.* 88(11):6294–306.
108. WHO (World Health Organization). 2012. The history of Zika. http://www.who.int/emergencies/zika-virus/timeline/en/.
109. Duffy, M. R., T. H. Chen, T. Hancock, et al. 2009. Zika virus outbreak on Yap Island, Federate States of Micronesia. *NEJM* 360:2536–43.
110. Cao-Lormeau, V., C. Roche, A. Teissier, et al. 2014. Zika virus, French Polynesia, South Pacific, 2013. *Emerg. Infect. Dis.* 20(6):1084–6.
111. Rodriguez-Faria, N., A. R. do Socorro da Silva, M. U. G. Kraemer, et al. 2016. Zika virus in the Americas: Early epidemiology and genetic findings. *Science* 352(6283):345–9.

112. CDC (Centers for Disease Control and Prevention). 2016. Geographic distribution 2016. https://www.cdc.gov/chikungunya/geo/.
113. Ikejezie, J, C. N. Shapiro, J. Kim, et al. 2017. Zika virus transmission: Region of the Americas, May 15–December 15, 2016. *MMWR* 66(12):329–34.
114. CDC (Centers for Disease Control and Prevention). 2017. Zika virus. https://www.cdc.gov/zika/symptoms/symptoms.html.
115. Cunningham, A. 2018. Zika may harm nearly 1 in 7 babies exposed to the virus in the womb. *Science News*, August 7.
116. Petersen, L. R., D. J. Jamieson, A. M. Powers, and M. A. Honein. 2016. Zika virus. *NEJM* 374:1552–63.
117. WHO (World Health Organization). 2017. Zika virus and complications: 2016 Public health emergency of international concern. http://www.who.int/emergencies/zika-virus/en/.
118. Rasmussen, S. A., D. J. Jamieson, M. A. Honein, L. R. Petersen. 2016. Zika virus and birth defects – Reviewing the evidence for causality. *N. Engl. J. Med.* 374:1981–7. DOI: 10.1056/NEJMsr1604338.
119. Villar, L., G. H. Dayan, J. L. Arredondo-Garcia, et al. 2015. Efficacy of a tetravalent dengue vaccine in children in Latin America. *NEJM* 372:113–23.
120. Smalley, C., J. H. Erasmus, C. B. Chesson, D. W. C. Beasley. 2016. Status of research and development of vaccines for chikungunya. *Vaccine* 34(26):2976–81.
121. NIH (National Institutes of Health). 2017. Phase 2 Zika vaccine trial begins in U.S., Central and South America. https://www.nih.gov/news-events/news-releases/phase-2-zika-vaccine-trial-begins-us-central-south-america.
122. Barrett, A. D. T. 2018. Current status of Zika vaccine development: Zika vaccines advance into clinical evaluation. *NPJ Vaccines* 3:24. DOI: 10.1038/s41541-018-0061-9, PMCID: PMC5995964.
123. Abad-Franch, F., E. Zamora-Perea, G. Ferraz, et al. 2015. Mosquito-disseminated pyriproxyfen yields high breeding-site coverage and boosts juvenile mosquito mortality at the neighborhood scale. *PLoS Negl. Trop. Dis.* 9(4):e0003702. DOI: 10.1371/journal.pntd.0003702. eCollection 2015 Apr.
124. Lowe, R., C. Barcellos, P. Brasil, et al. 2018. The Zika virus epidemic in Brazil: from discovery to future implications. *Int. J. Environ. Res. Public Health* 15(1):96. Published online 2018 Jan 9. DOI: 10.3390/ijerph15010096.
125. CDC (Centers for Disease Control and Prevention). 2019. Eastern equine encephalitis. https://www.cdc.gov/easternequineencephalitis/index.html.
126. Nuclo, C., R. Polasky, and P. Del Rio. 2019. 21 towns impacted by EEE, communities taking precautions. *WFSB*, September 27.
127. Soper, F. L. 1963. The elimination of urban yellow fever in the Americas through the eradication of Aedes aegypti. *Am. J. Public Health Nations Health* 53(1):7–16.
128. The World Health Report. 1999. Rolling back malaria. http://www.who.int/whr/1999/en/whr99_ch4_en.pdf.
129. CDC (Centers for Disease Control and Prevention). 2017. Tick-borne diseases. https://www.cdc.gov/niosh/topics/tick-borne/.
130. CDC (Centers for Disease Control and Prevention). 2017. Powassan virus. https://www.cdc.gov/powassan/.
131. CDC (Centers for Disease Control and Prevention). 2015. Tick geographic distribution. https://www.cdc.gov/ticks/geographic_distribution.html.
132. Rettner, R. 2018. Scary 'new tick' has US officials worried. *Live Science*, November 29.
133. Gander, K. 2019. Untreatable form of Lyme disease could hit 2 million Americans by 2020, scientists warn. *Newsweek*, April 23.

134. EPA (Environmental Protection Agency). 2016. Reducing the risk of tick-borne diseases through smart, safe and sustainable pest control. https://www.epa.gov/pesp/reducing-risk-tick-borne-diseases-through-smart-safe-and-sustainable-pest-control.
135. CDC (Centers for Disease Control and Prevention). 2010. Diseases directly transmitted by rodents. https://www.cdc.gov/rodents/diseases/direct.html.
136. CDC (Centers for Disease Control and Prevention). 2012. Hantavirus. https://www.cdc.gov/hantavirus/hps/symptoms.html.
137. Benedictow, O. J. 2008. The black death: The greatest catastrophe ever. *History Today* 55(3):1–15.
138. CDC (Centers for Disease Control and Prevention). 2016. Plague. https://www.cdc.gov/plague/maps/.
139. Castro, J. 2013. Bats host more than 60 human-infecting viruses. *Live Science*, February 6.
140. Cara, E. 2018. What is Nipah, the virus spread by bats that's killing people in India? *Gizmodo*, June 4.
141. WHO (World Health Organization). 2019. Soil-transmitted helminth infections. https://www.who.int/news-room/fact-sheets/detail/soil-transmitted-helminth-infections.
142. Pinkington, E. 2017. Hookworm, a disease of extreme poverty, is thriving in the US south. Why? *The Guardian*, September 5.
143. Carter Center. 2019. Guinea worm case totals. https://www.cartercenter.org/health/guinea_worm/case-totals.html.
144. Roberts, L. 2019. Exclusive: Battle to wipe out debilitating Guinea worm parasite hits 10 year delay. *Nature*, September 30.
145. CDC (Centers for Disease Control and Prevention). 2019. Parasites – American trypanosomiasis (also known as Chagas Disease). https://www.cdc.gov/parasites/chagas/epi.html.
146. WHO (World Health Organization). 2019. Chagas disease (American trypanosomiasis). https://www.who.int/news-room/fact-sheets/detail/chagas-disease-(american-trypanosomiasis).
147. Miller, K. 2015. Superbugs: What they are and how you get them. *WebMD News Archives*, April 17.
148. CDC (Centers for Disease Control and Prevention). 2019. Antibiotic resistance/antimicrobial resistance. https://www.cdc.gov/drugresistance/index.html.
149. ETH Zurich. 2019. Antimicrobial resistance is drastically rising. *Science Daily*, September 19.
150. Germanos, A. 2019. 'Superbugs' could kill 10 million annually without urgent action, warns new report. *Common Dreams*, April 29.
151. Staff. 2018. Superbugs to 'kill millions' by 2050 unless countries act. *The Straits Times*, November 7.
152. CDC (Centers for Disease Control and Prevention). 2019. Antibiotic resistance/antimicrobial resistance: Biggest threats and data. https://www.cdc.gov/drugresistance/biggest-threats.html.
153. WHO (World Health Organization). 2019. *In the face of slow progress, WHO offers a new tool and sets a target to accelerate action against antimicrobial resistance.* Geneva: Office of Media Affairs.
154. Jacobs, A. 2020. W.H.O. warns that pipelines for new antibiotics is running dry. *The New York Times*, January 17.
155. McMaster University. 2020. Antibiotics discovered that kill bacteria in a new way. *Science Daily*, February 12.
156. Trafton, A. 2020. A deep-learning model identifies a powerful new drug that can kill many species of antibiotic-resistant bacteria. *MIT News Office*, February 26.

157. Hamers, L. 2018. A new antibiotic uses sneaky tactics to kill drug-resistant superbugs. *Science News*, September 12.
158. Sandoiu, A. 2018. 'Game changing' antibiotic can kill superbugs. *Medical News Today*, March 27.
159. Lin, D. M., B. Koskella, and H. C. Lin. 2017. Phage therapy: An alternative to antibiotics in the age of multi-drug resistance. *World J. Gastrointest. Pharmacol. Ther.* 8(3):162–73.
160. University of Virginia. 2019. How E. coli knows how to cause the worst possible infection. *Science Daily*, August 16.
160a. Fuller-Wright, L. 2020. Princeton team develops 'poisoned arrow' to defeat antibiotic-resistant bacteria. *Princeton University Office of Communications*, June 3.
161. Healthline. 2019. What are nosocomial infections? https://www.healthline.com/health/hospital-acquired-nosocomial-infections#outlook.
162. CDC (Centers for Disease Control and Prevention). 2019. General information about Candida auris. https://www.cdc.gov/fungal/candida-auris/candida-auris-qanda.html.
163. Canton, N. 2019. Deadly drug-resistant superbug fungus now in 14 states, including Georgia. *Atlanta Journal-Constitution*, November 18.
164. Gallagher, J. 2019. Drug-resistant superbug spreading in Europe's hospitals. *BBC News*, July 29.
165. WebMD. 2019. Necrotizing fasciitis (flesh-eating bacteria). https://www.webmd.com/skin-problems-and-treatments/necrotizing-fasciitis-flesh-eating-bacteria#1.
166. Duncan, C. 2019. Pool parasites sickening swimmers are on the rise, CDC says. Here's how to stay safe. *The Charlotte Observer*, July 1.
167. CDC (Centers for Disease Control and Prevention). 2019. Legionella (Legionnaires' disease and Pontiac fever). https://www.cdc.gov/legionella/fastfacts.html.
168. Sullivan, B. 2019. Why the brain-eating amoeba is so deadly. *Live Science*, August 2.
169. CDC (Centers for Disease Control and Prevention). 2019. Parasites – Naegleria fowleri – Primary Amebic Meningoencephalitis (PAM) – Amebic Encephalitis. https://www.cdc.gov/parasites/naegleria/infection-sources.html.
170. Piesman, J. and C. B. Beard. 2012. Prevention of tick-borne diseases. *J. Environ. Health* 74(10):30–5.
171. EPA (Environmental Protection Agency). 2016. Reducing the risk of tick-borne diseases through smart, safe and sustainable pest control. https://www.epa.gov/pesp/reducing-risk-tick-borne-diseases-through-smart-safe-and-sustainable-pest-control.
172. McVeigh, K. 2018. Tropical disease outbreaks are growing threat in Europe as temperatures rise. *The Guardian*, August 23.
173. NASA (National Aeronautics and Space Agency). 2017. Global climate change. https://climate.nasa.gov/.
174. Caminade, C., J. Turner, S. Metelmann, et al. 2016. Global risk model for vector-borne transmission of Zika virus reveals the role of El Niño 2015. *Proc. Natl. Acad. Sci. U.S.A.* 114(1):119–24.
175. Ryan, S. J., C. J. Carlson, E. A. Mordecai, and L. R. Johnson. 2019. Global expansion and redistribution of Aedes-borne virus transmission risk with climate change. *PLoS Negl. Trop. Dis.* 13(3):e0007213.

2 Solid and Hazardous Waste

2.1 INTRODUCTION

This chapter is about waste generation and its management, with mismanagement of waste constituting a hazard to human and ecological health, as related herein. Moreover, as an alarming human and ecological health portent, the World Bank forecasts that global waste is on pace to triple by the year 2100 [1]. The definition of "waste" used in this work is simply "material that is discarded." Other more elaborate definitions can be found in dictionaries and elsewhere, but this definition is preferred due to its simplicity. This definition leaves open the reality that one person's waste could become another person's desirable resource, i.e., the idea of recycling of "waste."

As related in this chapter, waste in general requires careful management, lest human, and ecological consequences ensue. While waste was undeniably left by our primordial ancestors, as humankind grew in population size, social complexity, and outreach, one factor became strongly associated with the nature and volume of waste generated, especially in the context of waste's impact on the human environment. It is asserted here that technology has been a prime factor in humankind's waste generation and its management. For instance, the Industrial Revolution of the 17th and 18th centuries began the development, manufacture, and distribution of goods and services that in turn generated waste. The wastes came from such factors as inefficiencies in manufacturing of goods, too easily available sinks (e.g., rivers) for waste disposal, and social acceptance of wastes as a sign of industrialization and economic growth. Irrespective of these and other factors, the outcome was the same: materials were discarded into air, water, and land. As industrialization continued to increase across national borders, waste concomitantly grew in volume and character.

> As an alarming human and ecological health portent, the World Bank forecasts that global waste is on pace to triple by the year 2100.

Two world wars of the early 20th century fueled further growth of industrialization, followed by what could be called the Chemicals Age, which continues today. The globalization of the production of various chemicals, ranging from cosmetics to pharmaceuticals to weapons, has been accompanied by wastes of special character and challenge. As described in this chapter, these chemical wastes can have properties that are hazardous to human and ecological health. These kinds of chemical wastes are in addition to the solid and liquid wastes produced daily by human populations, as subsequently described herein.

Of particular note is the problem of contemporary sources and three kinds of waste: plastics, food, and electronics products (e-waste). Illustrated in Figure 2.1 is an example

DOI: 10.1201/9781003212621-2

FIGURE 2.1 Plastics waste pollution in the Pacific Ocean. (NOAA, Marine Debris Program, 2016.)

of plastics waste floating in the Pacific Ocean. As described in this chapter, plastics are a major threat to marine life and coastal ecosystems. Plastics waste is the second major threat to ocean life, with climate change causing the most serious threat. Food waste represents a lost opportunity to feed hungry people and other living creatures. And e-waste is a threat to both human and ecological health because of both its sheer volume and for its content of hazardous materials such as carcinogenic metals like cadmium. Because of the importance of these three forms of waste, the chapter begins with a description of their prevalence and consequences to human and ecological health.

The policies attending the management of municipal solid waste, uncontrolled hazardous waste, and oil waste will also be described in this chapter. Of particular note will be a presentation of issues of recycling of waste materials.

> Plastics waste is the second major threat to ocean life, with climate change causing the most serious threat.

2.2 CONTEMPORARY CHALLENGES TO WASTE MANAGEMENT

The generation and management of wastes are of global import and concern. All societies are faced with similar challenges: what to do with wastes produced by human activities such as industrialization, agriculture, and food acquisition. These challenges have given rise to regional, national, state/provincial, and local waste management policies that are tailored to waste conditions that present themselves to policymakers. On a global scale, three forms of waste are of contemporary importance and impactful on human and ecosystem health: plastics, food, and electronics. All three will be described in this section.

2.2.1 Plastics Waste

Plastics are omnipresent in modern society. They are used to contain food, beverages, and retail products in general; plastic bags are provided by merchants in which to carry purchases, and plastic straws are provided by food services. As a contribution to public health, meat and

> The total amount of plastics produced since the mid-20th century is approximately 5 billion tons, which is projected to rise to 30 billion tons by the end of the 21st century.

other food products sealed in plastic wrappers inhibit contaminants from entering the food. Plastics are also used in the construction of products as diverse as clothing and aircraft parts. They have become companions to humankind, but at a cost of their safe disposal. The total amount of plastics produced since the mid-20th century is approximately 5 billion tons (5.51 billion tons), which is projected to rise to 30 billion tons (33 billion tons) by the end of the 21st century [2]. The sheer volume of plastic material is a challenge to waste managers and a threat to human and ecosystem health. As background, plastics is the general common term for a wide range of synthetic or semi-synthetic materials used in a huge, and growing, range of applications. Plastics are organic, the same as wood, paper, or wool. The raw materials for plastics production include cellulose, coal, natural gas, salt, and crude oil. This section will describe some of the environmental health implications of plastics in waste streams.

2.2.1.1 Impacts of Plastics Waste on Human and Ecological Health

Any implications of plastics waste per se affecting human health seem undocumented. In contrast, the effects of plastics waste on ecosystems, particularly in regard to aquatic systems and marine life, are consequential and have led to considerable policy efforts to prevent or reduce releases of plastics waste into environmental media. A sample of investigations on the effects of microparticles on marine life is given in Table 2.1 While the presence of microparticles in tissues of fish and other marine

TABLE 2.1
Effects of Microparticles on Marine Life

Effect of Exposure	Reference
Fish eggs had a lower hatch rate; fish were smaller and less physically active; predator avoidance impaired.	[3]
Stunted growth of creatures near the bottom of the food chain such as worms, plankton, mussels and oysters.	[4]
Exposed oysters produced fewer larvae, which grew at a slower rate, and ultimately reached a smaller size than larvae unexposed to microplastics.	[5]
Microplastics found in the skin of both farmed and wild fish; unknown effect.	[6]
Strong inflammatory response in blue mussels demonstrated by the formation of granulocytomas and lysosomal membrane destabilization.	[7]
Beached dead whale in the Philippines found with 88 pounds of plastics trash inside body; one of several examples of such beached dead whales.	[8]
More than 70% of deep-sea fish found to have ingested plastics.	[9]

animals is well documented, the health consequences are a subject of research, with the investigations shown in the table raising concern for the implications.

Another ecological health impact of plastics waste is the discovery that their decomposition in environmental media contributes to climate change [10]. Researchers from the University of Hawaii discovered that several greenhouse gases are emitted as common plastics degrade in the environment. The researchers reported the unexpected discovery that the most common plastics when exposed to sunlight produced the greenhouse gases methane and ethylene. The science team tested polycarbonate, acrylic, polypropylene, polyethylene terephthalate, polystyrene, high-density polyethylene, and low-density polyethylene (LDPE), which are materials used to make food storage, textiles, construction materials, and various plastics goods. Polyethylene, used in shopping bags, is the most produced and discarded synthetic polymer globally and was found to be the most prolific emitter of both greenhouse gases. Additionally, the team found that the emission rate of the gases from virgin pellets of LDPE increased during a 212-day experiment and that LDPE debris found in the ocean also emitted greenhouse gases when exposed to sunlight. Once initiated by solar radiation, the emission of these gases continued in the dark. Research is continuing to better ascertain the contribution of the disintegration of plastics waste to climate change greenhouse emissions.

> Another ecological health impact of plastics waste is the discovery that their decomposition in environmental media contributes to climate change.

2.2.1.2 Plastics as Solid Waste

Plastics in the solid waste stream constitute a major challenge to waste managers. Two plastic products, beverage bottles and bags used in retailing, are the principal challenges, given the sheer volume of both plastic products, together with the longevity of plastics in general. Although recycling of plastics is a commercial enterprise in some areas of the world, much plastics remains for dumping in landfills or by incineration. The volume of plastics waste, particularly bottles and bags, has stimulated some policymakers to limit or ban their use. This section will detail the plastics waste challenge presented by plastic bottles and bags.

> Enough plastics are discarded annually to circle Earth four times.

According to one nongovernment organization (NGO), the following items provide a picture of the impact of plastics waste [11]:

- Enough plastics are discarded annually to circle Earth four times.
- Only 5% of the plastics produced are currently recovered.
- The average American discards approximately 185 pounds of plastics per year.
- Plastics accounts for around 10% of the total waste generated.
- Americans discard 35 billion plastic water bottles annually.
- Approximately 500 billion plastic bags are used annually worldwide. More than 1 million bags are used every minute.

Solid and Hazardous Waste

- **It takes 500–1,000 years for plastics to degrade.**
- Virtually every piece of plastics that was ever made still exists in some shape or form (with the exception of the small amount that has been incinerated).

These statistics indicate that the presence of plastic products, especially bottles and bags, is significant in terms of waste disposal. The volume and variety of plastic products pose a monumental challenge for their waste management. One major waste concern is the presence of plastics waste in the oceans and seas of planet Earth.

2.2.1.3 Plastics in Oceans

There are two major threats to the security of oceans and the marine life they contain. One threat is climate change, as described in Volume 1, Chapter 1. Climate change is warming ocean waters and contributing to their acidification, resulting in ecosystem damage that might be irreversible. For instance, some coral reefs are dying and some fish populations are threatened by warmer ocean waters. The second major threat to ocean security is plastics waste, as described in this section.

> Plastics constitute approximately 90% of all trash floating on the oceans' surface, with 46,000 pieces of plastics per square mile.

As preface, according to one NGO, the following items provide an overview of the impact of plastics waste in ocean waters [11]:

- Billions of pounds of plastics can be found in swirling convergences in the oceans making up about 40% of Earth's ocean surfaces. Eighty percent of pollution enters the ocean from the land.
- Plastics constitute approximately 90% of all trash floating on the oceans' surface, with 46,000 pieces of plastics per square mile.
- Plastics in the ocean decays into such small segments that pieces of plastics from a 1-liter bottle could end up on every mile of beach throughout the world.
- 46% of plastics float (Environmental Protection Agency (EPA) 2006) and they can drift for years before eventually concentrating in the ocean gyres.
- It takes 500–1,000 years for plastics to degrade.
- The Great Pacific Garbage Patch is located in the North Pacific Gyre off the coast of California and is the largest ocean garbage site in the world. This floating mass of plastics is about four times the area of California and comprises an estimated 1.8 trillion pieces of plastics rubbish, an estimated 87,000 tons. The plastics pieces eventually disintegrate into tiny particles that often get eaten by fish and may ultimately enter the human food chain [12].

> Plastics constitute approximately 90% of all trash floating on the oceans' surface, with 46,000 pieces of plastics per square mile.

- One million sea birds and 100,000 marine mammals are killed annually from plastics in Earth's oceans. As one of many examples of marine life laden with plastics waste, illustrated in Figure 2.2, is a dead young sea turtle

FIGURE 2.2 Baby sea turtle with ingested plastics waste. (Prendergast, R. Melbourne Zoo, EPA, 2021, https://blog.epa.gov/tag/sea-turtle/.)

that washed ashore in Boca Raton, Florida [13]. Dozens of tiny pieces of plastic dot the counter top next to the dead baby sea turtle. The 104 remnants, which range from a wrapper to a twist tie, are used in trash bags.
- 44% of all seabird species, 22% of cetaceans, all sea turtle species, and a growing list of fish species have been documented with plastics in or around their bodies.
- In samples collected in Lake Erie, 85% of the plastics particles were smaller than 0.2 inch, and much of that was microscopic. Researchers found 1,500 and 1.7 million of these particles per square mile [11].

The quantity and prevalence of ocean plastics were estimated by a team of researchers from six nations who conducted 24 garbage-collecting ocean expeditions between 2007 and 2013. Based on their calculations, at least 5.25 trillion pieces of plastic – weighing nearly 269,000 tons – are currently present in Earth's oceans [14]. A different study linked worldwide data on solid waste, population density, and economic status in order to estimate the mass of land-based plastics waste entering the ocean. Investigators calculated that 275 million metric tons (MT) of plastics waste were generated in 192 coastal countries in 2010, with 4.8–12.7 million metric tons entering the ocean. The research also lists the world's 20 worst plastics polluters. According to the estimate, China tops the list, producing as much as 3.5 million metric tons of marine debris annually. The U.S., which generates as much as 110,000 metric tons of marine debris a year, ranked 20th [15].

2.2.1.4 Microbeads in Waste

Microbeads are a special form of plastics waste that is a hazard to the health of ecosystems. These are tiny bits of plastics used as ingredients in exfoliating body washes and facial scrubs. Since their introduction in 1972, they have made their way into more than 100 personal care products. Microbeads are used in soaps because exfoliating products need small,

> Since their introduction in 1972, microbeads have made their way into more than 100 personal care products.

hard particles to rub debris from the skin [16]. Microbeads range in diameter from 5 µm to 1 mm. They are made from synthetic polymers including polyethylene, polylactic acid, polypropylene polystyrene, or polyethylene terephthalate.

One group of researchers calculated that 8 trillion microbeads per day are emitted into aquatic habitats in the U.S. Further, the investigators estimate that the 8 trillion beads that make it into aquatic habitats are only 1% of the total release, which would amount to 800 trillion microbeads ending up in the sludgy runoff from sewage plants [17]. A study by Japanese investigators found that the concentration of microplastics, defined as plastic particles up to 5 mm in diameter, is higher in sediment than in sea water, raising concerns that it could affect organisms living in and on the bottom of the ocean. The study examined samples of sediment collected from the sea floor and elsewhere in areas that included Tokyo, Southeast Asia, and Africa [18]. As will be described in the next section, microbeads present a serious threat to the health of ecosystems.

Another study that illustrates the global swath of plastic microbeads is an investigation of microplastics in Arctic ice [19]. German polar ice researchers in 2018 found higher amounts of microplastics in Arctic sea ice than ever before. However, the majority of particles were microscopically small. The ice samples from five regions throughout the Arctic Ocean contained up to 12,000 microplastic particles per liter of sea ice. The researchers found a total of 17 different types of plastics in the sea ice, including packaging materials like polyethylene and polypropylene, but also paints, nylon, polyester, and cellulose acetate, the latter is primarily used in the manufacture of cigarette filters. Taken together, these six materials accounted for about half of all the microplastic particles detected.

> German polar ice researchers in 2018 found ice samples from five regions throughout the Arctic Ocean contained up to 12,000 microplastic particles per liter of sea ice.

The global omnipresence of microplastics stimulated the WHO in 2019 to issue a health policy statement concerning their presence in the environment and potential impact on public health. According to WHO's analysis, which summarized the available knowledge on microplastics in drinking water, microplastics larger than 150 µm are not likely to be absorbed in the human body and uptake of smaller particles is expected to be limited. However, the agency called for a further assessment of microplastics in the environment and their potential impacts on human health. The WHO also called for a reduction in plastic pollution to benefit the environment and reduce human exposure [20]. Supporting this WHO concern was a study by UK scientists who calculated that the top 200 m of the ocean alone contains up to 21 million metric tons of plastic in the form of microplastic produced principally by water's grinding action on larger pieces of plastic such as discarded shopping bags [21].

The production of new plastic stock produces tiny plastic pellets. Those small spheres, sometimes known as "nurdles," are a massive source of plastic pollution, with about 22,000 nurdles per pound of plastic. Nurdles are about the size of a lentil. Nurdles can escape into the environment as microplastics before they have a chance to be molded into a useful shape. Environmental groups and scientists are worried about the lack of regulation to specifically address plastic pellet pollution. California is the only state with regulations to specifically control for plastic pellet pollution, although enforcement of the state's regulations is difficult, given the pellets' size and wide distribution [22].

2.2.1.5 Microbead-Free Waters Act, 2015

Because of concerns that microbeads can deleteriously affect aquatic life and ecosystems in general, several policymakers have enacted policies to prevent microbeads from entering environmental media. Of particular note, the U.S. Congress enacted the Microbead-Free Waters Act of 2015, which was signed into law by President Barack Obama (D-IL) on December 28, 2015. The act requires the manufacturing of products containing microbeads to end by July 1, 2017, and their sale to cease by July 1, 2018 [23]. This action by Congress was stimulated by concern that microbeads were polluting the Great Lakes. Similarly, the UK banned microbeads used in cosmetics and cleaning products effective in 2017 [24]. Further, the federal government of Canada has announced that it intends to ban microbeads used in personal care products. The Canadian federal government intends to develop regulations to prohibit the manufacture, import, and sale of microbead-containing care products [25].

2.2.1.6 Recycling of Plastics Waste

One business source notes that nearly 80% of the world's plastic waste ends up in landfills or gets thrown away as trash. That plastic can linger in the environment for more than 400 years before it degrades, clogging up oceans and seeping into our air and soil. Environmentalists now consider plastic pollution a threat to human health as well as marine life [26].

Each year, at least 8 million tons of plastics find their way into the ocean, the equivalent of dumping the contents of one garbage truck into the ocean every minute, according to the World Economic Forum [27]. Further, the source notes that only 14% of plastics packaging is collected for recycling, while most plastics packaging is used only once. Overall, the plastic recycling rate in the U.S. peaked in 2014 at 9.5%. In addition, 95% of the value of plastics packaging material, worth between $80 billion and $120 billion annually, is lost to the economy [26]. What can be done with plastics waste? The problem has several elements of possibility. First, what to do with the plastics waste that has already been left in the environment? And, second, what to do in support of preventing more intractable plastics waste?

> Each year, at least 8 million tons of plastics find their way into the ocean, the equivalent of dumping the contents of one garbage truck into the ocean every minute.

Regarding plastics waste already generated, can more of it be recycled? Plastics recycling refers to the process of recovering waste or scrap plastics and reprocessing the materials into functional and useful products. A reviewer of recycling of plastics cites the following statistics:

- "Every hour, Americans use 2.5 million plastic bottles, most of which are thrown away.
- About 9.1% of plastics production was recycled in the U.S. during 2015, varying by product category. Plastics packaging was recycled at 14.6%, plastics durable goods at 6.6%, and other nondurable goods at 2.2%.

- Currently, 25% of plastics waste is recycled in Europe.
- Americans recycled 3.14 million tons of plastics in 2015, down from 3.17 million in 2014.
- Recycling plastics takes 88% less energy than producing plastics from new raw materials.
- Currently, around 50% of plastics we use are thrown away just after a single use.
- Plastics account for 10% of total global waste generation.
- Plastics can take hundreds of years to degrade.
- The plastics that end up in the oceans break down into small pieces and every year around 100,000 marine mammals and 1 million seabirds get killed eating those small pieces of plastics.
- The energy saved from recycling just a single plastic bottle can power a 100 watt light bulb for nearly an hour" [28].

> Every hour, Americans use 2.5 million plastic bottles, most of which are thrown away

The relatively low percentage of plastics waste that gets recycled is due in part to the commercial factors of supply and demand. Recycled plastics have a market if their cost to a potential purchaser is less than the cost of producing a plastics product using nonrecycled material. Some U.S. states have enacted policies that require recycling of specified plastic products, sometimes subsidizing a portion of the cost of recycling. A number of U.S. states, including California, Connecticut, Delaware, Hawaii, Iowa, Maine, Massachusetts, Michigan, New York, Oregon, and Vermont, have passed laws that establish deposits or refund values by encouraging reusing and recycling of beverage containers, including plastic bottles [28]. This is known as a Bottle bill or container deposit law. Most states refund five cents per can or bottle, but some can refund up to 10 cents.

There are many examples of recycling a specific plastics waste into a useful alternative. For illustration, when Indian fishermen in India's southern state of Kerala drag their nets through the water, they end up scooping up huge amounts of plastics along with the fish. Once all the plastics waste caught by the Keralan fishermen reaches the shore, it is collected by people from the local fishing community and fed into a plastics shredding machine. The shredded plastics are then converted into material that is used for road surfacing in India [29]. As a second example, the UK's University of Chester is now using gasification, a process that involves heating plastics waste with air or steam, to produce electricity. Unlike most recycling methods, the technology does not require plastics to be sorted or washed beforehand. The researchers believe the process could one day be scaled to power cars, homes and, eventually, entire cities and towns [26].

Considerable research has been undertaken for the purpose of developing biodegradable plastics. Two examples will illustrate the thrust of this form of research. In one research endeavor, the U.S. Department of Energy (DOE) has designed a type of plastic that can be recycled any number of times without any loss of performance. The DOE's Lawrence Berkeley National Laboratory has produced a new kind of polymer that can be broken down and reassembled into a different shape, texture, and color without impairing its quality. The new material is called poly(diketoenamine), or PDK [30]. The researchers believe that their new plastic could be a good alternative to many nonrecyclable plastics in use today.

A second example comes from research in Indonesia, where plastics waste is a significant problem for that country. A company in Indonesia has created a plastic bag so ecofriendly you can eat it. It is made of cassava, the vegetable root which is a staple in the diets of many in Africa, Latin America, and Asia, but which can also be used in manufacturing. The bag looks and feels like plastic, but is completely degradable and compostable [31].

> A company in Indonesia has created a plastic bag so eco-friendly you can eat it. It's made of cassava, a vegetable root.

As the last example, some researchers have turned to biology as a solution to dealing with plastics waste. Several species of fungi and bacteria can do the job, but only slowly. Researchers at Stanford University have found that bacteria in the guts of mealworms can break down polymers much more quickly. Their research found that mealworms could digest both forms of popular plastics, polystyrene and polyethylene, producing only CO_2 as the byproduct, a greenhouse gas that would need to be trapped [32]. Further research is ongoing.

2.2.1.7 Bans on Single-Use Plastics

The ecological problem of plastics waste grew gradually over time. As reports of plastics waste in the oceans mounted, policymakers, and the general public in Europe and North America became under increasing pressure to act in ways that would help interdict the ecological problems being caused by plastics waste. News media and social networks reports on large floating islands of plastics in the Pacific Ocean prompted policymakers to commence action. Single-use plastic products became a primary target of policymakers and the general public in Europe and North America. These products became a focus of attention due to their familiarity to the general public and the rather easy rationale for their discontinuance, that is, how does a person justify the presence of an ecological hazard that is only going to be used once? Policymakers understood this kind of political dynamic, which formed the genesis of actions to ban some single-use plastic products.

2.2.1.7.1 Government Bans

Pressure from environmental organizations and concerned members of the public have brought actions by different levels of government in several countries to ban single-use plastic products. These bans were sometimes specific to a particular single-use product, e.g., plastic straws, and other times were more sweeping in coverage, e.g., plastics used in restaurant service. Policymakers in several countries have instituted bans or other controls on the production and use of plastic bags and bottles. For example, California is the first U.S. state to ban single-use

> News media and social networks reports of large floating islands of plastics in the Pacific Ocean prompted policymakers to commence action.

plastic bags. Large grocery stores and smaller businesses, like liquor and convenience stores, are prohibited from using plastic bags. Businesses are permitted to charge customers for use of paper bags [33]. Likewise, beginning March 1, 2020, single-use plastic carryout bags will be banned in New York State.

> France has apparently become the first country in the world to ban plastic plates, cups, and utensils, passing a law that will go into effect in 2020.

The law allows a number of exceptions, from the expected (dry cleaning, food delivery) to the niche (bulk-buying live insects), but these exceptions may be decreased over time [34].

Another U.S. state, New Jersey, enacted in November 2020 the strongest single-use bag ban in the U.S., effectively prohibiting the use of single-use plastic and paper bags in all stores and food service establishments across the state [35]. The ban goes into effect in May 2022. Food service businesses will be allowed to offer single-use plastic straws upon request beginning in November 2021. Penalties for violating the new law are scaled from a warning on the first offense to up to $5,000 in fines for a third violation.

France has apparently become the first country in the world to ban plastic plates, cups, and utensils, passing a law that will go into effect in 2020. The new law is a part of the country's Energy Transition for Green Growth Act, the same legislation that outlawed plastic bags in grocery stores and markets beginning in July 2016 [36]. In 2019, the Canadian government announced that it will ban single-use plastics as early as 2021. Plastic straws, cotton swabs, drink stirrers, plates, cutlery, and balloon sticks are just some of the single-use plastics that will be banned in Canada [37]. Similarly, in September 2020, South Australia became the first Australian state to introduce laws banning some single-use plastics including cutlery, straws, and stirrers [38].

Several cities globally have implemented policies to ban the sale of plastic water bottles. As four examples, in 2009 residents of Bundanoon, a town of 2,500 residents about 100 miles south of Sydney, Australia, voted to ban the sale of bottled water, in part due to litter concerns. They were possibly the first community in the world to take such a step [39]. In 2013, Concord, Massachusetts, became the first U.S. city to enact a bylaw prohibiting the use of plastic water bottles [40]. Similarly, in

> In 2013 Concord, Massachusetts, became the first U.S. city to enact a bylaw prohibiting the use of plastic water bottles.

2014 San Francisco banned the sale of plastic water bottles. In 2011, the National Park Service of the U.S. issued Policy Memorandum 11-03 regarding the "recycling and reduction of disposable plastic bottles in parks." Regional directors of national parks could review and approve "a disposable plastic water bottle recycling and reduction policy, with an option to eliminate sales on a park-by-park basis." The ban on bottled water in 23 U.S. national parks prevented up to 2 million plastic bottles from being used and discarded every year, according to a May 2017 study by the National Park Service. In September 2017 U.S. President Donald Trump's (R-NY) Department of Interior vacated the ban on bottle water at the behest of the bottled water industry [41].

Policymakers in some jurisdictions have taken alternative routes to outright banning of plastic bags and bottles for the purpose of reducing the prevalence of plastics waste. For example, some jurisdictions such as New York City have enacted a bill that requires certain retailers to collect a 5-cent fee on each carryout bag, paper or plastic, with some exceptions [42]. Similarly, since 2014 retailers in Great Britain must charge customers five pence (eight U.S. cents) for plastic bags in an effort to wean the population off plastic shopping bags. The fee is estimated to reduce the garbage cleaning cost by £60 million ($80 million) over the next 10 years. In Wales, since 2001 when a similar measure was instituted, the use of plastic bags has decreased 79%. Rwanda banned plastic bags in 2008, and Bangladesh removed them in 2002 [43]. All these bans or restrictions on use of plastic bags are policies implemented for the reduction of plastics waste in landfills, incinerators, and water sources.

Concerned about the environmental impact of single-use plastic products, the United Nations Environ Programme (UNEP) released a report that had surveyed the global problem. The report noted, "More than 60 countries have introduced bans and levies to curb single-use plastics waste. Plastic bags and, to a certain extent, foamed plastic products like Styrofoam have been the main focus of government action so far. This is understandable. These plastic products are often the most visible forms of plastics pollution. It is estimated that 1 to 5 trillion plastic bags are consumed worldwide each year. Five trillion is almost 10 million plastic bags per minute. If tied together, all these plastic bags could be wrapped around the world seven times every hour" [44].

> The UNEP noted that more than 60 countries have introduced bans and levies to curb single-use plastics waste.

Following are examples of governments' actions to ban single-use plastic products.

Governments around the world are increasingly awake to the scale of plastics pollution. In May 2018 the European Commission (EC) proposed a total ban on some single-use plastic products and measures to drastically cut the consumption of others for the purpose of reducing carbon emissions and marine litter threatening its seas [45]. The ban would apply to plastic cotton buds, cutlery, plates, straws, drink stirrers, and sticks for balloons. In October 2018, the European Parliament approved the EC proposal. Member states would be required to reduce the use of plastic food containers and drink cups by prohibiting their free-of-charge distribution, with anticipation that the ban will be fully in place by 2021 [46].

In addition to the collective action of the countries comprising the EU, several individual countries have enacted similar policies to ban or restrict the use of single-use plastic products. For instance, in January 2019, the government of South Korea banned all single-use plastic bags in its major supermarkets. Instead, stores will instead be required to offer alternatives such as paper or cloth bags, which can be recycled [47]. Canada plans to ban single-use plastics by 2021 [48], Iran has banned plastic water bottles [49], Kenya banned plastic bags [50], New Zealand has banned plastic bags [51], and in 2019 Panama became the first Central American nation to ban plastic bags [52]. Of note, in 2020 the New Zealand Ministry of the Environment reported that banning single-use plastic bags from supermarkets and other retail stores had resulted in more than 1 billion bags kept out of the nation's waste stream [53]. The U.S. has no federal plans to ban single-use plastics, although, as noted below, some U.S. states and cities have taken that action.

California, Hawaii, and New York have acted on banning the distribution of single-use plastic bags statewide, as have several U.S. cities banned single-use straws or other plastics product. Regarding plastic straws, according to one source, about 500 million plastic straws are used in the U.S. every day, with most of them ending up in the environment [54]. A partial list of U.S. cities that have banned plastic straws include New York City, Seattle, Miami Beach, Washington, D.C., and Oakland, California. A larger list of smaller population U.S. cities has also acted on banning plastic straws.

2.2.1.7.2 Commercial Sector Bans

While governments globally at federal through municipal levels have acted to ban some forms of single-use plastics, the private commercial sector has also acted for the same ecological purpose. In one particularly noteworthy action, in 2018 the Ellen MacArthur Foundation launched its New Plastics Economy Global Commitment in collaboration with the United Nations Environment Programme (UNEP) [55]. Two hundred and fifty organizations pledged to cut all plastics waste from their operations in what the UN called the most ambitious effort yet to fight plastics pollution. The pledge included many of the world's biggest packaging producers, leading consumer brands, retailers and recyclers, as well as governments and NGOs. Signatories promised to eliminate single-use and unnecessary plastics and to innovate so that all packaging could be recycled, with targets to be reviewed regularly and updates posted on their progress to fulfill their pledge.

Additionally, several U.S. food service companies have pledged to forego some single-use plastics. For instance, Kroger, the largest U.S. grocery chain, plans to quit using plastic bags in their stores by 2025, Starbucks will eliminate plastic straws by 2020, and McDonald's has ceased using plastic straws in the UK and Ireland. And in 2018 the Hyatt hotel chain quit using plastic straws and drink picks in their hotels. As a matter of environmental policy, these actions by private sector entities occurred not by government regulation, but rather by recognition of customers' pressure and expectations.

> New York City, Seattle, Miami Beach, Washington, D.C., and Oakland, California, have banned plastic straws.

Perspective: Single-use plastic products such as straws and bags have served important commercial purposes for making food services and shopping experiences more satisfactory. But with that commercial benefit came the challenge of what to do with their waste. Gradually, as public awareness grew as concerns the huge global burden of plastics waste, social support for single-use plastics diminished. That change brought policymakers both in government and private sector entities to realize that change was needed, as exemplified by the policies to ban single-use plastics. The policy lesson here is that public awareness, conveyed and broadcast by social media, affected a positive environmental outcome, banning single-use plastics.

> Kroger, the largest U.S. grocery chain, plans to quit using plastic bags in their stores by 2025. Starbucks will eliminate plastic straws by 2020.

2.2.2 Food Waste

As noted in Volume 1, Chapter 4, an important aspect of food security is food wastage. The seminal study on this subject was conducted in 2013 by the United Nations' Food and Agriculture Organization (FAO), which estimated that annually, approximately one-third of all food produced globally for human consumption is lost or wasted. This food wastage represents a missed opportunity to improve global food security, and also to mitigate environmental impacts [56].

On a national scale, researchers estimated that annually about 40% of all food in the U.S. goes uneaten, which represented the equivalent of $165 billion each year from food wastage. For comparison, this amount is approximately equal to the 2016–2017 total budget for the State of California. The top three food groups in terms of the value of food loss at these levels were: meat, poultry, and fish (41%); vegetables (17%); and dairy products (14%). These estimates at the U.S. consumer-level translate into almost 124 kg (273 lbs.) of food lost from human consumption, per capita, in 2008 at an estimated retail price of $390/capita/year [57].

> On a national scale, researchers estimated that annually about 40% of all food in the U.S. goes uneaten.

Stating this figure differently, U.S. consumers throw away $29 billion worth of good food annually. Much of it rots in landfills, producing methane, a potent greenhouse gas [58]. Some food is wasted because of consumers' confusion with food labels. In an interesting policy move, in 2017 some of the world's biggest food manufacturers and retailers agreed they would simplify food labels to only two labels: "use by" date on perishable items that indicates when a food is no longer safe to eat and "best if used by" date that indicates when a food bypasses its optimal quality [58].

> U.S. consumers throw away $29 billion worth of good food annually. Much of it rots in landfills, producing methane.

Similarly, a study by the UK government found that UK households waste 7 million tons of food – worth £700 per family – every year at a total cost of £12.5 billion. This is just less than half the 15 million tons of food wasted annually in the UK – the rest by supermarkets, restaurants, and elsewhere in the supply chain. Within these totals, UK householders throw away 34,000 tons of beef annually, wasting about £260 million worth of raw and cooked beef annually [59].

The aforecited numbers relating to food wastage must be understood in terms of the policy issue of lack of an international standard on how to define food waste. As observed by one group of researchers, agreement on how to define food wastage is lacking on a global scale. For instance, some consider any food not consumed by humans as being waste; however, some other sources do not consider discarded food that is fed to livestock (e.g., swine) as food wastage. In recognition of the lack of an international standard, an expert group has put forth a proposed set of guidelines that national and other organizations can use when accounting for food wastage, but adoption of the guidelines awaits [60].

Because food is one of the three necessities of life (air, water, food), food wastage is a misuse of a vital resource. One might propose two paths to reduce food waste: a) prevention of waste generation and/or b) reuse/recycle generated food waste. Stated in different words, to reduce food wastage, simply don't generate food waste, or if it is generated, find ways to make use of the waste. For several reasons, the former path is preferable to the latter path. For instance, food waste not generated does not require energy and economic resources necessary for dealing with generated food waste. Some exempts of each path follow.

> Because food is one of the three necessities of life (air, water, food), food wastage is a misuse of a vital resource.

Denmark has become a hallmark country in terms of reducing food waste. Celebrity chefs proselytize on entertainment media the virtues and methods to reduce food waste. Contemporaneously, social media campaigns also promote the message, "don't waste food." These efforts have resulted in Danes' increasing willingness to buy and consume items like just-expired dairy products and produce. According to a recent report from the Danish government, Danes now discard 25% less food than they did 5 years ago. Danes today discard 104 pounds of food per year on average compared to an estimated 273 pounds per person per year in the U.S. [61]. A similar resolve to reduce food waste is exemplified by efforts in South Korea. Seoul's city government requires residents to deposit their food waste in bins, where the amount of food they discard is weighed per household using a key-card system. Dispose of too much food and a fee is charged by municipal officials. Trial districts in Seoul have succeeded in reducing household food waste by 30% and restaurant food waste by 40%. Such programs are now underway in 90 localities nationwide in South Korea [62].

Another path to waste reduction is reuse/recycling of generated food waste. Three examples will suffice. In Colorado, a facility takes wasted food from Colorado's most populous areas and turns it into electricity. Through anaerobic digestion, spoiled milk, old pet food, and vats of grease combine with helpful bacteria in massive tanks to generate methane gas. The gas is transferred via a pipeline as fuel to use with electric power generators [63].

In California, a company has developed a process that transforms truckloads of supermarket food waste into farm-ready fertilizer. The company developed an aerobic digestion method that accelerates the composting process, turning food waste into liquid fertilizer in three hours.

> Greenhouse gases, especially methane, released from landfills containing food waste are a significant contributor to climate change.

The company claims that since its launch in 2012, it has diverted more than 2.2 million pounds of food waste from a landfill, preventing the emissions of 3.2 million pounds of greenhouse gases and preventing the need for more than 1.1 million pounds of nitrogen fertilizers [64].

In Hong Kong, scientists at the City University of Hong Kong have found a method to turn coffee grounds and stale bakery goods into a sugary solution that can be used to manufacture plastic. The food waste is mixed with bacteria and fermented to produce succinic acid, a substance usually made from petrochemicals, and which can be found in a range of fibers, fabrics, and plastics [65].

Food waste is an environmental hazard and is a threat to the health of ecosystems. In a 2007 report, the FAO estimated that the global carbon footprint of wasted food was about 3.3 billion tons of CO2-equivalents, which equates to about 7% of all global emissions in 2007. As a perspective, this is more carbon than most countries emit in a year, with only China and the U.S. exceeding this amount in nationwide carbon emissions in 2007 [66]. Greenhouse gases, especially methane, released from landfills containing food waste are a significant contributor to climate change, as noted in the FAO study. Further, food waste can serve to unbalance the mix of wildlife within an ecosystem. And a growing body of evidence suggests that food waste may be reshaping the way the natural world functions globally, inadvertently food subsidizing some opportunistic predators and thus contributing to the decline of other species. For example, gulls with access to food waste in landfills have increased in numbers, resulting in fewer juvenile fish, a consequence of increased populations of gulls. Similarly, bears, coyotes, and wolves with access to food waste can increase in numbers and then pose a threat to domestic animals such as dogs and cats. Access to food waste serves to unbalance the normal mix of predators and prey [67].

2.2.3 ELECTRONIC PRODUCTS WASTE (E-WASTE)

Consumer products that contain electronics now abound globally. These products include vehicles with engine computers, kitchen equipment, entertainment devices such as flat screen TVs, computers, telephones, and many more. Electronics represent the global digital world upon which social media and modern day commerce rely. The variety and volume of electronic products is enormous. But as with other products of popular commerce, popularity equates to volume of production, accompanied by challenges of waste disposal when the products become passé or inoperative. Modern societies find themselves facing the nitty-gritty reality of what has become known as electronic waste or e-waste, which is defined as any device with an electric cord or battery – from refrigerators to smart phones [68]. And as with other environmental health issues, e-waste is a global problem.

A study conducted by researchers at the United Nations University found that the U.S. and China contributed most to record mountains of electronic waste in 2016 and worldwide less than a fifth was recycled [68]. Overall, 44.7 million metric tons of e-waste were discarded globally in 2016, with a 17% increase in e-waste globally predicted by 2021 [69]. The researchers estimated that the discarded materials, including Au, Ag, Fe, and Cu, were worth some $52 billion in 2014. Small equipment (e.g., cell phones, tables) is the largest category of e-waste [69].

> Overall, 44.7 million metric tons of e-waste were discarded globally in 2016, with a 17% increase in e-waste globally predicted by 2021.

The U.S led e-waste dumping with 7.1 million tons in 2014, ahead of China at 6.0 million, followed by Japan, Germany, and India. The U.S., where individual U.S. states enforce e-waste laws, reported collection of 1 million tons for 2012 while China

Solid and Hazardous Waste

said it collected 1.3 million tons of e-waste in 2013. Norway led per capita waste generation, with 28 kg (62 lbs.) dumped per inhabitant, followed by Switzerland, Iceland, Denmark, and Britain. On a per capita ranking, the U.S. was ninth and China was not among a list of the top 40 [68].

However, on a positive note, the amount of electronic waste generated in the U.S. has shrunk 10% since 2015. The decline is due to the phasing out of bulky products, such as large cathode-ray tube televisions and computer monitors, according to Rochester Institute of Technology researchers [70].

Well, given the huge volume of e-waste produced each year globally, where does it go? Who gets it? What do they do with it? According to a UNEP study of the fate of e-waste, 60%–90% of e-waste – worth nearly $19 billion – is illegally traded or dumped, often with the involvement of transnational criminal gangs [71]. West Africa has been reported by the UN Office on Drugs and Crime (UNODC) to be a major destination for electronic waste, while some Asian countries are also recipients of millions of tons of e-waste, sometimes as part of so-called trade-free agreements with Western countries.

> According to a UNEP study, 60%–90% of e-waste – worth nearly $19 billion – is illegally traded or dumped, often with the involvement of transnational criminal gangs.

Nigeria is an example of a country that receives a large volume of illegal e-waste. A study, led by the Basel Convention Coordinating Centre for Africa and United Nations University, found 66,000 tons of used electronics were shipped to Nigeria during 2015 and 2016. About 16,900 tons did not work – which qualifies it as waste and made shipping it illegal. European Union countries had sent about 77% of the used electronics. The U.S. accounted for 7%. TV monitors made up about one-third of the waste, followed by photocopiers and refrigerators. In none of the inspected cases were there any consequences, neither for the exporters nor for the importers, according to the researchers [72].

There can be serious health consequences to workers who process e-waste. This reality was investigated by a research team from the University of Benin who studied Nigerian male waste workers [73]. This cross-sectional study which was carried out in Benin City, South-South Nigeria recruited male e-waste workers and apparently healthy non-e-waste workers, as exposed and unexposed participants, respectively. Results showed Nigerian men who work with electronic waste had much lower levels of crucial fertility hormones than men unexposed to the waste. The research, conducted in Benin City in southern Nigeria, suggests toxic metals and other contaminants in e-waste are reducing levels of key hormones in men who process e-waste and could contribute to reduced sperm levels, and cause damage to the sperm's DNA, stability, and ability to fertilize eggs.

The huge volume of e-waste has implications for human and ecosystem health. As noted by the WHO, recycling of e-waste to retrieve metals, for example, Cu and Ag, has become a source of income in developing or emerging industrialized countries [74]. E-waste-connected health

> E-waste-connected health risks may result from direct contact with harmful materials such as Pb, Cd, Cr, brominated flame retardants, or PCBs, from inhalation of toxic fume.

risks may result from direct contact with harmful materials such as Pb, Cd, Cr, brominated flame retardants, or PCBs, from inhalation of toxic fumes (e.g., from burning electric cable to reclaiming Cu), as well as from accumulation of chemicals in soil, water, and food.

As pervasive and polluting as cigarette butts are, however, the e-waste from e-cigarettes presents an even more significant health hazard [75]. The legacy of cigarette butts imparts a dark story. An estimated two-thirds of cigarette butts are littered, clogging sewer drains, blighting city parks, and contributing to estimated cleanup costs of $11 billion yearly for U.S. litter alone. Cigarettes are environmentally irresponsible, and yet e-cigarettes pose an environmental threat of considerably greater proportions. Instead of merely being thrown away, these complex devices present simultaneously a biohazard risk with potential high quantities of leftover or residual nicotine and an environmental health threat as littered electronic waste. A 2014 study found that none of the surveyed e-cigarette packages contained disposal instructions.

Children are especially vulnerable to the health risks that may result from e-waste exposure and, therefore, need more specific protection. As they are still growing, children's intake of air, water, and food in proportion to their weight is significantly increased compared to adults [75]. Journalists who visited two of the world's largest e-waste dumps reported children were at risk of exposure to e-waste contaminants during recycling processes. At one e-waste dump, many of the workers at the Agbogbloshie site in Nigeria were middle-school-age boys who smash electronics to get their metals, thereby exposing themselves to toxic metals such as Cu, Cd, and Cr [76]. Similarly, the Chinese city of Guiyu, Guangdong Province, receives some 15,000 metric tons of waste every day. Most of the city's residents work in the recycling industry. Research by a Chinese university found abnormally high levels of lead in the city's children's blood [77].

2.3 U.S. SOLID WASTE MANAGEMENT POLICIES

As a preface, involvement of the U.S. federal government in the regulation of solid and hazardous waste was not part of the environmental movement of the early 1960s [78]. Environmentalists had given priority to supporting legislation that would improve air and water quality. Moreover, U.S. states, territories, and municipalities had long had the responsibility for managing municipal waste collection, waste dumps, and sanitary landfills. At an earlier age, during the years of Colonial America and the agrarian period that followed, farmers disposed of their own solid wastes, much of which was recycled as fertilizer for soil and crop enrichment. Towns and cities during this period continued the longstanding practice of creating open waste dumps, usually located at a distance from occupied areas. Human wastes were disposed of in privies and some cities established rudimentary sewage management facilities. These were local responsibilities; the federal government simply was not involved until early in the 20th century.

Perhaps the earliest federal involvement in solid waste management is found in the Public Health Service Act (PHSAct), which in 1913 stated, "The Public Health Service may study and investigate the diseases of man and conditions influencing the propagation and spread thereof, including sanitation and sewage [..]." [79]. While this authority led to research on waste disposal, almost two decades were to pass before there appeared federal legislation specific to management of solid and hazardous waste. In 1965, Congress enacted the Solid Waste Disposal Act. It was the

> In 1965, Congress enacted the Solid Waste Disposal Act. It was the first federal statute focused solely on waste management.

first federal statute focused solely on waste management. Congress had found "[t]hat the problem presented by solid waste disposal was national in scope and necessitated federal action in assistance and leadership" [80]. However, the act also stated that the collection and disposal of solid waste should continue to be primarily the function of state, regional, and local agencies. Under the Solid Waste Disposal Act, funds were made available for research on solid waste disposal. The PHSAct, in effect, continued, but more directly focused, research on waste disposal already authorized in the PHSAct.

The proper disposal of hazardous wastes became a concern of Congress commencing in the 1970s. Described in this chapter are the Solid Waste Disposal Act, the RCRAct, two acts that deal with the permitted disposal of solid and hazardous wastes, and the CERCLAct, enacted by Congress to address the environmental and human health problems caused by uncontrolled hazardous waste, particularly abandoned hazardous waste sites. Also described are federal statutes for controlling dumping of waste into oceans and for preventing environmental pollution from oil spills. The chapter includes a discussion of the Pollution Prevention Act (PPAct), a statute that focuses expressly on the prevention of pollution through means of recycling and improved waste management.

2.3.1 Impacts of Solid Waste on Human Health

The human health consequences of permitted landfills and incinerators have not been the subject of any sustained program of research. However, one study in 2001 of adverse birth outcomes in populations residing near landfill sites found small excess risks of congenital anomalies (neural tube defects, hypospadias, abdominal wall defects) and low to very low birth weight babies. The landfills in the study included some hazardous waste sites [81]. Concerning incinerators, the National Research Council reported in 2000 that few studies have tried to establish a link between an incinerator and illness in the surrounding area, and that most studies found no adverse health effects [82]. In contrast, some studies have shown that municipal incinerator workers have been exposed to high concentrations of dioxins and metals, but any adverse health effects have not been pursued in follow-up studies [83].

Medical waste incinerators are of public health concern because highly toxic dioxins are formed as a byproduct of incinerated plastics materials. EPA has issued standards to reduce emissions from waste incinerators, based on a standard of "maximum achievable control technology" (BACT), and emissions should decrease over time. As public health policy, emissions from incinerators merit scrutiny by state environmental departments in order to assure that harmful emissions are not occurring.

2.3.2 Impacts of Solid Waste on Ecosystem Health

The associations between solid waste and ecosystem health are largely confined to issues of landfills. Incinerated solid waste has not been a subject of ecosystem investigation, given little evidence that a problem exists. Solid waste, especially plastics, that reach oceans and other bodies of water are discussed in a subsequent section of this chapter. Therefore, the ensuing material focuses on the impacts of solid waste landfills on ecosystems. A review of landfills and ecosystem impacts, conducted by the Illinois Chicago Metropolitan Agency for Planning (CMAP) follows [84]:

Hazardous gas emissions: In 1987, EPA estimated that the nation's 7,124 landfills emitted 15 million tons of methane per year and 300,000 tons of other gases like toluene and methylene chloride. As mentioned in Volume 1, Chapter 1, methane is a powerful greenhouse gas and landfills contributed 23% of total emissions in 2006. In addition to its effect in the ozone layer, methane is also a highly combustible gas that might be responsible for various explosion hazards in and around landfills.

Water quality/contamination: There is no expert consensus about the impact of municipal solid waste (MSW) on surface and groundwater sources. Some argue that even common MSW items such as newspaper pose a significant risk to water quality, while others argue that the effect of landfills on groundwater would be negligible if hazardous materials (e.g., motor oil, paint, chemicals, incinerator ashes) were prohibited from the sites. Experts also argue that while leachate is a clear environmental liability, the frequency and severity of leachate-related problems is uncertain and can be minimized through proper siting and sealing measures. However, if leachate does seep into groundwater, it can be the source of many contaminants, specifically organic compounds that may decrease the oxidation-reduction potential and increase the mobility of toxic metals. Locally, some solid waste managers catch errant leachate and pump it back into the landfill. This process helps keep it from seeping away and actually hastens the decomposition of the landfill contents [85].

Energy consumption: As a community's tolerance for landfills decreases, they are moved farther from densely populated areas, requiring collection trucks to drive farther distances to unload. Also, the complexity of collection routes can affect energy consumption. This frequent and lengthy travel by fossil fuel-consuming vehicles is also detrimental to air quality and results in increased greenhouse gases.

Natural habitat degradation: As land is claimed for landfills, it is no longer hospitable to many plants and wildlife. Often, this fertility cannot be completely reclaimed, even after the landfill is capped.

Biodegradation: Responsibly sited and managed landfills are often preferred over other waste disposal methods, such as incineration, because, aside from being more economical, they allow most waste to decay safely and naturally. Conversely, the positive effects of biodegradation are often overstated when, in reality, landfills tend to mummify their contents, severely prolonging oxidation and natural breakdown processes [85].

2.3.3 Solid Waste Disposal Act, 1965

In the early 1960s, cities and towns across the U.S. practiced open air burning of trash. In response, Congress passed the Solid Waste Disposal Act (SWDAct) in 1965

Solid and Hazardous Waste

as part of the amendments to the Clean Air Act. This was the first federal law that required environmentally sound methods for the disposal of household, municipal, commercial, and industrial waste. The SWDAct became law on October 20, 1965. The act declared, "Congress hereby declares it to be the national policy of the United States that, wherever feasible, the generation of hazardous waste is to be reduced or eliminated as expeditiously as possible. Waste that is nevertheless generated should be treated, stored, or disposed of so as to minimize the present and future threat to human health and the environment" [86].

In its original form, it was a broad attempt to address the solid waste problems confronting the U.S. through a series of research projects, investigations, experiments, training, demonstrations, surveys, and studies. The decade following its passage revealed that the act was not sufficiently structured to resolve the growing mountain of waste disposal issues facing the U.S. [87]. As a result, significant amendments were made to the act with the passage of the Resource Conservation and Recovery Act of 1976 (RCRA), which became law on October 21, 1976. The SWDA as amended in 1976 is more commonly known as the RCRA.

2.3.4 Resource Conservation and Recovery Act, 1976

The RCRAct amends earlier legislation, the Solid Waste Disposal Act of 1965, but the amendments were so comprehensive that the act is commonly called RCRA rather than by its official title [88]. This act is the principal solid waste management statute of the U.S. federal government.

> The Solid Waste Disposal Act and the Resource Conservation and Recovery Act provide regulation of solid and hazardous wastes [67].

2.3.4.1 History

The RCRAct established the federal program that regulates solid and hazardous waste management. As an overview, the RCRAct defines solid and hazardous waste, authorizes EPA to set standards for facilities that generate or manage hazardous waste, and establishes a permit program for hazardous waste treatment, storage, and disposal facilities. As a policy, controlling waste releases through a permitting system for individual waste generators emulates the permitting system in the Clean Water Act (CWAct). Amendments to the RCRAct have set deadlines for permit issuance, prohibited the land disposal of many types of hazardous waste without prior treatment, required the use of specific technologies at land disposal facilities, and established a new program regulating underground storage tanks. EPA is also given authority to inspect hazardous waste facilities coverable under the RCRAct and is given enforcement powers to ensure compliance with federal RCRAct requirements [88].

The amounts of waste generated in the U.S. are huge and as such bring challenges to waste managers. As characterized by the National Research Council, the three categories of waste are municipal solid waste, medical wastes, and hazardous waste [82]. An appreciation of the generated amounts and composition of these wastes is useful for public health considerations.

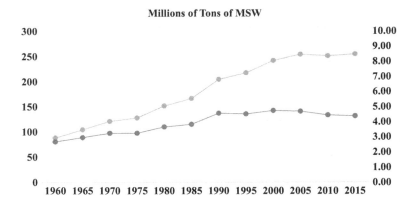

FIGURE 2.3 Municipal solid waste (MSW) generation rates in the U.S., 1960–2013. Top curve: Total MSW; Bottom: per-capital generation. (EPA [83].)

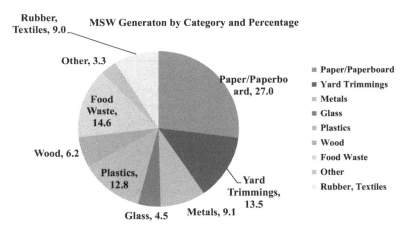

FIGURE 2.4 Total MSW generation, 254 million tons (by material), 2013. (EPA [83].)

Municipal Solid Waste (MSW): This category of waste is defined as "[t]he solid portion of the waste (not classified as hazardous or toxic) generated by households, commercial establishments, public and private institutions, government agencies, and other sources" [82]. The volume of MSW has steadily increased in the U.S., as shown in EPA data in Figure 2.3, although the per capital rate has begun to decrease, commencing in the year 2000, which suggests that programs of recycling have occurred [83]. Shown in Figure 2.4 are the top eight generators in the U.S. of municipal solid waste (MSW) in 2013 with a total volume of 254 million tons. Researchers estimated that more than 2.1 billion tons of MSW are generated globally each year – enough to fill 822,000 Olympic-size swimming pools, which would stretch approximately the length

EPA: About 50% of municipal solid waste is paper and paperboard products.

Solid and Hazardous Waste

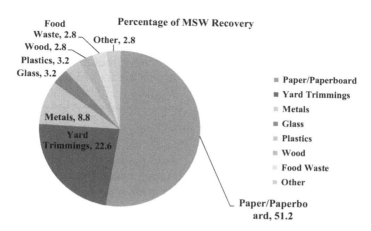

FIGURE 2.5 Percentage of total MSW recovery, 87 million tons (by material), 2012. (EPA [85].)

of Earth's circumstance (24,901 miles). However, only 16% (323 million tons) of this is recycled each year, while 46% (950 million tons) is disposed of unsustainably. The U.S. is a major contributor to the global generation of MSW [89]. The three major contributors, China, India, and the U.S., each has a large population and consumer economies.

Recycling of MSW represents potential economic benefits as well as environmental advantages (i.e., lesser volume of waste disposal). Of the 2013 MSW volume, EPA estimated that the U.S. recycled and composted about 87 million tons of this material, equivalent to a 34.3% recycling rate, a rate that has shown a steady increase. Commending in 1985, MWS recycling rates in the U.S. have increased [85].[1] Shown in Figure 2.5, are seven groups of MSW materials that are recovered through recycling or other means. Noteworthy in Figure 2.5 is that about 50% of this total represents paper and paperboard products. The environmental health policy implication is that recycling of paper and paperboard can significantly reduce the amount of municipal solid waste that is taken to permitted landfills, thereby decreasing landfills' volume and area.

Medical waste: These wastes constitute a particularly important hazard to human health. Medical wastes are generated throughout the U.S. healthcare system. Hospitals, in particular, produce the greatest volume of medical waste, generating, according to one source, about 26 pounds of waste per bed per day. An estimate of the annual volume of generated medical wastes is unknown. A crude division of medical wastes consists of that part that contains infectious pathogens (e.g., HIV) and wastes that do not pose an infectious hazard. Under the provisions of the RCRAct, infectious medical wastes must be incinerated or otherwise handled by permitted waste disposal facilities. Noninfectious medical wastes under the RCRAct can be handled as municipal solid waste and taken to permitted landfills [82].

---

[1] Solid waste refers here to household and industrial wastes, not bodily wastes.

Hazardous waste: Under the provisions of the RCRAct, hazardous waste is a waste material that can be categorized as potentially dangerous to human health or ecosystems. According to EPA, the five greatest forms of hazardous waste are: chemicals (52%), metals (19%), petroleum/coal (13%), metals products (10%), and plastics (6%) [90]. As inferred from the figure, the chemical and allied industries are the top generators. According to the EPA, more than 20,000 hazardous waste producers annually produce more than 40 million tons of hazardous waste regulated under the RCRAct [85]. Of this amount, about 4% is hazardous waste produced in households, i.e., more than 1.6 million tons. Leftover household products that contain corrosive, toxic, ignitable, or reactive ingredients are considered by EPA to be household hazardous waste [90]. Products such as paints, cleaners, oils, batteries, pest poisons, and pesticides contain potentially hazardous substances that require special care for proper waste disposal.

> According to the EPA, more than 20,000 hazardous waste producers annually produce more than 40 million tons of hazardous waste regulated under the RCRAct.

As previously described, the amount of municipal, medical, and hazardous wastes produced in the U.S. is enormous. If one accepts the proposition that waste generation is fundamentally wasteful, having important consequences such as environmental quality (e.g., air pollution, landfills), economic burdens (e.g., cost of waste disposal), health impacts (e.g., effects of air pollution on children's health), and social disruption (e.g., disputes on where to site landfills), what can be done to lessen these impacts? Some waste will always be inevitable, but a policy of waste reduction and minimization comports with good public health practice.

> The RCRAct declares as national policy that wherever feasible, the generation of hazardous waste is to be reduced or eliminated as expeditiously as possible.

EPA promotes the three Rs of waste reduction: Reduce, Reuse, and Recycle [91], to which the authors of this book have added Redesign. The following material elaborates on EPA's three Rs, to which a fourth R is added, Redesign:

- **Redesign**: All products and devices must go through a design phase. This occurs for new products as well as for existing products or devices that already exist, but are being considered for update or revision. At this stage of development, consideration should be given to whether a redesign should occur in order to reduce or eliminate waste.
- **Reduce**: Source reduction, often called waste prevention, means consuming and throwing away less. Source reduction includes purchasing durable, long-lasting goods and seeking products and packaging that are as free of toxics as possible. It can be as complex as redesigning a product to use less raw material in production, have a longer life, or be used again after its original use is completed. Because source reduction actually prevents the generation of waste in the first place. It is the most preferable method of waste management and goes a long way toward protecting the environment [91].

- **Reuse**: Reusing items by repairing them, donating them to charity and community groups, or selling them also reduces waste. Use a product more than once, either for the same purpose or for a different purpose. Reusing, when possible, is preferable to recycling because the item does not need to be reprocessed before it can be used again [91].
- **Recycle**: Recycling turns material that would otherwise become waste into valuable resources and generates a host of environmental, financial, and social benefits. After collection, materials (e.g., glass, metals, plastics, and paper) are separated and sent to facilities that can process them into new materials or products [91].

According to the EPA, recycling is one of the best environmental success stories of the late 20th century. By the agency's assessment, recycling, including composting, diverted 68 million tons of material away from landfills and incinerators in 2001, an increase of 34 million tons from that in 1990 [91]. Further, EPA credited curbside waste recycling programs with producing a diversion of about 30% of the U.S.'s solid waste in 2001.

The RCRAct contains a statement of national environmental health policy, "The Congress hereby declares it to be the national policy of the U.S. that, wherever feasible, the generation of hazardous waste is to be reduced or eliminated as expeditiously as possible. Waste that is nevertheless generated should be treated, stored, or disposed of so as to minimize the present and future threat to human health and the environment" [92]. This pollution prevention policy set a course for subsequent federal and state regulatory action.

The RCRAct is a regulatory statute designed to provide 'cradle-to-grave' control of hazardous waste by imposing management requirements on generators of hazardous waste and transporters, and upon owners and operators of treatment, storage, and disposal facilities. The statute applies principally to operating waste management facilities. The RCRAct deals both with hazardous waste and nonhazardous waste, although the main emphasis in the act is on the former. More than 500,000 companies and individuals in the U.S. who generate more than 172 million metric tons of hazardous waste each year are covered under the RCRAct regulatory programs [87,88]. The RCRAct, as amended, represents a significant challenge to the regulated community. In particular, the industry is challenged to find new ways to minimize, treat, and dispose of hazardous waste. The use of innovative technologies, like bioremediation, to reduce waste is the subject of active research and development.

Wastes covered under the RCRAct are defined in the statute. The RCRAct, Subtitle A, defines *solid waste* as being "any garbage; refuse; sludge from a waste treatment plant, water supply treatment plant, or air pollution control facility; and other discarded material including solid, liquid, semisolid, or contained gaseous material resulting from industrial, commercial, mining, and agricultural operations, and from community activities; but does not include solid or dissolved material in domestic sewage, or solid or dissolved materials in irrigation return flows or industrial discharges which are point sources subject to permits under §402 of the Federal Water Pollution Control Act, as amended, or source, special nuclear, or byproduct material as defined by the Atomic Energy Act of 1954, as amended" [92].

Under the RCRAct, Subtitle A, hazardous waste "means a solid waste, or combination of solid wastes, which because of its quantity, concentration, or physical, chemical, or infectious characteristics may either cause, or significantly contributed to an increase in mortality or an increase in serious irreversible or incapacitating reversible illness; or pose a substantial present or potential hazard to human health or the environment when improperly treated, stored, transported, or disposed of, or otherwise managed" [92].

> The SWAct Amendments of 1980 banned open waste dumps, thereby eliminating a public health hazard that had existed since antiquity.

The RCRAct therefore applies to almost any waste regardless of its physical form. EPA regulations have further clarified the definition of wastes coverable under the RCRAct. It is a complex and comprehensive statute, implemented by EPA through a set of extensive regulations.

2.3.4.2 Amendments to the RCRAct

Commencing in 1980, Congress has enacted seven amendments to the RCRAct [88]. As subsequently elaborated, the amendments of 1980 and 1984 were substantive and made major changes in how solid waste is managed in the U.S. The other amendments were largely technical adjustments to existing legislation. The amendments address specific areas of solid and hazardous waste management but have not changed the basic thrust of the RCRAct's basic principles. Following are descriptions of the major amendments to the RCRAct.

The Used Oil Recycling Act of 1980 amended the SWDAct by defining the terms *used oil*, *recycled oil*, *lubricating oil*, and *re-refined oil*. The EPA Administrator was directed to promulgate regulations to establish performance standards and other requirements necessary to protect the public health and environment from the hazards of recycled oil. Moreover, EPA was authorized to provide grants to states with approved solid waste plans that (1) encourages the use of recycled oil, (2) discourages uses hazardous to the public's health and environment, (3) calls for informing the public of the uses of recycled oil, and (4) establishes a program for the collection and disposal of used oil in a safe manner [88].

Also, in 1980, the Solid Waste Disposal Act Amendments were substantive and gave EPA broader powers to deal with illegal disposal of hazardous waste. Two provisions were of special import. One important provision prohibited open dumping of solid waste and hazardous waste. This prohibition brought to close a human practice that dates to antiquity. Moreover, banning of open dumps was a major public

> **ENFORCEMENT EXAMPLE – RCRA**
>
> (Washington, D.C. – January 18, 2017) On this date a Cedar Falls, Iowa man was sentenced to 2 years in federal prison for knowingly storing hazardous waste without a permit. Cedar Valley Electroplating (CVE), was a large quantity hazardous waste generator, producing more than 1,000 kg of hazardous waste per month. Neither the operator nor CVE had a permit to treat, store, or dispose of hazardous waste under federal law [93].

health contribution. Gone were the open dumps that were rife with disease-carrying vermin and which provided human access to areas that contained decomposing food, hazardous chemicals, and physical hazards. Further, standards were developed for the sanitary disposal of solid waste in dump sites that are designed to prevent the releases of hazardous substances into ambient air and underground aquifers [88].

The other important provision authorized the EPA Administrator to issue orders requiring individual facility operators to do monitoring, testing, analysis, and reporting necessary to abate hazards to human health and the environment. Other key changes included (1) transferred from EPA to the Department of the Interior all responsibilities for managing coal mining wastes, (2) expanded EPA's standards applicable to generators of hazardous waste and their responsibility for the arrival of wastes at waste management facilities, (3) set forth criminal and civil penalties for failures to comply with waste management permits, and (4) directed each state to submit to EPA an inventory of hazardous waste storage and disposal sites.

An even more significant set of amendments to the SWDAct was the Hazardous and Solid Waste Amendments of 1984, comprising six titles and accompanying subtitles [90]. Title I revised findings and objectives of the act to include minimizing the generation and the land disposal of hazardous waste and declared it to be the national policy that, wherever feasible, the generation of hazardous waste is to be reduced or eliminated as expeditiously as possible, but without stating how. This title states, "Waste that is nevertheless generated should be treated, stored, or disposed of so as to minimize the present and future threat to human health and the environment."

One of the most significant provisions of the 1984 amendments to the RCRAct is the prohibition of land disposal of hazardous wastes [90]. In a phased approach, the act bans the disposal in landfills of bulk or noncontainerized liquid hazardous wastes, and hazardous wastes containing free liquids. The act then required EPA to determine whether to ban, in whole or in part, the disposal of all the RCRAct hazardous wastes in land disposal facilities. At the same time, EPA must establish treatment standards for each restricted waste based on the Best Demonstrated Available Technologies (BDAT). If the restricted waste is first treated to BDAT levels, the treated waste or residue can then be placed in land disposal facilities.

The Federal Facility Compliance Act of 1992 amended the Solid Waste Disposal Act to: (1) waive the sovereign immunity of the U.S. for purposes of enforcing federal, state, interstate, and local requirements with respect to solid and hazardous waste management, (2) make federal employees subject to criminal sanctions under such laws, (3) prohibit federal agencies from being subject to such sanctions, (4) require the Secretary of Energy to develop a treatment capacity and technology plan for each facility at which DOE generates or stores mixed wastes, (5) direct the Administrator to promulgate regulations identifying when military munitions become hazardous waste and providing for the safe transportation and storage of such waste, (6) exclude from the definition of "solid waste" solid or dissolved material in domestic sewage, and (7) require the Administrator to establish a program to assist small communities in planning and financing environmental facilities and compliance activities [88].

The Land Disposal Program Flexibility Act of 1996 exempted small U.S. landfills located in arid or remote areas from groundwater monitoring requirements, provided there is no evidence of groundwater contamination. This act also exempts hazardous waste from the RCRAct regulation if it is treated to a point where it no longer exhibits the characteristics that made it hazardous and is subsequently disposed in a facility regulated under the CWAct or SDWAct [94].

2.3.4.3 Key Provisions of the RCRAct Relevant to Public Health

The subtitles of the RCRAct, as amended with particular relevance for public health policies and practices, are discussed in the following sections.

> The Land Disposal Program Flexibility Act of 1996 exempted small U.S. landfills located in arid or remote areas from groundwater monitoring requirements.

> EPA compiles listings of hazardous wastes and develops standards applicable to: generators, transporters, owners and operators of hazardous waste treatment, storage, and disposal facilities.

Subtitle B – *Office of Solid Waste; Authorities of the EPA Administrator* – Establishes the EPA Office of Solid Waste and the authorities of the EPA Administrator in carrying out the provisions of the Act.

Subtitle C – *Hazardous Waste Management* – This part of the act is specific to the management of hazardous waste. §3001: requires that the Administrator develop and promulgate criteria for identifying the characteristics of hazardous waste. §§3002,3001(d), 3003, 3004: require EPA to compile listings of hazardous wastes and to develop standards applicable to: generators of hazardous waste; transporters of hazardous waste; owners and operators of hazardous waste treatment, storage, and disposal facilities. Permits for treatment, storage, and disposal of hazardous waste are required. EPA is given authority to inspect hazardous waste facilities coverable under the RCRAct and is given enforcement powers to ensure compliance with federal RCRAct requirements. Each U.S. state is required to submit to EPA a continuing inventory that describes the location of each site at which hazardous waste has at any time been stored or disposed of. Similarly, each federal agency must provide the same kind of information to EPA. §3004(u): authorizes EPA or a state to require corrective action for all releases of hazardous waste or constituents from any solid waste management unit at a TSDF seeking a permit under Subtitle C, regardless of the time at which the waste was placed in the unit. Under §3008(h) EPA is authorized to assess a civil penalty to any interim status facility that has released hazardous waste into the environment. §3005: requires that each application for a final determination about a permit for a landfill or surface impoundment shall be accompanied by information reasonably ascertainable by the owner or operator on the potential for the public to be exposed to hazardous wastes or hazardous constituents through releases related to the unit. The exposure information is to be provided to EPA, which in turn, shall make it available to ATSDR for public health purposes. When EPA or a state determines that a particular landfill or surface impoundment poses a substantial potential risk to public health, they may request that ATSDR conduct a health assessment of the population at potential risk. However, ATSDR can conduct the requested health assessment "[…] If funds are provided in connection with such request the Administrator of such Agency [i.e., ATSDR] shall conduct such health assessment."

Solid and Hazardous Waste

§3017: sets forth requirements on the export of hazardous waste to other countries. In general, the exporter must furnish EPA with information about the nature and amount of the waste, the country of destination, the ports of entry, the manner of transport, and the name and address of the ultimate treatment, storage, or disposal facility. Following receipt of the export information, EPA must request the Secretary of State to contact the receiving country to obtain that country's written consent to receive the exported hazardous waste.

Subtitle D – *State or Regional Solid Waste Plans* – Regulation of nonhazardous waste, under the RCRAct, is the responsibility of the states. The federal involvement is limited to establishing minimum criteria that prescribe the best practicable controls and monitoring requirements for solid waste facilities. Disposal of solid waste in open dumps is prohibited, but the RCRAct provides EPA with no enforcement authority for banning open dumps. (EPA's enforcement authority under the RCRAct covers only hazardous waste.)

> Disposal of solid waste in open dumps is prohibited, but the RCRAct provides EPA with no enforcement authority for banning open dumps.

Subtitle E – *Federal Responsibilities* – Federal statutes sometimes exempt federal agencies from an act's coverage. This may be for reasons of national security, economic factors, or political reasons. However, the RCRAct holds each department, agency, and instrumentality of the executive, legislative, and judicial branches of federal government having jurisdiction over any solid waste management facility or disposal site, or engaged in any activity resulting, or which may result, in the disposal or management of solid waste or hazardous waste to the same expectations and requirements "[a]s any person is subject to such requirements." Only the President can exempt a department's solid waste management facility if it is in the Nation's paramount interest.

Subtitle F – *Research, Development, Demonstration, and Information* – EPA is given authority to conduct research and studies on a range of areas that include: adverse health and welfare effects of solid waste releases; resource conservation systems [...].

Subtitle G – *Regulation of Underground Storage Tanks* – EPA is directed to develop a comprehensive regulatory program for "underground storage tanks" [95]. The RCRAct directs EPA to promulgate release, detection, prevention, and correction regulation applicable to all owners and operators of underground storage tanks (UST), as may be necessary to protect human health and the environment. EPA estimates there are 700,000 UST facilities with about 2,000,000 tanks covered by this regulation.

2.3.4.4 Coal Ash Regulations under RCRA

Coal has been the fuel of choice for electric power generation for much of the 20th century and continues that status into the 21st century, but in diminishing volumes, as was discussed in Volume 1, Chapter 1 (Climate Change). Indeed, coal has been the fossil energy for industrial development throughout the industrialized world. Coal was relatively abundant and inexpensive. Industrial growth in the U.S. and other countries would have either been slower or even absent were it not for coal. But with coal's utilitarian purpose for power generation came latent legacies. One such legacy is coal ash.

> When coal is burned to produce electricity, toxic chemicals become concentrated in the waste product – coal ash.

A review of the environmental and health consequences of coal ash noted that coal contains a long list of toxic chemicals, including arsenic, radium, and other carcinogens, several metals that can impair children's developing brains, and multiple chemicals that are toxic to aquatic life [96]. When coal is burned to produce electricity, these toxic chemicals become concentrated in the waste product – coal ash. Coal-fired power plants in the U.S. produce around 100 million tons of coal ash every year, with at least 2 billion tons stored in pits of varying quality. For much of the last century, power companies dumped this waste into unlined landfills and waste ponds, where the lack of a barrier between the coal ash and groundwater left them vulnerable to leaks and contamination of underground water supplies. The landfills and waste ponds in essence became uncontrolled hazardous waste sites, the principal environmental consequence being contamination of groundwater and surface bodies of water.

As with many other environmental problems with public health and ecological consequences, hazard interventions and policies did not occur until a disaster brought the problem to the attention of the public and policymakers. In the instance of coal ash, a 2008 coal ash spill in Kingston, Tennessee, led to the release of more than 5 million cubic yards of coal ash, destroying dozens of homes and allegedly contributing to the illness or deaths of scores of cleanup workers [96]. Yet the most enduring legacy of coal ash disposal will undoubtedly be groundwater pollution.

In 2015 U.S. President Barack Obama's (D-IL) administration put in place the first national regulations for coal-burning wastes, favoring dry storage in landfills over wet storage in ponds. Called the "Coal Ash Rule," the rules require utilities to conduct groundwater monitoring at ponds and landfills, close leaking ash ponds, clean up polluted groundwater, and disclose their studies and actions on their websites. These regulations were eased, however, by U.S. President Donald Trump's (R-NY) administration, when in July 2019, the EPA extended by 18 months the time that industry can use unlined coal ash ponds for dumping. On July 30, 2020, the EPA announced it had finalized a new regulation for the more than 400 coal ash pits across the nation, where coal residue is mixed with liquid and stored in open air,

> The groundwater at a majority of coal plants (52%) has unsafe levels of arsenic, which is known to cause multiple types of cancer.

often in unlined ponds. With those extensions, coal ash ponds that were supposed to stop receiving waste by 2021 can keep receiving sludge for two to seven more years. Including the additional time for closing ponds, which allows some pits to stay open as late as 2038 [97].

Separately, using the power companies' data submitted under the Coal Ash Rule, a coalition of nonprofit organizations conducted an extensive analysis of the data [96]. The nonprofit Environmental Integrity Project (EIP), in collaboration with Earthjustice, the Sierra Club, Prairie Rivers Network, and other organizations, obtained and analyzed all of the groundwater monitoring data that power companies posted on their websites in 2018. "The data cover 265 coal plants or offsite coal ash disposal areas, including more than 550 individual coal ash ponds and landfills that are monitored by more than 4,600 groundwater monitoring wells. This represents about three-quarters of the coal power plants across the U.S. The rest of the coal plants have not posted groundwater data either because they closed their ash dumps before the Coal Ash Rule took effect in 2015, or because they were eligible for an extension or exemption. After comparing monitoring data to health-based EPA standards and advisories, our analysis confirmed that groundwater beneath virtually all coal plants is contaminated:

> **CONSENT DECREE**
>
> Harrisburg, PA – July 31, 2019
>
> In a consent decree with four environmental groups, a large central Pennsylvania power plant has agreed to stop tainted water in its coal ash disposal sites from leaking into the Susquehanna River. The Brunner Island Generating Station has agreed to close and excavate one of its active coal ash landfills and address leaks at seven other sites. The plant also will be fined $1 million by the state Department of Environmental Protection. The fine is the largest ever involving coal ash disposal in Pennsylvania [98].

- 91% of coal plants have unsafe levels of one or more coal ash constituents in groundwater, even after we set aside contamination that may naturally occurring or coming from other sources.
- The groundwater at a majority of coal plants (52%) has unsafe levels of arsenic, which is known to cause multiple types of cancer. Arsenic is also a neurotoxin, and, much like lead, can impair the brains of developing children. Of note, the Trump administration in July 2019 proposed scrapping restrictions on arsenic-laden waste from coal-fired power plants. The latest proposal from the EPA would eliminate restrictions from 2015 that limited coal ash use to 12,400 tons per site. The Trump administration proposal would allow projects to use as much coal ash as they want but users would have to demonstrate how the project won't cause harm if it is close to certain features like groundwater or wetlands [99].
- The majority of coal plants (60 percent) also have unsafe levels of lithium, a chemical associated with multiple health risks, including neurological damage.
- The contamination at a given site typically involves multiple chemicals. A majority of coal plants have unsafe levels of at least four toxic constituents of coal ash" [96].

The researchers observed that the threat to groundwater comes from both coal ash ponds and dry coal ash landfills. Monitoring data examined for this report revealed unsafe levels of contamination at 92% of ash ponds and 76% of ash landfills.

The extent of the environmental problem caused by releases of toxic substances from utilities" coal ash ponds has revealed serious groundwater contamination in at least 22 states. At dozens of power plants across the U.S., including many in the Southeast, utilities have found coal-ash pollution severe enough to force them to propose cleanup plans [100]. Additionally, some U.S. states have undertaken coal ash cleanups in the absence of EPA's involvement. For example, the governor of Virginia in January 2019 called for the complete removal of coal ash ponds within the Chesapeake Bay watershed [100]. Some other coal ash mitigations have come by way of litigation, as exemplified by a superior court judge in Raleigh in 2016 ordering Duke Energy to remove coal ash from three plants in eastern North Carolina. The ruling came in lawsuits filed by environmentalists in 2013 [100].

Although several coal ash spills have occurred in the U.S., one merits citation as an example of the vastness of ecological damage caused by releases of toxicants into environmental media. The Kingston Fossil Plant, operated by the Tennessee Valley Authority (TVA), spill of coal fly ash slurry was an environmental and industrial disaster that occurred on December 22, 2008 [101]. A dike ruptured at a coal ash pond at the TVA's Kingston Fossil Plant in Roane County, Tennessee. This event released 1.1 billion U.S. gallons of coal fly ash slurry. The coal-fired power plant, located across the Clinch River from the city of Kingston, Tennessee, used ponds to dewater the fly ash, a byproduct of coal combustion, which is then stored in ponds. The spill released a slurry of fly ash and water, which traveled across the Emory River and its Swan Pond embayment, onto the opposite shore, covering up to 300 acres of the surrounding land.

> The TVA's Kingston Fossil Plant in Roane County, Tennessee, released 1.1 billion U.S. gallons of coal fly ash slurry into environmental media.

The spill damaged multiple homes and flowed into nearby waterways, including the Emory River and Clinch River (tributaries of the Tennessee River). It was the largest fly ash release and worst coal ash-related disaster in U.S. history. Cleaning up the ash spill cost TVA $1.2 billion. The TVA razed 100 of the homes it purchased and transformed the neighborhood near Swan Pond Road into public parks with trails, fishing docks, and sports fields. In November 2018, a federal jury found that the engineering firm hired to conduct clean-up activities had broken its contract with TVA and could be held responsible for the illnesses of 300 workers ranging from high blood pressure to cancer [101].

Perspective: Coal ash is a significant hazardous waste problem in the U.S. Releases from ash ponds and landfills can convey toxic substances such as carcinogenic metals into environmental media such as groundwater. The remediation of coal ash sites could be subject to EPA's actions under the CERCLAct authorities for some sites. In the absence of EPA's

ENFORCEMENT SETTLEMENT

Knoxville, June 13, 2019:
The Tennessee Department of Environment and Conservation and Tennessee Attorney General announced that the state reached a settlement with TVA over coal ash contamination at the Gallatin plant and nearby waterways. The Tennessee Valley Authority has agreed to dig up 12 million tons of coal ash stored in unlined pits at its Gallatin Fossil Plant in Middle Tennessee and clean up contamination from it [102].

Solid and Hazardous Waste

actions, U.S. states will find themselves subject to the challenges of coal ash site remediation, usually a costly proposition.

2.3.5 U.S. State's Solid Waste Act Example

As previously noted, the Solid Waste Act (SWAct), as amended, places solid waste management as primarily the responsibility of the U.S. states. This responsibility is effectuated by enactment of state laws, regulations, and solid waste codes. While state laws vary in content according to specific needs and circumstances, all state laws contain provisions that require permits to manage solid waste. This provision reflects requirements found in the SWAct, as amended. Other provisions are illustrated in one state's solid waste law. Following are excerpted provisions of the State of Georgia's solid waste code [103].

Permits: "(a)[2] No person shall engage in solid waste or special solid waste handling in Georgia or construct or operate a solid waste handling facility in Georgia [...] without first obtaining a permit from the director authorizing such activity.

(b) No permit for a biomedical waste thermal treatment technology facility shall be issued by the director unless the applicant for such facility demonstrates to the director that a need exists for the facility for waste generated in Georgia by showing that there is not presently in existence within the state sufficient disposal facilities for biomedical waste being generated or expected to be generated within the state [...].

(c) On or after March 30, 1990, any permit for the transportation of municipal solid waste from a jurisdiction generating solid waste to a municipal solid waste disposal facility located in another county shall be conditioned upon the jurisdiction generating solid waste developing and being actively involved in, by July 1, 1992, a strategy for meeting the state-wide goal of waste reduction by July 1, 1996.

Permit revocation: (e) (1) The director may suspend, modify, or revoke any permit issued pursuant to this Code section if the holder of the permit is found to be in violation of any of the permit conditions or any order of the director or fails to perform solid waste handling in accordance with this part or rules promulgated under this part [...].

Site modification: (2) Prior to the granting of any major modification of an existing solid waste handling permit by the director, a public hearing shall be held by the governing authority of the county or municipality in which the municipal solid waste facility or special solid waste handling facility requesting the modification is located [...].

(3) Except as otherwise provided in this part, major modifications shall meet the siting and design standards applicable to new permit applications in effect on the date the modification is approved by the director [...]

Site inspection: (j) The director or his designee is authorized to inspect any generator in Georgia to determine whether that generator's solid waste is acceptable for the intended handling facility [...]."

Perspective: A bit of reflection on the Georgia code shows several environmental health policies. The core policy is the permitting of waste managers, with provisions

---

[2] These headings were added for purpose of enhancing clarity. They do not appear in the cited Georgia code.

for permit revocation. State and federal permits for the management of environmental hazards are a feature of many statutes. Permits provide a legal means for application of the command and control policy that leads to regulations and actions that are intended to ensure accountability of solid waste managers.

Other policies of note in the Georgia code include: (1) provisions for upgrading solid waste facilities when they undergo major modifications, with the intention of keeping such facilities in compliance with current management practices, and (2) provisions for on-site inspections by state inspectors. In a sense, both provisions are an expression of the public health policy of prevention of disease and disability.

2.3.6 Solid Waste Policy Issues

The SWAct, as amended, is an example of federalism policy. The states work with a federal agency, EPA, to implement a federal environmental statute. As such, states bear considerable responsibilities for conducting their duties under the Act. For example, federal funding is not within states' control and, therefore, some states must supplement their solid waste program's funding via state funds. On a different policy plane, states must issue permits and conduct inspections of facilities that manage solid waste. These kinds of state policies can lead to variability between states, given differences in sociopolitical conditions and funding.

2.4 U.S. HAZARDOUS WASTE POLICIES

The management of hazardous waste presents challenges to waste managers. By their very nature and definition, hazardous waste conveys a potential risk to human and ecological health. Because of this potential risk, environmental health policies have evolved for the purpose of avoidance of adverse effects of contact with hazardous waste. Of special importance is the matter of **uncontrolled** (emphasis added) hazardous waste. This is distinguished from the **control** (emphasis added) of hazard waste under the authorities of EPA in the RCRAct. As a reminder, under Subtitle C – Hazardous Waste Management – the EPA Administrator must develop and promulgate criteria for identifying the characteristics of hazardous waste, requires EPA to compile listings of hazardous wastes and to develop standards applicable to: generators of hazardous waste; transporters of hazardous waste; owners and operators of hazardous waste treatment, storage, and disposal facilities.

As will be discussed in this section, two forms of uncontrolled hazardous waste led to action by Congress. These forms were uncontrolled (i.e., abandoned or otherwise not appropriately managed) hazardous waste sites and the unintentional release of hazardous substances via spills and accidental releases.

2.4.1 Impacts of Uncontrolled Hazardous Waste Sites on Human Health

The consequences to the health of persons who reside near uncontrolled hazardous waste sites (HWS) are both feared and real. Communities located near these kinds of sites often create grassroots groups that actively express their fears that excess cancer rates and reproductive health problems are caused by substances released from HWS into their midst. Although relating a specific community's health problems to a given HWS is investigatively very challenging, there is, nonetheless, a compelling body of epidemiological and toxicological data that associates adverse effects on community health with residential proximity to some HWS.

The effects on human reproductive outcomes from exposure to hydrocarbon solvents such as trichloroethylene released from HWS constitute the strongest evidence for an HWS–human health association. According to one source who comprehensively reviewed relevant epidemiological literature, residential proximity to HWS is associated with lower birth weight and an increased risk of congenital malformations that include defects of the heart, neural tube, and oral palate [104]. Particularly compelling were findings from two investigations of two HWS: Love Canal, New York, and Lipari, New Jersey [105,106]. For both sites, average birth weights decreased during the span of time when documented releases of hazardous substances were migrating from the waste sites. For both HWS, when the releases of substances were interdicted, mean birth weights in the geographic areas returned to normal. In 2002, a study of congenital anomalies and residence near hazardous waste landfill sites compared 245 cases of chromosomal anomalies to 2,412 controls that lived near 23 landfills in Europe [107]. After adjusting for confounders, a higher risk of chromosomal anomalies was found in people who lived close to sites (0–3 km) than in persons who lived farther away.

> The effects on human reproductive outcomes from exposure to hydrocarbon solvents released from HWS constitute the strongest evidence for an HWS–human health association.

Any association between residential proximity to HWS and elevated cancer rates is less well substantiated than for adverse reproductive effects. A review of the epidemiological literature indicates elevated rates of certain cancers, primarily those of the urinary bladder and gastrointestinal tract, in counties that contain HWS and for which groundwater contamination was either documented or assumed (e.g., [108,109]). There also exists published work that associates increased rates of childhood leukemia with the presence of trichloroethylene (TCE) in municipal wells that supplied segments of Woburn, Massachusetts, with residential water (e.g., [110,111]).

There are also toxicological data that have importance for cancer rates in communities impacted by releases of substances from HWS. One source examined the most frequently occurring substances released into groundwater supplies and noted that of these 30 chemicals, 18 were known or reasonably anticipated to be human carcinogens [104]. Given the long latency associated with most cancers, whether these toxicological observations portend any increase in cancer rates will not be known for many years. But generally, clusters of cancer in persons residing near HWS have not been identified. This outcome could be a consequence of investigating any kind of disease clusters, given the required rigor needed for epidemiological associations between outcome and potential causal factors. It is also possible that actual exposure to carcinogens released from HWS did not occur.

In general, there is sufficient published scientific data to designate some uncontrolled hazardous waste sites as a hazard to the public's health. Given this knowledge, and as a matter of public health, remediation of HWS is an example of primary disease prevention in action. That is, adverse health effects are prevented by the elimination of the causal hazard.

2.4.2 Impacts of Uncontrolled Hazardous Waste Sites on Ecosystem Health

As discussed, exposure to uncontrolled hazardous waste can be harmful to human health, depending on the circumstances of substances' toxicity and receptor characteristics. Additionally, exposure to hazardous waste can be extremely harmful to plants and animals. Hazardous waste stunts plant growth, much of which is useful to humans for consumption or manufacturing. In addition, the elimination of plant life reduces the natural food supply for feral and domesticated animals. Similarly, hazardous waste can harm fish and other animals in bodies of water contaminated by hazardous waste.

2.4.3 Nuclear Waste Policies and Issues

A major environmental concern related to nuclear power is the creation of radioactive wastes such as uranium mill tailings, spent (used) reactor fuel, and other radioactive wastes. These materials can remain radioactive and dangerous to human health for thousands of years. The U.S. has more than 90,000 metric tons of nuclear waste that requires disposal according to a study by the Government Accountability Office (GAO) [112]. The U.S. commercial power industry alone has generated more waste (nuclear fuel that is "spent" and is no longer efficient at generating power) than any other country – nearly 80,000 metric tons. This spent nuclear fuel, which can pose serious risks to humans and the environment, is enough to fill a football field about 20 m deep. The U.S. government's nuclear weapons program has generated spent nuclear fuel as well as high-level radioactive waste and accounts for most of the rest of the total at about 14,000 metric tons, according to the Department of Energy (DOE). For the most part, this waste is stored where it was generated – at 80 sites in 35 states.

The amount of nuclear waste is expected to increase in the U.S. to about 140,000 metric tons over the next several decades. However, there is still no disposal site in the U.S. After spending decades and billions of dollars to research potential sites for a permanent disposal site, including at the Yucca Mountain site in Nevada that has a license application pending to authorize construction of a nuclear waste repository, the future prospects for permanent disposal remain unclear. In 2018, President Trump's fiscal year 2019 budget request included $120 million for the resumption of the license review for the repository

> The amount of nuclear waste is expected to increase in the U.S. to about 140,000 metric tons over the next several decades. However, there is still no disposal site in the U.S.

at Yucca Mountain and for interim storage of nuclear waste. As of February 2019, Congress has not directed funding for the license application to resume. It remains a fact that the U.S. and all other countries that have accumulated nuclear waste have no policy on how to safely manage it.

2.4.4 Comprehensive Environmental Response, Compensation, and Liability Act, 1980

As commented by political pundits, nothing motivates policymakers to act more quickly and resolutely than the occurrence of a catastrophe. No legislator or other policymaker wants to be characterized as having turned a blind eye and deaf ear to dealing with a disaster. The Federal Meat Inspection Act (Volume 1, Chapter 4) is but one of several examples of environmental health policies that could be termed "disaster reaction" legislation. Another example, as will be evident from its history, is the CERCLAct of 1980.

2.4.4.1 History

The CERCLAct (aka Superfund) was enacted in 1980 and was reauthorized by the Superfund Amendments and Reauthorization Act of 1986 [113]. The statute was a direct consequence of the discoveries of releases of hazardous substances from abandoned landfills into community residences, in particular, the community of Love Canal, a suburb of Niagara Falls, New York, which was evacuated following the discovery that it overlays an abandoned chemical dump. Decades earlier, a chemical company had used an abandoned canal as a dump for chemical waste stored in metal barrels. The dump was eventually covered with layers of soil and a housing development built over what had become a forgotten chemical waste dump. As the years passed, some metal barrels rusted and their stored chemicals leached into the soil and vapors seeped into homes build over the chemical dump.

The story of Love Canal captured the public's attention because of intense news-media coverage. Rarely did a day pass without national newsmedia interviewing Love Canal residents, who expressed their concerns about the health of their children and future generations of children. They associated health problems in the community's children with the release of noxious chemicals that had seeped into their homes. As will be subsequently described in this chapter, health investigations confirmed some of the residents' fears that adverse health outcomes had occurred.

The drama of Love Canal occurred during the waning months of U.S. President Jimmy Carter's (D-GA) administration. In 1979, the federal government offered to buy the homes of Love Canal residents and assisting with their relocation elsewhere. Approximately 950 families were evacuated from the Love Canal area [114]. Over the next 20 years, more than $400 million was spent to remediate the Love Canal area. While this is an impressive expenditure, it is noteworthy that the 21,000 tons of chemical soup that characterized the Love Canal site are still there. To remediate the site, EPA capped it with a thick layer of clay, installed pumps and drains to control runoff of chemicals from the site, and replaced miles of contaminated sewer pipe. The chemicals themselves were left in the contained site and the area was surrounded by a fence [115].

As a policy, leaving hazardous substances in place, but interdicting human contact with them has evolved into a risk management decision by EPA, states, and some private sector entities. The theory is that interdiction of contact between hazardous substances and humans and ecological systems will prevent adverse effects. The costs of containment are generally less than for the removal of contaminated soil or water. In theory, the cost savings could be used to remediate more sites that require cleanup. On the other side of this argument is the problem that failure to remove hazardous contamination can simply prolong the life of a hazardous waste site and there are no guarantees that future cleanup actions will ever occur. Moreover, even contained sites, using current best available technology, will in time deteriorate. Therefore, the costs of site maintenance and upkeep are passed along to future generations.

In March 2004, EPA removed the Love Canal site from its list of most significant uncontrolled hazardous waste sites, ending 21 years of government and community concern that uncontrolled hazardous waste was a threat to public health. Today the former Love Canal neighborhood is called Black Creek Village, a neighborhood constructed largely of new houses [115].

The CERCLAct was therefore the product of great public concern that toxic materials could invade private homes and cause harm to children and future generations. The intent of the law is stated to be, "To provide for liability, compensation, and emergency response for hazardous substances released into the environment and the cleanup of inactive hazardous waste disposal sites." The CERCLAct's basic purposes are to provide funding and enforcement authority for remediating (i.e., cleaning up) uncontrolled hazardous waste sites, and for responding to hazardous substance spills. The statute includes provisions for remediating waste sites, responding to public health concerns, enforcement authorities to identify "potential responsible parties," and emergency removal of chemical spills [116].

> The CERCLAct, as revised, requires EPA, in cooperation with states, to identify and remediate uncontrolled hazardous waste sites.

The Hazardous Substance Trust Fund was created by the CERCLAct, which was intended by Congress to be a source of funds for site remediation when other sources were unavailable. Until 1995 the trust fund was financed primarily by a tax on crude oil and certain chemicals and an environmental tax on select corporations [117]. The trust fund also receives revenue from interest accrued on the unexpended balance, recovery of cleanup costs from responsible parties, and collections of fines and penalties [117]. The authority for these taxes expired in December 1995 and has not been reauthorized by Congress. Neither the U.S. President Bill Clinton (D-AR) administration nor subsequent administrations sought reauthorization of the Superfund taxes. This trust fund fulfills part of the Act's philosophy of "the polluter pays" for environmental cleanup. (The other part of this philosophy is the authority given to EPA to identify polluting parties and force them to bear the cost of site remediation.)

The CERCLAct places special emphasis on those uncontrolled hazardous waste sites (HWS) ranked by EPA to pose the greatest hazard to human health and natural resources damage. The worst of the HWS comprise what is called the National Priorities List (NPL). As a policy, ranking the worst HWS provides decision-makers at EPA and states with a means to prioritize sites and remediate first those sites posing the potential greatest risk to human and ecological health. Under the CERCLAct, *Potentially Responsible Parties* (PRPs) must be identified by EPA and costs recovered from them to pay for site remediation. Further, the CERCLAct stipulates that all NPL sites must receive a health assessment conducted by ATSDR for the purpose of ascertaining any adverse effects in persons impacted by HWS. Uncontrolled waste storage/treatment facilities and former manufacturing facilities constitute about 75% of NPL sites. Both categories represent industrial operations that operated in the past and then went out of operation, leaving a legacy of hazardous waste in the environment.

> Under the CERCLAct, PRPs must be identified by EPA and costs recovered from them to pay for site remediation.

Sites can be placed on the NPL by three mechanisms: (1) EPA's Hazard Ranking System (HRS), (2) states or territories designate one top-priority site regardless of HRS score, or (3) ATSDR has issued a health advisory that recommends relocating people from the site [118]. The HRS uses a structured analysis for scoring and ranking sites. This approach assigns numerical values to factors that relate to risk, based on conditions at the site under consideration. Sites can be deleted from the NPL if EPA determines that no further action is required to protect human health or the environment.

At the heart of the CERCLAct process is the process of identification, inspection, remediation, and closure of NPL sites. The process used by EPA to effectuate this process is complex in details, but fairly straightforward overall if considered as a step-by-step program.

The CERCLAct's central philosophy is to require parties, called the PRPs, to bear the costs of remediating sites to which the parties had contributed wastes and for costs attending environmental and health problems created by releases of substances from the waste sites. PRPs include the past and current owners and/or operators of a site, those who arranged for the transportation of hazardous substances to the site, and those who arranged for the treatment or disposal of the substances, and they are subject to retroactive liability. This is established under the legal concept of *retroactive joint and several liabilities* for parties whose wastes had contributed to environmental degradation at the waste sites. The concept means that a company or other accountable entities that long ago had disposed of wastes could now be held responsible for all of a site's remediation costs unless other responsible parties can be identified and costs shared. Needless to say, retroactive joint and several liabilities have often led to litigation as to who pays what portion of a site's remediation [117]. The large ligation that accompanies identifying parties to pay for waste site cleanups led some to half-jokingly call the CERCLAct the "Lawyers Full Employment Act!"

In addition to the identification, ranking, remediation, and cost recovery provisions of the CERCLAct that pertain to uncontrolled releases of hazardous substances from sites, the CERCLAct contains other provisions of relevance to public health [119]:

- Removal actions are conducted by EPA in instances where a short-term, limited response to a manageable environmental release is indicated (e.g., spills of hazardous substances from transportation mishaps), rather than a long-term remedy (i.e., remediation) is indicated, e.g., for an NPL site.

- The CERCLAct of 1980 created ATSDR within the U.S. Public Health Service for the purpose of investigating public health implications of hazardous substances in the community environment, with emphasis on those released from NPL sites.

> Removal actions are conducted by EPA under CERCLA in instances where a short-term, limited response to a manageable environmental release is indicated.

ATSDR conducts public health assessments of all NPL sites, develops Toxicological Profiles for priority hazardous substances, conducts epidemiological and other applied research, provides medical education for physicians and other health professionals, responds to emergency releases of hazardous substances (e.g., through transportation spills), and maintains a national registry of persons exposed to some specific hazardous substances known to have been released from NPL sites [119].

Sometimes confusion arises about the differences between the CERCLAct and the RCRAct. For example, what does one law cover that the other law does not? Is there a difference between the authorities of EPA and the states? Do the laws have different purposes? The key differences between the CERCLAct and the RCRAct are summarized in Table 2.2. As shown in the table, states have the primary responsibility for the RCRAct facilities, whereas the federal government (i.e., EPA) has primacy on CERCLAct sites. It should be noted that most states also have their own programs to remediate those uncontrolled hazardous waste sites that were not designated as NPL sites by EPA.

2.4.4.2 Key Provision of the CERCLAct Relevant to Public Health

The CERCLAct, as amended in 1986, contains four titles, under which are found the various sections that constitute the statute. Several standard references contain all sections in the statute (e.g., [117]). The following excerpts are relevant to the CERCLAct's public health authorities.

Title I – *Provisions Relating Primarily to Response and Liability:*

§101 –- Definitions – (14) The term *hazardous substance* means (A) any substance designated pursuant to §11(b)(2)(A) of the Federal Water Pollution Control Act [...], (B) any element, compound, mixture, solution, or substance designated pursuant to §9602 of this title, (C) any hazardous waste having the characteristics identified under or listed pursuant to §3001 of the Solid Waste Disposal Act [..]. The terms "remove" or

TABLE 2.2
Comparison of key differences between RCRA and CERCLA

RCRA	CERCLA
Focus is on *controlled* facilities that treat, store, and destroy solid and hazardous waste	Focus is on *uncontrolled* hazardous waste sites
States have primacy in setting emission standards and enforcement	Federal government has primacy in setting cleanup standards and enforcement

"removal" mean the cleanup or removal of released hazardous substances from the environment, [o]r the taking of such other actions as may be necessary to prevent, minimize, or mitigate damage to the public health or welfare or to the environment [...]. (24) The terms *remedy* or *remedial action* means instead of or in addition to removal actions in the event of a release or threatened release of a hazardous substance into the environment, to prevent or minimize the release of hazardous substances so that they do not migrate to cause substantial danger to present or future public health or welfare or the environment. [...]

§104 – (i) Agency for Toxic Substances and Disease Registry; establishment, functions, etc. (1) There is hereby established within the Public Health Service an agency, to be known as the Agency for Toxic Substances and Disease Registry, which shall report directly to the Surgeon General of the U.S. Health officials effectuate and implement the health-related authorities of this chapter. [...].

§120 – Federal facilities – (a) Application of chapter to Federal Government: (1) In general – Each department, agency, and instrumentality of the U.S. (including the executive, legislative, and judicial branches of government) shall be subject to, and comply with, this chapter in the same manner and to the same extent, both procedurally and substantively, as any nongovernmental entity, including liability under §9607 of this title.

§121- Cleanup standards – (b) General rules – [T]he President shall select a remedial action that is protective of human health and the environment, that is cost-effective, and that utilizes permanent solutions and alternative treatment technologies or resource recovery technologies to the maximum extent practicable [...].

Title IV – *Radon Gas and Indoor Air Quality Research* – includes sections on findings (§402); radon gas and indoor air quality research programs (§403), authorizations (§405).

> **ENFORCEMENT EXAMPLE - CERCLA**
> Washington, D.C. – On August 23, 2017, a consent decree between the U.S., state of Wisconsin, and settling defendants NCR Corporation (NCR) and Appvion, Inc., was approved by the U.S. District Court for the Eastern District of Wisconsin. Under the consent decree, NCR will perform and fund the remaining contaminated sediment cleanup work at the Fox River Superfund site in Green Bay, Wisconsin at an estimated cost of more than $200 million [120].

2.4.5 HEALTH SIGNIFICANCE OF NPL SITE REMEDIATION

The National Priorities List (NPL) sites are the uncontrolled hazardous waste sites of greatest importance under the terms of the CERCLAct. The NPL list provides EPA with a priority for which sites to remediate. In a sense, NPL sites are the worst amongst the worst. The remediation of NPL sites is a substantive

> Businesses covered by CERCLA Title III must notify state and local emergency planning entities of the presence and amounts in inventory of hazardous materials on their premises.

TABLE 2.3
Status of NPL Sites, Fiscal Years 2018–2013 [120a]

Action	2018	2017	2016	2015	2014	2013
Proposed NPL sites	53	49	53	53	49	54
NPL sites	1,338	1,342	1,337	1,324	1,322	1,315
Deleted NPL sites	412	394	392	390	384	370
NPL sites with construction completions	1,205	1,195	1,187	1,177	1,164	1,156
NPL sites with partial deletions	67	65	62	62	61	58

contribution to human and ecological health. Further, remediation also translates into improved economic benefits as property unfit for development is converted into property of worth. Remediated NPL sites remove the hazardous materials that pose a health hazard to persons residing in the vicinity of sites.

The number and status of EPA's listed NPL sites is shown in Table 2.3. The number of active NPL sites in 2019 was 1,335, with 42 sites removed from the NPL in 2019. While these numbers may seem great or small, depending on one's perspective, it must be borne in mind that remediation of NPL sites is a time-consuming and costly effort, ranging in years and millions of dollars.

2.4.6 Trump EPA Superfund Sites & Climate Change

As elaborated in Volume 1, Chapter 1, the Trump administration adopted a policy of climate change denial, with the president announcing climate change was a "hoax." The incoming Biden administration was confronted in 2021 with 34 Superfund sites nationwide for which there was no reliable cleanup funding – the largest backlog of "unfunded" sites in 15 years. Of perhaps even greater urgency, a GAO analysis in 2019 found that 945 Superfund sites are vulnerable to extreme weather events that are intensifying because of human-caused climate change, including hurricanes, flooding, sea level rise, increased precipitation, or wildfires [121]. The warning of the potential impact of climate change outcomes on releases of toxic substances from Superfund sites was ignored by the Trump EPA. The Biden administration will face a challenge in locating funds to remediate Superfund sites that present the greatest health and ecological threats in conjunction with climate change impacts.

2.4.7 Emergency Planning and Community Right-to-Know Act, 1986

Title III – *Emergency Planning and Community Right-to-Know* – The Emergency Planning and Community Right-to-Know Act of 1986 was enacted as a freestanding provision of the Superfund Amendments and Reauthorization Act of 1986 [122], constituting Title III of the CERCLAct, as amended. Title III is a significant statement from Congress concerning the obligations of government and private industry to protect the public from releases of hazardous substances. Under Title III, state and local governments are required to develop emergency plans for responding to unanticipated environmental releases of acutely toxic materials [123]. (Title III required different kinds of notifications according to groups of chemical substances.) Additionally, businesses covered by Title III must notify state and local emergency planning entities of the presence and amounts in inventory of hazardous materials on their premises and to notify federal, state, and local authorities of planned and uncontrolled environmental releases of those substances. Regulatory agencies, in turn, are required to make available to the public the data on releases of substances in the environment.

Title III includes sections on the establishment of state commissions, planning districts, and local committees (§301); substances and facilities covered and notification (§302), comprehensive emergency response plans (§303), emergency notification (§304), emergency training and review of emergency systems (§305), material safety data sheets (§311), emergency and hazardous chemical inventory forms (§312), toxic chemical release forms (§313), trade secrets (§322), providing information to health professionals (§324); public availability of plans, data sheets, and follow-up notes (§324); enforcement (§325), and regulations (§328).

EPA has developed a reporting system for businesses to use when providing information required of them for placement in the Toxics Release Inventory (TRI).[3] The TRI database can be accessed through the EPA website (http://www.epa.gov). Citizen and environmental groups have used the TRI data to bring attention to the amounts of hazardous substances released by industry within geographic areas of concern. The impact has been to bring the weight of public opinion to bear on companies' reduction of emissions.

ENFORCEMENT EXAMPLE

Gibbsboro, NJ, May 30, 2019 – Sherwin-Williams Co. will pay an estimated $21 million to remove contaminated soil from a 19-acre Superfund site once used for paint disposal in Gibbsboro, according to a recent consent decree with the EPA. Separately, the firm has agreed to work with Camden County to dredge tainted sediment from Kirkwood Lake in Voorhees [124].

Another use of TRI data is for research purposes. As an example of one organization's use of TRI data, the Greater Boston Physicians for Social Responsibility found that of the 20 TRI chemicals with the greatest total releases into the environment, about 75% were known or suspected neurotoxicants [125]. On a more positive note, EPA reported in 2005 that TRI data showed that the amount of toxic substances released into the U.S. environment had declined 42% between the years 1999 and 2003 [126]. It is unclear if this decline is a product of relaxed changes in how industry reports emissions data to EPA, or actual reductions due to public pressure, or reductions as a consequence of federal, state, or local emission standards.

2.4.8 EPA's Brownfields Program

Brownfields is an EPA program akin in concept to the CERCLAct program of site remediation. A brownfield is a property, the expansion, redevelopment, or reuse of

---

[3] As described in this chapter, the Pollution Prevention Act of 1990 requires that industrial facilities also report data on recycling of wastes and other information on pollution prevention.

which may be complicated by the presence or potential presence of a hazardous substance, pollutant, or contaminant [127]. Brownfields are often urban properties located in underserved communities. Since its inception in 1995, EPA's Brownfields Program is designed to empower states, communities, and other

> A brownfield is a property, the expansion, redevelopment, or reuse of which may be complicated by the presence or potential presence of a hazardous substance, pollutant, or contaminant.

stakeholders in economic redevelopment to work together in a timely manner to prevent, assess, safely clean up, and sustainably reuse brownfields. It is estimated that there are more than 450,000 brownfields in the U.S. Remediating and reinvesting in these properties increases local tax bases, facilitates job growth, utilizes existing infrastructure, takes development pressures off of undeveloped, open land, and both improves and protects the environment.

Beginning in the mid-1990s, EPA provided small amounts of seed money to local governments that launched hundreds of two-year Brownfields "pilot" projects and developed guidance and tools to help states, communities, and other stakeholders in the cleanup and redevelopment of Brownfields sites. The 2002 Small Business Liability Relief and Brownfields Revitalization Act (the "Brownfields Law") codified many of EPA's practices, policies, and guidance. The Brownfields Law expanded EPA's assistance by providing new tools for the public and private sectors to promote sustainable Brownfields cleanup and reuse [127].

EPA asserted in 2018, "Through fiscal year 2018, on average, $16.86 was leveraged for each EPA Brownfields dollar and 8.6 jobs leveraged per $100,000 of EPA brownfields funds expended on assessment, cleanup, and revolving loan fund cooperative agreements. A 2017 study concluded that cleaning up brownfield properties led to residential property value increases of 5–15.2% within 1.29 miles of the sites. Analyzing data near 48 of those brownfields, another study found an estimated $29 to $97 million in additional tax revenue for local governments in a single year after cleanup – 2 to 7 times more than the $12.4 million EPA contributed to the cleanup of those brownfields. Initial anecdotal surveys indicate a reduction in crime in recently revitalized brownfields areas" [128]. Shown in Table 2.4 is a cumulative summary of EPA's Brownfields Program accomplishments. Perhaps the most noteworthy statistic

TABLE 2.4
Brownfields Program Accomplishments as of February 1, 2019 [128]

Performance Measure	Accomplishments
Properties assessed	28,697
Properties cleaned up	1,816
Jobs leveraged	144,800
Dollars leveraged	$27,527 Billion
Properties made ready for reuse	7,262
Acres made ready for anticipated reuse	80,952

in the table is the 7,262 properties made ready for reuse. These properties can be used for parks, greenspace, playgrounds and other uses that can promote improved community well-being and public health.

2.5 KEY GLOBAL HAZARDOUS WASTE POLICY

The generation and management challenge of hazardous waste is a global problem. Previously discussed were the U.S. policies regarding the management of hazardous waste. In this section is discussed a key international policy developed in regard to managing hazardous waste. The Basel Convention on the Control of Transboundary Movements of Hazardous Wastes and their Disposal (commonly referred to as the Basel Convention) was adopted on March 22, 1989, in Basel, Switzerland [129]. This international covenant was in response to a public outcry following the discovery, in the 1980s, in Africa and other parts of the developing world of deposits of toxic wastes imported from abroad. The Convention entered into force in 1992. The objective of the Convention is stated as: "to protect human health and the environment against the adverse effects of hazardous wastes."

The Convention's scope of application covers a wide range of wastes defined as "hazardous wastes" based on their origin and/or composition and their characteristics, as well as two types of wastes defined as "other wastes" – household waste and incinerator ash.

The provisions of the Convention center around the following principal aims:

- The reduction of hazardous waste generation and the promotion of environmentally sound management of hazardous wastes, wherever the place of disposal.
- The restriction of transboundary movements of hazardous wastes except where it is perceived to be in accordance with the principles of environmentally sound management.
- And a regulatory system applying to cases where transboundary movements are permissible.

The regulatory system is the cornerstone of the Basel Convention as originally adopted. Based on the concept of prior informed consent, it requires that, before an export may take place, the authorities of the State of export notify the authorities of the prospective States of import and transit, providing them with detailed information on the intended movement. The movement may only proceed if and when all States concerned have given their written consent. The Basel Convention also provides for cooperation between parties, ranging from the exchange of information on issues relevant to the implementation of the Convention to technical assistance, particularly to developing countries. In the event of a transboundary movement of hazardous wastes having been carried out illegally, the Convention attributes responsibility to one or more of the States involved, and imposes the duty to ensure safe disposal, either by re-import into the State of generation or otherwise [129].

An example of the Basel Convention's work occurred in 2019 when almost all the world's countries agreed to restrict shipments of hard-to-recycle plastic waste to poorer countries [130]. Exporting countries – including the U.S. – will now have to obtain consent from countries receiving contaminated, mixed or unrecyclable plastic waste. Currently, the U.S. and other countries can send lower-quality plastic waste to private entities in developing countries without getting approval from their governments. The U.S. is not a signatory to the Basel Convention.

2.6 U.S. OIL POLLUTION POLICIES

Discussed in this chapter to this point are matters of solid waste, including descriptions of municipal solid waste and both controlled and uncontrolled hazardous waste. Liquid waste, exemplified by oil spills, is also a matter of environmental and health concern. As a consequence of increased vast amounts of oil being shipped in huge oil tankers, oil spills occurred when tankers have ruptured. Further, the U.S. possesses a large network of oil pipelines that are used to move oil supplies from points of access to points of refining. When pipelines rupture, oil spills occur, resulting in environmental contamination. As a consequence of these kinds of oil spills, U.S. federal legislation was enacted, as described in this section.

2.6.1 OIL POLLUTION ACT, 1990

Environmental disasters can serve as impetus for federal and state/provincial policymaking. Legislators are loathe to be seen as ignoring catastrophes that have drawn public attention. A case in point is the U.S. Oil Pollution Act of 1990, which followed a massive oil spill in Alaska [131].

The OPAct was enacted by Congress in reaction to the large number of oil spills occurring annually in the U.S., and in particular, the legislation was markedly influenced by the 1989 oil spill in Alaska from the ruptured vessel *Exxon Valdez* [132]. The act is a comprehensive amendment to the CWAct, §311. The OPAct is designed to enhance oil spill prevention, preparedness, and response capabilities and authorities of government agencies. The act established a new liability and compensation regime for oil pollution incidents in the aquatic environment and provided the resources necessary for the removal of discharged oil.

The Oil Spill Liability Trust Fund, which is similar in statutory concept to the CERCLAct's Hazardous Substance Trust Fund, created a $1 billion fund to be used to respond to, and provide compensation for damages caused by, discharge of oil. Similar to authorities in the CERCLAct, the OPAct provides that the responsible party for a vessel or a facility from which oil is discharged, or which poses a substantial threat of a discharge, is liable for certain damages and costs of

ENFORCEMENT EXAMPLE – CWACT

Under provisions of the Clean Water Act, the pipeline company Enbridge was fined $61 million as part of a $177 million settlement between the company and EPA for a massive 2010 oil spill into Michigan's Kalamazoo River. The spill cost more than a billion dollars to remediate. The settlement requires the company to pay for cleanup costs and to spend at least $110 million on sp6ll prevention safeguards [133].

removal of oil. In addition, the OPAct provides new requirements for contingency planning both by government and industry and establishes requirements on construction (e.g., §4115 of the act mandates newly constructed tank vessels must be equipped with double hulls, with the exception of vessels used only to respond to discharges of oil or hazardous substances), manning, and licensing for tank vessels.

When oil spills occur, the authorities and resources of the National Contingency Plan are used to quickly respond to emergency conditions that involve the release of oil or hazardous substances. The NCP, because it is based on law, brings together the coordination between federal government agencies and others in order to protect the public's health and the well-being of ecosystems. As a policy, yoking government agencies in a legally binding rope of coordination well serves the public, since such cooperation is often difficult to achieve in the absence of law.

The most severe oil spill that ever occurred was on April 20, 2010, in the Gulf of Mexico when the BP (British Petroleum) *Deepwater Horizon* oil rig exploded 50 miles off the Louisiana coast, killing 11 oil rig workers. An estimated 210 million gallons of oil and 1.8 million gallons of chemical dispersants seeped into the deep ocean, polluting the coast and the seafloor. This oil pollution has been characterized by some as the most devastating environmental disaster in U.S. history [134]. It soon became the largest and most expensive marine oil spill in history, with an estimated $17.2 billion in damages to properties, fisheries, and tourism across the Gulf Coast. In July of 2015, BP agreed to pay a $20 billion settlement over 15 years. However, further litigation against BP resulted in additional charges against the oil company. A Court Supervised Settlement Program was established in response to more than 400,000 individual claims against BP. As of 2018, the BP Deepwater Horizon costs had risen to $65 billion [134].

A full report on the extent of the oil impacts revealed that hundreds of thousands of sea turtles, marine mammals, and birds died, along with trillions of larval fish. An area 20 times the size of Manhattan is still polluted and buried on the Gulf seafloor.

In response to the BP oil spill, U.S. President Barack Obama (D-IL) signed an executive order creating a commission to study the spill. The commission recommended new safety rules, accountability standards, and environmental regulations for drilling in U.S. waters. Obama then signed another executive order to promote environmental stewardship of the ocean, coasts, and the Great Lakes in light of the oil spill. In June 2018, President Trump revoked Obama's stewardship directive, replacing it with a new executive order giving more responsibility to states for offshore oil and gas drilling, as well as prioritizing business interests ahead of the environment [135].

> The *Deepwater Horizon* oil spill of April 2010 has been characterized as the most devastating environmental disaster in U.S. history.

2.6.2 National Oil and Hazardous Substances Pollution Contingency Plan

The National Oil and Hazardous Substances Pollution Contingency Plan, more commonly called the National Contingency Plan or NCP, is the federal government's blueprint for responding to both oil spills and hazardous substance releases [136].

The NCP is the result of efforts to develop a national response capability and promote coordination among the hierarchy of responders and contingency plans.

The first NCP was developed and published in 1968 in response to a massive oil spill from the oil tanker *Torrey Canyon* off the coast of England. More than 37 million gallons of crude oil spilled into the water, causing massive environmental damage. To avoid the problems faced by response officials involved in this incident, U.S. officials developed a coordinated approach to cope with potential spills in U.S. waters. The 1968 plan provided the first comprehensive system of accident reporting, spill containment, and cleanup. The plan also established a response headquarters, a national reaction team and regional reaction teams (precursors to the current National Response Team and Regional Response Teams).

Congress has broadened the scope of the NCP over the years. As required by the Clean Water Act of 1972, the NCP was revised to include a framework for responding to hazardous substance releases, as well as oil spills. Following the passage of CERCLA legislation in 1980, the NCP was broadened to cover releases at hazardous waste sites requiring emergency removal actions. Over the years, additional revisions have been made to the NCP to keep pace with the enactment of legislation. The latest revisions to the NCP were finalized in 1994 to reflect the oil spill provisions of the Oil Pollution Act of 1990.

2.7 U.S. POLLUTION PREVENTION ACT, 1990

Much of U.S. federal environmental health policymaking has in a sense been focused on controlling ongoing releases of pollutants into various environmental media such as air and water. This is understandable in the context of legislative awareness of trying to respect economic interests but contemporaneously attempting to protect society against harm from environmental hazards. In other words, most policies have been directed at managing ongoing pollution, rather than preventing it from occurring. An exception to this theme is the PPAct, which is described in this section.

The PPAct of 1990 is an example of a federal environmental statute whose genesis derives substantively from EPA initiative [137]. While the Executive Branch of the U.S. government has the authority, indeed, the responsibility, to propose legislation to the Congress, in fact, the main body of federal environmental law has generally been the outcome of Congressional interest and action. With the election of U.S. President George H.W. Bush (R-TX) came the appointment of William Reilly as EPA Administrator. Reilly was at the time the president of the World Wildlife Federation, a respected and venerable international environmental organization. He is the only EPA Administrator to have served as the leader of a major environmental organization.

One of Reilly's priorities was pollution prevention, which was thought to be preferable to dealing with "after-the-fact" pollution management. This prevention orientation is, of course, identical to the central thesis of public health – prevention of disease and disability. Reilly's pollution prevention priority found favor in Congress, contributing to the enactment of the PPAct.

> The PPAct seeks to prevent pollution through reduced generation of pollutants at their point of origin.

Solid and Hazardous Waste

The PPAct declares "[i]t to be the national policy of the United States that pollution should be prevented or reduced at the source whenever feasible, pollution that cannot be prevented should be recycled in an environmentally safe manner, whenever feasible; pollution that cannot be prevented or recycled should be treated in an environmentally safe manner whenever feasible; and disposal or other release into the environment should be employed only as a last resort and should be conducted in an environmentally safe manner [138]." This act encourages voluntary reduction of pollution and is, therefore, a complement to the command and control regulatory approach whereby pollution control is mandated of polluting sources.

As a matter of policy, it should be noted that federal environmental laws generally exempt U.S. federal government agencies from responsibilities required of the private sector (e.g., businesses and corporations), unless specifically specified by statute. Another example of an environmental statute that specifies accountability required of federal agencies is found in the CERCLAct, where uncontrolled hazardous waste sites under the control of federal agencies must be remediated, as is the situation with private sector responsibility. Presidential Executive Orders, such as one on pollution prevention, is one way to bring federal agencies into compliance with environmental laws directed at the private sector [139].

2.8 U.S. OCEAN WASTE POLLUTION POLICIES

Earth's oceans are vital for humankind's well-being in myriad ways. Oceans provide food, transportation, security, recreation, and contributions to climate stability, as but a few examples. Yet, in spite of this vital importance, humans have proceeded to damage our oceans by harmful mismanage of wastes, especially toxic waste. According to the MarineBio Conservation Society, a nonprofit marine conservation organization, "The most toxic waste material dumped into the ocean includes dredged material, industrial waste, sewage sludge, and radioactive waste.

Dredging contributes about 80% of all waste dumped into the ocean, adding up to several million tons of material dumped each year. Rivers, canals, and harbors are dredged to remove silt and sand buildup or to establish new waterways. About 20–22% of dredged material is dumped into the ocean. The remainder is dumped into other waters or landfills and some are used for development. About 10% of all dredged material is polluted with heavy metals such as cadmium, mercury, and chromium, hydrocarbons such as heavy oils, nutrients including phosphorous and nitrogen, and organochlorines from pesticides. Waterways and, therefore, silt and sand accumulate these toxins from land runoff, shipping practices, industrial and community waste, and other sources. When these materials find their way into the ocean, marine organisms suffer toxic effects and seafood is often contaminated" [140].

> Dredging contributes about 80% of all waste dumped into the ocean, adding up to several million tons of material dumped each year.

In addition to dredged materials being dumped into oceans, the volume of plastics waste has become a major factor in marine pollution. According to one source, in 2010, coastal countries dumped 8 million tons of plastics trash in the ocean, which was much greater than the total that has been measured floating on the surface in the ocean's "garbage patches." The study also identified the major sources of plastics debris and named the top 20 countries generating the greatest amount of ocean-bound trash. China ranked first.

> According to one source, in 2010, coastal countries dumped 8 million tons of plastics trash in the ocean. China ranked first.

After China were 11 other Asian countries, Turkey, five African countries, and Brazil. The U.S. was 20th. Researchers estimated that the tonnage is on target to increase 10-fold in the next decade unless waste collection and management policies are improved globally [141].

The variety and extend of ocean debris are suggested by the variety and volume of waste that washes up on coastal beaches. An NGO organization, Ocean Conservancy, conducts periodic collection by volunteers of coastal marine trash. In 2018, one million volunteers across 122 countries collected more than 23 million pounds of trash [142]. Noteworthy is the distribution of items found on beaches. Cigarette butts were the most prevalent trash until 2019, when as itemized by the Conservancy, plastic food wrappers supplanted them in prevalence [143]. Moreover, plastics items dominated the trash collection, including items such as food wrappers, straws/stirrers, plastic food ware, and beverage bottles. As will be subsequently discussed, plastics waste in Earth's oceans is a major global environmental health problem. In summary, cigarette butts and plastic items were the most prevalent items constituting coastal trash.

Regarding the problem of cigarette butts as environmental waste, environmental advocacy groups have spurred increased public education about the environmental impacts and pushed for the installation of more bins to safely dispose of butts [144]. Some cities have put restrictions on where people can smoke or instituted additional fees on cigarettes to fund clean-up costs. Butt pollution continues. Further, some legislators are trying a different approach – producer responsibility. New legislation in several states, including a bill in California to ban products with single-use filters, could force cigarette manufacturers to take responsibility for the environmental impacts of their products. How policies help solve the environmental pollution problem of cigarette butts remains to be seen.

> The ODAct focuses on the regulation of intentional disposal of materials into ocean waters and authorizing related research.

U.S. federal policies on controlling ocean pollution consist of two statutes, the Ocean Dumping Act (ODAct) and the Act to Prevent Pollution from Ships [145]. Both acts are intended to reduce the release of waste into ocean waters. The older of the two acts, the ODAct, is discussed first.

2.8.1 OCEAN DUMPING ACT, 1972

The U.S. has derived enormous benefits from a geography positioning the nation next to three major oceans (Atlantic, Arctic, and Pacific) and one ocean basin (Gulf of Mexico). These bodies of water are sources of seafood, transportation, revenue, and protection for the U.S. Regrettably, the U.S. and other coastal nations have over

Solid and Hazardous Waste

their histories chosen to use these oceans as dumping basins for waste. As related in this section, recognition of the impact of ocean dumping has led to both international and domestic policies meant to control ocean dumping. The Marine Protection, Research, and Sanctuaries Act, also known as the Ocean Dumping Act, prohibits dumping material into the ocean that would unreasonably degrade or endanger human health or the marine environment [146,147]. The ODAct implements policies to control waste dumped into ocean waters.

2.8.2 Act to Prevent Pollution from Ships, 1980

Ocean pollution has doubtless occurred from the time that humans first found ways to traverse great bodies of water. Whether human waste of mariners, or damaged goods being dumped, or oil from leaking vessels, the oceans and its resident creatures have continuously been victims of pollution from ships.

As ocean pollution became a concern to the global environmental protection community, international maritime treaties emerged to regulate and control pollution from ships [148]. These treaties are in addition to laws enacted by individual nations, such as the U.S. Ocean Dumping Act. In concert with other nations, the U.S. has enacted policies to restrict waste from being released from ships that bear U.S. credentials. The Act to Prevent Pollution from Ships implements the international MARPOL Convention. The act restricts pollution from ships, establishes record keeping of materials released from ships and provides penalties for violations of the act [149].

2.8.3 Obama's Oceans Policy, 2010

In July 2010 the Obama (D-IL) administration issued executive order 13547 --Stewardship of the Ocean, Our Coasts, and the Great Lakes [150]. The overall intent of the order was to solidify administrative corporation between government and private entities for the purpose of protection and conservation of marine resources. The implementation plan, issued in April 2013 states, "Today, the Obama Administration released its final plan for translating the National Ocean Policy into on-the-ground actions to benefit the American people. With significant public input from a wide spectrum of individuals and interests, the final Implementation Plan focuses on improving coordination to speed Federal permitting decisions; better manage the ocean, coastal, and Great Lakes resources that drive so much of our economy; develop and disseminate sound scientific information that local communities, industries, and decision-makers can use; and collaborate more effectively with State, Tribal, and local partners, marine industries, and other stakeholders. Without creating any new regulations or authorities, the plan will ensure the many Federal agencies involved in ocean management work together to reduce duplication and red tape and use taxpayer dollars more efficiently" [151].

This order created a framework for regional marine spatial planning and an Ocean Council, a cabinet-level forum to coordinate the nation's regulatory and diplomatic responsibilities for the oceans. At the end of the president's second term, Obama approved rules implementing recommendations of the ocean council to combat global illegal fishing and seafood fraud through import monitoring and certification requirements.

In June 2018 the U.S. President Donald Trump (R-NY) administration replaced the Obama National Ocean Policy with a different policy. The order formally revokes the 2010 oceans policy issued by then-President Barack Obama and replaces it with a markedly different template for what the government should focus on in managing the nation's oceans, coastal waters, and Great Lakes. The Trump policy removed the Obama policy's emphasis on conservation and climate, replacing them with an emphasis on economic and security concerns [152].

2.9 INTERNATIONAL POLICIES ON MARITIME POLLUTION

Because maritime pollution is a global issue, as related previously by a description of ocean pollution, international treaties and actions by the UN have been implemented.

2.9.1 LONDON CONVENTION AND PROTOCOL, 1972

The Convention on the Prevention of Marine Pollution by Dumping of Wastes and Other Matter of 1972, known as the London Convention, is one of the first global conventions to protect the marine environment from human activities [153]. The Convention has been in force since 1975. The Convention is an international treaty and is administered by the International Marine Organization (IMO). The objective of the Convention is to promote the effective control of all sources of marine pollution and to take all practicable steps to prevent pollution of the sea by dumping of wastes and other matter [154].

In 1996, the "London Protocol" was accepted in order to further modernize the Convention and, eventually, replace it. Under the Protocol, all dumping is prohibited, except for possibly acceptable wastes on the so-called "reverse list." This list includes the following: 1: dredged material; 2: sewage sludge; 3: fish wastes; 4: vessels and platforms; 5: inert, inorganic geological material (e.g., mining wastes); 6: organic material of natural origin; 7: bulky items primarily comprising iron, steel and concrete; and 8: CO_2 streams from CO_2 capture processes for sequestration.

The London Protocol entered into force on March 24, 2006. According to the IMO, Under the unregulated dumping and incineration activities that developed in the late 1960s and early 1970s have been halted. Parties to the Convention agreed to control dumping by implementing regulatory programmers to assess the need for, and the potential impact of, dumping. They eliminated dumping of certain types of waste and, gradually, made this regime more restrictive by promoting sound waste management and pollution prevention. Prohibitions are in force for dumping of industrial and radioactive wastes, as well as for incineration at sea of industrial waste and sewage sludge. Under the Protocol, all dumping is prohibited, except for the so-called "reverse list" [154]. The U.S., Russia, Brazil, and other South American countries are Parties to

> The London Convention is one of the first global conventions to protect the marine environment from human activities. The Convention has been in force since 1975.

the Convention, but not Parties to the Protocol, which indicates that they accept the general principles of the Convention, but reserve the right to determine how to implement the Convention's protocol.

2.9.2 INTERNATIONAL MARITIME ORGANIZATION (IMO)

On an international scale, the IMO, an agency of the UN, has led the development and implementation of treaties to control ocean pollution from ships. Created in 1948, the IMO's purposes include, "[t]o provide machinery for cooperation among Governments in the field of governmental regulation and practices relating to technical matters of all kinds affecting shipping engaged in international trade; to encourage and facilitate the general adoption of the highest practicable standards in matters concerning maritime safety, efficiency of navigation and prevention and control of maritime pollution from ships" [154]. Concerning marine pollution from ships, in 1973 the IMO issued the International Convention for the Prevention of Pollution from Ships, modified by the Protocol of 1978, both being referred to as MARPOL 73/78. The MARPOL Convention covers pollution by chemicals, goods in packaged form, sewage, garbage, and air pollution, as well as accidental and operational oil pollution (i.e., oil pollution from the operation of a ship).

As an international treaty, adoption of the MARPOL Convention requires national legislation for its implementation. In the U.S., the Act to Prevent Pollution from Ships implements the provisions of the Convention. As stated in the Act, it "[a]pplies to all U.S. flag ships anywhere in the world and to all foreign flag vessels operating in the navigable waters of the U.S. or while at a port or terminal under the jurisdiction of the U.S. The oil and noxious liquid substances provisions apply only to seagoing ships. The regulations implementing Annex I and Annex II of MARPOL limit discharges of oil and noxious substances, establish report requirements for discharges, and establish specific requirements for monitoring equipment and record keeping aboard vessels" [155]. The act contains provisions for both criminal and civil penalties for violations.

> The International Maritime Organization, an agency of the UN, has led the development and implementation of treaties to control ocean pollution from ships.

An example of IMO's actions on setting global regulations applicable to maritime commerce is the establishment of limits on sulfur emissions from ships [156]. In October 2016, the IMO set global regulations to limit the amount of sulfur emissions from vessels. The new requirements will result in sulfur emissions decreasing from the current maximum of 3.5% of fuel content to 0.5%. The regulation will come into force after the year 2020. Analysts estimate the additional costs for the container shipping sector alone could be $35–$40 billion [156]. The goal of the IMO regulation is to reduce air emissions of sulfur compounds released from vessels' diesel engines.

2.9.3 EU MARITIME POLLUTION POLICIES

Many of the Member States of the EU have historic ties to Earth's oceans. These nations continue to derive multiple benefits from their seafaring endeavors. As such, the quality of ocean waters is of concern. One concern is the nature and volume of litter deposited in oceans. In response, the EU is developing a marine litter program that will be constructed around a directive issued by the European Commission (EC). Within the context of the European Marine Strategy Framework Directive [157], the Task Group for Marine Litter recommended that the overriding objective should be a measurable and significant decrease of the total amount of marine litter in comparison with initial baseline values by 2020. In support of developing the directive, the EC has provided important data on the extent of marine litter that impacts the EU, stated as follows [158]:

- Approximately 80% of marine litter is land-based.
- In 2004, marine water samples contained 6 times more plastic than plankton, i.e., out of 7 kilo, 6 kilos of plastics vs. 1 kilo of plankton.
- Cruise ships: 95,000 m³ of sewage from toilets and 5,420,000 m³ of sewage from sinks, galleys, and showers are released into the oceans each day.
- 250,000 kg of waste are removed from the North Sea yearly.
- Marine litter can cause serious economic damage: losses for coastal communities, tourism, shipping, and fishing.
- Potential cost across the EU for coastal and beach cleaning was assessed at almost €630 million per year, while the cost to the fishing industry could amount to almost €60 million, which would represent approximately 1% of total revenues of the EU fishing fleet in 2010. Taking into account its accumulation and dissemination, marine litter may be one of the fastest growing threats to the health of the world's oceans.

> EU: Cruise ships release 95,000 m³ of sewage from toilets and 5,420,000 m³ of sewage from sinks, galleys, and showers into the oceans each day.

The 7th Environment Action Programme calls for the development of an EU-wide "quantitative reduction headline target for marine litter, supported by source-based measures and taking into account marine strategies established by Member States." Efforts in that direction are ongoing with discussions also to be held in the context of the Circular Economy review [159].

2.10 WASTE REDUCTION: RECYCLING OF WASTE

What to do about waste has been a perplexing challenge to humankind from the time that small societies began to form. As human experience accrued, generation of waste, especially that produced by industrialization of national economies, became a societal issue. Landfills and waste dumps became sources of vermin, environmental pollution, and a burdensome economic impact. Gradually, 20th century policymakers, urban planners, environmental groups, and public health advocates came to the realization that waste reduction should become a societal policy. The importance of waste reduction is fueled by the knowledge that the volume of waste generated globally will continue to increase according to World Bank researchers, who estimated

Solid and Hazardous Waste

that global solid waste generation would rise from more than 3.5 million tonnes per day in 2010 to more than 6 million tons per day in 2025. The researchers further predicted that by 2100 solid waste generation rates will exceed 11 million tons per day – more than three times the 2010 rate. Although these predictions come with uncertainty due to data limitations, the researchers aver that waste generation will significantly increase unless policy interventions are implemented [160].

One way to limit the impact of increased volumes of waste is to implement waste reduction strategies, as exemplified by the four Rs in Table 2.2 [161]. Recycling, one of the four Rs, is the process of collecting and processing materials that would otherwise be thrown away as trash and then turning the materials into new products. EPA cites several benefits of waste recycling [161]:

- reduces the amount of waste sent to landfills and incinerators,
- conserves natural resources such as timber, water, and minerals,
- prevents pollution by reducing the need to collect new raw materials,
- saves energy,
- reduces greenhouse gas emissions that contribute to global climate change,
- helps sustain the environment for future generations,
- helps create new jobs in the recycling and manufacturing industries.

Americans recycle 34% of all the waste they create, according to EPA. In total, Americans generated 254 million tons of trash in 2013, which is about 4.4 pounds per person per day [162]. Many European countries have developed more successful recycling programs, with Austria and Germany executing the highest recycling rates at 63% and 62%, respectively [163]. There are numerous examples of recycling of waste materials that are familiar to residents of industrialized countries. Perhaps the premier example is recycling of aluminum products, particularly containers of beverages. In 2016, nearly $1 billion worth of aluminum cans are discarded annually in the U.S. The aluminum industry spends more than $800 million annually for recycled cans. The U.S. aluminum can recycling rate is approximately 67%, yielding nearly a billion dollars of recycling profit annually [164].

> Americans recycle 34% of all the waste they create, according to the EPA.

2.10.1 Recycling Issues

There are some policy-relevant issues in recycling of waste. One issue is the fact that recycling itself is often a commercial enterprise, a desirable outcome whereby waste is converted into something useful and concurrently provides jobs and income for individuals and businesses. Additionally, recycling can be the genesis of creative processes for managing waste reduction. However, commercial recycling can suffer the same kinds of problems as any other business endeavor. For example, the creation of products that lack customer appeal can lead to loss of revenues and present business sustainability challenges. Recycling of plastic products is a case in point. In 2016, global oil prices decreased to new lows due in part to a glut of oil placed in global markets. One consequence was to hinder recycling of plastics materials, since it became less expensive to produce a plastics product using cheaper oil supplies than to purchase a recycled plastics product that costs more to manufacture [165].

> Once a profitable business for cities and private employers alike, the recycling industry in recent years has become a negative income business.

Further, commercial recycling is sometimes subsidized by governments anxious to comply with the ordinance or local policies for waste reduction. Recycling industry sources assert that 2,000 U.S. municipalities are paying to dispose of their recyclables instead of the other way around. Once a profitable business for cities and private employers alike, the recycling industry in recent years has become a negative income business due to multiple factors. These factors include decreased demand for commodities produced through recycling, a change in waste from paper to glass to plastics, both of which are less attractive for recycling, and more waste that requires sorting prior to recycling. The diminished profitability of waste recycling has resulted in more pressure on municipalities to increase their subsidies [166]. This kind of economic pressure ultimately translates into strain on decision-making by policymakers.

2.10.2 Innovative Technology for Waste Reduction

Economic and sociopolitical forces have contributed to a larger recognition of the benefits of waste reduction. Economic benefits can accrue from converting waste into commercial commodities. Sociopolitical forces include government policies developed for the purpose of waste control and environmental groups are active in promoting waste reduction policies. A consequence of these kinds of forces can be innovative efforts forged for waste reduction. Two examples of innovation in waste reduction are exemplified by recycling of cigarette butts and changes underway in the global garment industry.

Cigarette butts are the most numerous form of trash that volunteers collect from the world's beaches on the Ocean Conservancy's cleanup days. More than 2 million cigarette parts were recently collected in a single year worldwide. Other sources present the same data. For instance, New York State experiences an estimated 1.5 million tons of cigarette butts a year. In

> Cigarette butts are the most numerous form of trash that volunteers collect from the world's beaches.

Texas, butts account for about 13% of the litter accumulated on the state's highways, 130 million butts a year. However, cigarette filters are made from wood-based plastic fibers that can be recycled. There are now a handful of companies working to collect and recycle spent cigarette butts and recycle them into plastic lumber that can be used for benches, pallets, and other uses [167].

One U.S. state, California, in an effort to reduce litter, wildfire risk, and ocean pollution from cigarette butts, has enacted legislation to ban smoking on all of California's state beaches and in state parks. Starting January 1, 2020, it will be illegal to smoke cigarettes, cigars, pipes, vaping devices "or any other lighted or heated

Solid and Hazardous Waste

tobacco or plant product intended for inhalation" on any state beach or in any state park in California. Violators face fines of $25 [168].

As shown in Figure 2.5, textiles are a significant item in the waste stream. Some critics point out the damage being caused by a throwaway culture that is fueled by inexpensive clothing, which has seen a sharp rise in the number of garments sold annually around the world [169]. In a nascent effort to reduce the volume of textile waste, some fashion firms are turning to recycling of textile material for conversion into new fashions and for other uses. For instance, a company in Sweden launched a line of jeans containing recycled cotton and will offer an annual 1-million-euro ($1.16 million) prize for new techniques to recycle clothes, On a smaller scale, Finnish entrepreneurs have managed to produce sweatshirts from 100% recycled cotton after improving existing recycling techniques and by recycling offcuts from clothes factories. However, others believe that recycling is just a distraction from the real challenge of the fashion industry: persuading customers to keep wearing their clothes for longer. To that end, a British designer is offering a 30-year guarantee on a range of T-shirts [169].

> Finnish entrepreneurs have managed to produce sweatshirts from 100% recycled cotton after improving existing recycling techniques.

Of special promise are findings from a team of Japanese researchers who discovered a species of bacteria that can break the molecular bonds of one of the world's most-used plastics - polyethylene terephthalate, also known as PET or polyester. The Japanese research team sifted through hundreds of samples of PET pollution before finding a colony of organisms using the plastics as a food source. Further tests found the bacteria almost completely degraded low-quality plastics within 6 weeks. Use of this process to biodegrade plastics waste would be a most remarkable contribution to waste management [170]. Similarly, German university researchers scientists discovered in 2020 a strain of bacteria, the first of its kind, that can degrade the harmful compounds in polyurethane products. The bacterium Pseudomonas putida was discovered in the soil of a site covered in plastic waste [171]. These are promising findings that might portend a microbiological method for recycling of plastic products.

2.10.3 THE CIRCULAR ECONOMY

An economic model that promotes waste prevention is the circular economy, a generic term for an industrial economy that is producing no waste and pollution, by design or intention, and in which material flows are of two types: biological nutrients, designed to reenter the biosphere safely, and technical nutrients. The latter are designed to circulate at high quality in the production system without entering the biosphere as well as being restorative and regenerative by design. This is in contrast to a Linear Economy that is a "take, make, dispose" model of production [172].

As an example of adoption of the circular economy model, the EU has developed and begun to implement an enabling strategy [173]. In 2015, the European Commission adopted its Circular Economy

> A circular economy is a generic term for an industrial economy that is producing no waste and pollution.

Package, which includes revised legislative proposals on waste to stimulate Europe's transition towards a circular economy, which will boost global competitiveness, foster sustainable economic growth, and generate new jobs.

The revised EU legislative proposals on waste set clear targets for the reduction of waste and establish a long-term path for waste management and recycling. "Key elements of the revised waste proposal include the following:

- A common EU target for recycling 65% of municipal waste by 2030
- A common EU target for recycling 75% of packaging waste by 2030
- A binding landfill target to reduce landfill to a maximum of 10% of municipal waste by 2030
- A ban on landfilling of separately collected waste
- Promotion of economic instruments to discourage landfilling
- Simplified and improved definitions and harmonized calculation methods for recycling rates throughout the EU
- Concrete measures to promote re-use and stimulate industrial symbiosis – turning one industry's by-product into another industry's raw material
- Economic incentives for producers to put greener products on the market and support recovery and recycling schemes (e.g., for packaging, batteries, electric and electronic equipment, vehicles)" [173].

Some other countries, e.g., Australia, are considering or already implementing circular economy programs. It is worthy of mention that some persons prefer the term "regenerative economy," one in which, from beginning to end, the parts of a product can be reused over and over again. Regeneration is also about finding new ways of conceiving what a business, a building, or a product can be [174].

2.11 HAZARD INTERVENTIONS

This chapter is about waste generation and management. Some waste is inevitable, such as bodily wastes, agricultural and industrial wastes, and household wastes. And with waste generations must come the equally inevitable problem of waste management. Improperly managed waste disposal can be a hazard to human and ecosystem health, especially the latter. Interventions to mitigate or eliminate such hazards are therefore in the spirit of environmental health prevention policies. Some hazard interventions would include the following:

- Waste management policies at all levels of society, including individual members, must include elements of waste prevention and methods of proper waste management.
- Commercial products while in the design phase should include considerations of the implications of product waste. Waste management should be part of engineering education regardless of engineering specialty.
- Education of the public about waste management should be a subject of public service announcements on communication systems, e.g., television and social media.

- Commercial enterprises should be cognizant of the generation and management of waste during their operations and implement policies of waste minimization.
- Agricultural operations should implement waste minimization policies and methods, utilizing educational material available from local and national agricultural authorities.
- Research on ways to reduce or prevent waste products should be encouraged. For example, university researchers have discovered a cocktail of enzymes that degrade plastic, offering a new form of recycling that is faster and more affordable [175].
- Individuals can make a positive impact on reducing waste by adopting personal policies such as avoiding using plastic bags and bottles through the use of alternative products.
- Individuals should support policymakers whose agenda includes environmental health concerns.
- Individuals should encourage environmental and commercial organizations that promote waste reduction policies.
- The power of the purse is a forceful impetus in policymaking. In that vein, purchasing products made in whole or part from recycled materials will increase recycling of wastes.

2.12 SUMMARY

The two major statutes discussed in this chapter are the SWAct, which upon amendment became known as the Resource Conservation and Recovery Act (RCRAct), and the Compressive Environmental Response and Liability Act (CERCLAct, or Superfund). Both statutes contain a number of important environmental health policies of relevance to public health. From the RCRAct, open dumps were prohibited in the U.S., thereby eliminating places where disease-carrying vermin prospered, and placed upon states the responsibility for solid waste management, which is an example of federalism at work. The RCRAct also requires permits for those facilities that conduct hazardous waste management activities. The act also authorizes EPA to inspect hazardous waste facilities and issue corrective action orders when standards are not being met by facility operators. The policies of establishing national standards, issuance of permits for control of environmental hazards and inspection of facilities, in concept, are the same found in some other environmental statutes, e.g., the CAAct, as amended. These policies' purpose is to control releases of potentially hazardous environmental substances into the environment.

The CERCLAct also contains several important policies of relevance to public health. The act identifies uncontrolled hazardous waste sites and ranks those that are the most hazardous to human health and the environment (i.e., the NPL sites). The highest ranked sites are remediated preferentially before those of lesser hazard, a policy that comports with the public health practice of addressing first those hazards of the greatest threat to human health. The CERCLAct also contains the policy of "the polluter pays for the costs of their pollution," which is a statement of accountability to the public. Lastly, the CERCLAct is unique among federal environmental statutes in its creation of a federal public health agency, ATSDR, for the purpose of addressing a specific environmental hazard – in this case, human exposure to uncontrolled hazardous substances.

Other acts discussed in this chapter consisted of the Oil Pollution Act, the Ocean Dumping Act, Act to Prevent Pollution from Ships, and the PPAct. All four acts are intended to prevent or reduce the amount of pollution released into the environment. Reduced pollution levels are commensurate with improved public health protection since the opportunity for human exposure is lessened.

Mismanagement of waste is a global issue and challenge. Described in this chapter are environmental health problems associated with the disposal of plastics waste, electronics waste, and food waste. All three forms of waste are huge in volume and each represents a threat to ecosystem health. Global treaties have been developed, e.g., London Convention, for the purpose of controlling waste generation and its mismanagement.

REFERENCES

1. World Bank. 2013. Global waste on pace to triple by 2100. http://www.worldbank.org/en/news/feature/2013/10/30/global-waste-on-pace-to-triple.
2. McKie, R. 2016. Plastic now pollutes every corner of Earth. *The Guardian*, January 23.
3. Harvey, C. 2016. What tiny plastic particles are doing to tiny fish. *The Washington Post*, June 2.
4. Derbyshire, D. 2016. A poison worse than any spill: How a third of fish caught in the English Channel have microplastics in their guts. *Daily Mail*, August 28.
5. Anonymous. 2016. Oysters are munching on our microplastics. *Seeker*, February 1. http://www.seeker.com/oysters-are-munching-on-our-microplastics-1770821154.html.
6. Tang, W. J., et al. 2015. Chitin is endogenously produced in vertebrates. *Curr. Biol.* 25(7):897–900. DOI: 10.1016/j.cub.2015.01.058.
7. von Moos, N., P. Burkhardt-Holm, and A. Köhler. 2012. Uptake and effects of microplastics on cells and tissue of the blue mussel Mytilus edulis L. after an experimental exposure. *Environ. Sci. Technol.* 46(20):11327–35. DOI: 10.1021/es302332w.
8. Victor, D. 2019. Dead whale in the Philippines found with 88 pounds of plastic inside body. *The New York Times*, March 18.
9. O'Sullivan, K. 2018. Over 70% of deep-sea fish have ingested plastic, study finds. *The Irish Times*, February 18.
10. Royer, S-J., S. Ferrón, S. T. Wilson, D. M. Karl. 2018. Production of methane and ethylene from plastic in the environment. *PLoS One* 13(8): e0200574. DOI: 10.1371/journal.pone.0200574.
11. D'Alessandro, N. 2014. 22 facts about plastic pollution (and 10 things we can do about it). *EcoWatch*, April 7.
12. Albeck-Ripka, L. 2018. The 'Great Pacific Garbage Patch' is ballooning, 87,000 tons of plastic and counting. *The New York Times*, March 22.
13. Diaz, J. 2019. Baby sea turtle found dead in South Florida had 104 pieces of plastic in its body. *South Florida Sun Sentinel*, October 4.
14. Eriksen, M., et al. 2014. Plastic pollution in the world's oceans: More than 5 trillion plastic pieces weighing over 250,000 tons afloat at sea. *PLoS One* 9(12): e111913. DOI: 10.1371/journal.pone.0111913.
15. Jambeck, L., et al. 2015. Plastic waste inputs from land into the ocean. *Science* 347(6223):768–71.

16. Bushwick, S. 2015. What are microbeads and why are they illegal? *Popular Science*, December 22.
17. Rochman, C., et al. 2015. Scientific evidence supports a ban on microbeads. *Environ. Sci. Technol.* 49:10759–61. DOI: 10.1021/acs.est.5b03909.
18. Anonymous. 2016. Microplastic litter surging globally: Study. *Japan Times*, August 7.
19. Staff. 2018. Record concentration of microplastic in Arctic sea ice. *Science Daily*, April 24.
20. WHO (World Health Organization). 2019. WHO calls for more research into microplastics and a crackdown on plastic pollution. Geneva: News Room, August 22.
21. Simon, M. 2020. Wait, how much microplastic is swirling in the Atlantic? *Wired*, August 18.
22. Schlanger, Z. 2019. Virgin plastic pellets are the biggest pollution disaster you've never heard of. *Quartz*, August 19.
23. Henry, T. 2016. Ban on microbeads quietly becomes law. *The Toledo Blade*, January 4.
24. BBC. 2016. Plastic microbeads to be banned by 2017, UK government pledges. *British Broadcasting Company*, September 3.
25. Staff. 2015. Ottawa plans to ban microbeads over environmental concerns, *CTV News*, July 30.
26. Bendix, A. 2019. Cities could one day power homes and fuel cars with plastic, a giant step toward solving the pollution crisis. *Business Insider*, July 25.
27. Beresford, N, M. S. Jensen, G. Edgan, M. Sarglen. 2018. Our plastic problem is out of control. Here's how we can fight it. *World Economic Forum*, June 5.
28. Leblanc, R. 2019. An overview of plastic recycling. *Sustainable Businesses*, June 25.
29. McKenna, J. 2018. These Indian fishermen take plastic out of the sea and use it to build roads. *World Economic Forum*, June 28.
30. Micu, A. 2019. This infinitely-recyclable plastic might help us finally clean up landfills and oceans. *ZME Science*, May 8.
31. Gray, A. 2018. This plastic bag is 100% biodegradable. *World Economic Forum*, May 4.
32. Staff. 2018. Another way to recycle plastic. *The Economist*, May 24.
33. Steinmetz, K. 2014. California becomes first state to ban plastic bags. *Time*, September 30.
34. Quinn, A. 2020. Take one last look at the (many) plastic bags of New York. *The New York Times*, February 28.
35. Kelley, A. 2020. New Jersey signs strongest plastic and paper bag ban in US. *Changing America*, November 4.
36. McAuley, J. 2016. France becomes the first country to ban plastic plates and cutlery. *The Washington Post*, September 19.
37. Thibedeau, H. 2019. Government to ban single-use plastics as early as 2021: Source. *CBC News*, June 9.
38. Readfearn, G. 2020. South Australia's ban on single-use plastic cutlery and straws hailed as 'historic'. *The Guardian*, September 10.
39. Anonymous. 2009. Australian town bans bottled water. *The Guardian*, July 9.
40. Bloom, T. 2013. Where the plastic water bottle has been banned. *Truthdig*, January 9.
41. Glenza, J. 2017. National park ban saved 2m plastic bottles – and still Trump reversed it. *The Guardian*, September 26.
42. Goodman, J. D. 2015. 5¢ fee on plastic bags approved by New York City Council. *The New York Times*, May 5.
43. NCPA (National Center for Policy Analysis). 2015. Charge for plastic bags in Britain Draws mixed reactions. http://www.ncpa.org/sub/dpd/index.php?Article_ID=26172.
44. UNEP. 2018. *Single-use plastics: A roadmap for sustainability*. Nairobi: United Nations Environment Programme, Office of Director General.
45. Propina, L. 2018. EU proposes a total ban on plastic forks and other products. *Bloomburg News*, May 28.

46. Staff. 2018. Single-use plastics ban approved by European Parliament. *BBC News*, October 24.
47. Osborne, S. 2019. South Korea bans single-use plastic bags. *Independent*, January 1.
48. Griggs, M. B. 2019. Canada plans to ban single-use plastics by 2021. *The Verge*, June 10.
49. Staff. 2018. Iran's environmental body takes unprecedented step in fight against plastic pollution. *Tehran Times*, January 22.
50. Houreld, K. and J. Ndiso. 2017. Kenya imposes world's toughest law against plastic bags. *Reuters*, August 28.
51. Staff. 2019. *Single-use plastic shopping bags are banned in New Zealand*. Wellington: Ministry of the Environment.
52. Moreno, E. 2019. Panama becomes the first Central American nation to ban plastic bags. *Reuters*, July 20.
53. Staff. 2020. 'Extraordinary' effort from Kiwis sees plastic bags cut by a billion, one year on from ban – Ministry for the Environment. *1 News*, June 30.
54. Locker, M. 2018. Here are the U.S. cities that have banned plastic straws so far. *Fast Company*, June 1.
55. Gerretsen, I. 2018. Big brands pledge to turn tide on global plastic waste. *Reuters*, October 28.
56. FAO (UN Food and Agriculture Organization). 2017. Food lost and food waste. http://www.fao.org/food-loss-and-food-waste/en/.
57. Buzby, J. C. and J. Hyman. 2012. Total and per capita value of food loss in the United States. *Sci. Direct* 37(5):561–70.
58. Gustin, G. 2017. Better labels, less food waste: Companies agree to simplify 'use by' tags. *Inside Climate News*, September 21.
59. Smithers, R. 2016. UK households wasting 34,000 tonnes of beef each year. *The Guardian*, February 25.
60. Harvey, C. 2016. This is why the world can't stop wasting so much food. *The Washington Post*, June 7.
61. Jacobs, S. 2015. Denmark wages war on food waste. *National Public Radio*, September 2.
62. Chrobog, K. 2015. In South Korea, an innovative push to cut back on food waste. *Environment 360*, May 20.
63. Runyon, L. 2016. How Colorado is turning food waste into electricity. *NPR WABE 90.1.*, April 5.
64. Wong, K. 2015. Feeding farms with supermarket food waste. *Civil Eats*, September 21.
65. McEachran, R. 2015. Turning our mountains of food waste into graphene. *The Guardian*, February 16.
66. Harvey, C. 2016. The enormous carbon footprint of food that we never even eat. *The Washington Post*, March 28.
67. Conniff, R. 2016. Unnatural balance: How food waste impacts world's wildlife. *Environment 360*, January 6.
68. Doyle, A. 2015. U.S., China top dumping of electronic waste; little recycled. *Reuters*, April 18.
69. Bienkowski, B. 2016. E-waste grew 8 percent in just 2 years. Just one-fifth was recycled. *Environmental Health News*, December 14.
70. Staff. 2020. Amount of electronic waste generated in the U.S. is shrinking. *Yale Environment 360*, December 30.
71. Kamal, B. 2017. Where do 50 million tonnes a year of toxic e-waste go? *Inter Press Service*, September 27.
72. Bienkowski, B. 2018. The world is sending tons of illegal, electronic waste to Nigeria: Report. *Environmental Health News*, April 17.

73. Igharo, O. G., et al. 2018. Endocrine disrupting metals lead to alteration in the gonadal hormone levels in Nigerian e-waste workers. *Universa Medicina* 37:65–74. DOI: 10.18051/UnivMed.2018.v37.65-74.Journal.
74. WHO (World Health Organization). 2016. *Children's environmental health: Electronic waste*. Geneva: Office of Director-General, Media Centre.
75. Hendlin, Y. H. 2018. E-cigarettes and a new threat: How to dispose of them. *EcoWatch*, October 23.
76. Kirkpatrick, N. 2015. The children who make a living in the toxic world of discarded electronics. *The Washington Post*, April 15.
77. Sommer, J. 2015. The world's largest electronic-waste dump looks like a post-apocalyptic nightmare. *Business Insider*, July 15.
78. Landy, M. K., M. J. Roberts, and S. R. Thomas. 1994. *The environmental protection agency: Asking the wrong questions from Nixon to Clinton*. New York: Oxford University Press.
79. SGPHS (Surgeon General of the Public Health Service). 1913. Annual report of the Public Health Service of the United States. Washington, D.C.: U.S. Department of Health and Human Services, Office of Assistant Secretary for Health.
80. Sherrod, H. F. 1971. *Environmental law review* – 1971, 323. Albany: Sage Hill Publishers.
81. Elliott, P., et al. 2001. Risk of adverse birth outcomes in populations living near landfill sites. *Br. Med J.* 323(7309):363–8.
82. NRC (National Research Council). 2000. *Waste incineration and public health*. Washington, D.C.: National Academy Press.
83. EPA (Environmental Protection Agency). 2016. Municipal solid waste. https://archive.epa.gov/epawaste/nonhaz/municipal/web/html/.
84. CMAP. 2015. *Impacts of municipal solid waste*. Chicago, IL: Chicago Metropolitan Agency for Planning.
85. EPA (Environmental Protection Agency). 2016. *Municipal solid waste*. Washington, D.C.: Office of Land and Emergency Response.
86. U.S. House of Representatives. 1965. Solid Waste Disposal Act. https://legcounsel.house.gov/Comps/Solid%20Waste%20Disposal%20Act.pdf.
87. Robinson, E. H. Jr. 2019. Solid Waste Disposal Act (1965). https://www.encyclopedia.com/history/encyclopedias-almanacs-transcripts-and-maps/solid-waste-disposal-act-1965.
88. McCarthy, J. E. and M. Tiemann. 1999. Solid Waste Disposal Act/Resource Conservation and Recovery Act. Congressional Research Service. Report RL 30022. http://www.ncseonline.org/nle/crsreports/briefingbook/laws/h.cgm.
89. Smith, N. 2019. US tops list of countries fueling the waste crisis. *Verisk Maplecroft*, July 2.
90. Skinner, J., 1991. Hazardous waste treatment trends in the U.S. *Waste Manag. Res.* 9:55.
91. EPA (Environmental Protection Agency). 2005. *Municipal solid waste. Reduce, reuse, and recycle*. Washington, D.C.: Office of Solid Waste
92. EPA (Environmental Protection Agency). 2000. *RCRA, Superfund & EPCRA hotline training module*. EPA530-R-99-063. Washington, D.C.: Office of Land and Emergency Response.
93. EPA (Environmental Protection Agency). 2017. *Richard Delp (Cedar Valley Electroplating)*. Washington, D.C.: Office of Enforcement and Compliance Assurance.
94. EPA (Environmental Protection Agency). 2005. History of the Resource Conservation and Recovery Act (RCRA). 2016. https://www.epa.gov/rcra/history-resource-conservation-and-recovery-act-rcra.
95. Bosco, M. E. and R. V. Randle. 1991. Underground storage tanks. In *Environmental law handbook*, ed. J. G. Arbuckle, et al., 443–70, 444. Rockville, MD: Government Institutes, Inc.

96. Russ, A., C. Bernhardt, and L. Evans. 2019. *Coal's poisonous legacy. Groundwater contaminated by coal ash across the U.S.* Washington, D.C.: Environmental Integrity Project.
97. Milman, O. 2019. Most US coal plants are contaminating groundwater with toxins, analysis finds. *The Guardian*, March 4.
98. Crable, A. 2019. Pennsylvania power plant to stop coal ash pollution, pay $1 million fine. *Bay Journal*, July 31.
99. Beitsch, R. 2019. EPA proposal scraps limits on plant waste. *The Hill*, July 30.
100. Bruggers, J. 2019. Coal ash is contaminating groundwater in at least 22 states, utility reports show. *Climate Action News*, January 18.
101. Staff. 2018. Historic disaster: 10 years after the ash spill. Knoxville: *WBIR News*, December 21.
102. Satterfield, J. 2019. TVA agrees to remove 12 million tons of coal ash from Gallatin plant, clean contamination. *Knox News*, June 13.
103. State of Georgia. 2003. *Solid waste code 12-8-24*. Atlanta, GA: Department of Natural Resources.
104. Johnson, B. L. 1999. *Impact of hazardous waste on human health*. Boca Raton, FL: CRC Press, Lewis Publishers.
105. Vianna, W. K. and A. K. Polan. 1984. Incidence of low birth weight among Love Canal residents. *Science* 226:1217–9.
106. Berry, M. and F. Bove. 1997. Birth weight reduction associated with residence near a hazardous waste landfill. *Environ. Health Perspect.* 105:856–61.
107. Vrijheid, M., et al. 2002. Chromosomal congenital anomalies and residence near hazardous waste landfill sites. *Lancet* 359:320–2.
108. Najem, G. R., et al. 1985. Clusters of cancer mortality in New Jersey municipalities; with special reference to chemical toxic waste disposal sites and per capita income. *Int. J. Epidemiol.* 14:528–37.
109. Griffith, J., et al. 1989. Cancer mortality in U.S. counties with hazardous waste sites and ground water pollution. *Arch. Environ. Health.* 44:69–74.
110. Lagakos, S. W., B. J. Wessen, and M. Zelen. 1986. An analysis of contaminated well water and health effects in Woburn, Massachusetts. *J. Am. Stat. Assoc.* 81:583–96.
111. MDPH. 1996. *Woburn childhood leukemia follow-up study*. Boston, MA: Massachusetts Department of Public Health.
112. GAO. 1977. *Disposal of high-level nuclear waste*. Washington, D.C.: Government Accountability Office.
113. Reisch, M. 1999. Superfund. Summaries of environmental laws administered by the EPA. Congressional Research Service. Report RL 30022. http://www.ncseonline.org/nle/crsreports/briefingbooks/laws/j.cfm.
114. DePalma, A. 2004. Love Canal declared clean, ending toxic horror. *The New York Times*, March 18.
115. DePalma, A. 2004. Pollution and the slippery meaning of 'clean.' *The New York Times*, March 28.
116. GAO (General Accounting Office). 2003. *Superfund program: Current status and future fiscal challenges*. Washington, D.C.: Office of Comptroller General of the United States.
117. ELI (Environmental Law Institute). 1989. *Environmental law deskbook*, 166, 203, 221, 245. Washington, D.C.: Environmental Law Institute.
118. EPA (Environmental Protection Agency). 2003. How sites are placed on the NPL. http://www.epa.gov/superfund/programs/npl_hrs/nplon.htm.
119. Johnson, B. L. 1990. Implementation of Superfund's health-related provisions by the Agency for Toxic Substances and Disease Registry. *Environ. Law Reporter* 20: 10277.

120. EPA (Environmental Protection Agency). 2017. *Case summary: NCR Corporation Agrees to End Litigation and complete PCBs cleanup at Fox River Superfund site.* Washington, D.C.: Office of Enforcement and Compliance Assurance.
120a. EPA (Environmental Protection Agency). 2021. *Number of NPL sites of each status at the end of each fiscal year.* Washington, D.C.: Office of Land and Emergency Management.
121. Hasemyer, D. and L. Olsen. 2020. Biden faces pressure to tackle backlog of 'unfunded' toxic waste sites. *EcoWatch*, December 28.
122. Arbuckle, J. G., R. V. Randle, and P. A. J. Wilson. 1991. Emergency Planning and Community Right-to-know Act, (EPCRA). In *Environmental law handbook*, ed. J. G. Arbuckle, et al., 190. Rockville, MD: Government Institutes, Inc.
123. Arbuckle, J. G., et al., ed. 1991. *Environmental law handbook*, 471. Rockville, MD: Government Institutes, Inc.
124. Walsh, J. 2019. Sherwin-Williams to fund Superfund cleanups in Gibbsboro, Voorhees. *Cherry Hill Courier-Post*, May 30.
125. GBPSR. 2000. *In harm's way: Toxic threats to child development.* Boston, MA: Director, Boston Chapter Physicians for Social Responsibility.
126. EPA (Environmental Protection Agency). 2005. Press advisory: 2003 Toxics Release Inventory shows continued decline in chemical releases, May 11. Washington, D.C.: Office of Media Relations.
127. EPA (Environmental Protection Agency). 2016. *Brownfield overview and definition.* Washington, D.C.: Office of Land and Emergency Management.
128. EPA (Environmental Protection Agency). 2019. *Brownfields program accomplishments and benefits.* Washington, D.C.: Office of Land and Emergency Management.
129. UNEP (United Nations Environment Programme). 2019. Basel convention. http://www.basel.int/TheConvention/Overview/tabid/1271/Default.aspx.
130. Holden, E. 2019. Nearly all countries agree to stem flow of plastic waste into poor nations. *The Guardian*, May 10.
131. EPA (Environmental Protection Agency). 1991. *OPA Q's & A's: Overview of the Oil Pollution Act of 1990.* Washington, D.C.: Office of Land and Emergency Response.
132. Baldera, A. and R. Guillory. 2019. Remembering the 9th anniversary of the Deepwater Horizon oil disaster. *Ocean Conservancy Blog: Ocean Currents*, April 15.
133. Hasemyer, D. 2016. Enbridge's Kalamazoo spill saga ends in $177 million settlement. *Inside Climate News*, July 20.
134. Bousso, R. 2018. BP Deepwater Horizon costs balloon to $65 billion. *Reuters*, January 16.
135. Irfan, U. 2018. Deepwater Horizon led to new protections for US waters. Trump just repealed them. *Vox*, June 23.
136. EPA (Environmental Protection Agency). 1991. *National Oil and Hazardous Substances Pollution Contingency Plan (NCP) overview.* Washington, D.C.: Office of Land and Emergency Response.
137. Schierow, L. 1999. Pollution Prevention Act of 1990. Congressional Research Service. Report RL 30022. http://www.ncseonline.org/nle/crsreports/briefingbooks/laws/c.cfm.
138. O'Grady, M. J., ed. 1998. *Environmental statutes outline.* Rockville, MD: Environmental Law Institute.
139. White House. 1993. Presidential documents. Executive Order 12856. Federal compliance with right-to-know laws and pollution prevention requirements. *Federal Register* 58(150):12856.
140. Marinebio. 2015. Ocean pollution. http://marinebio.org/oceans/ocean-dumping/.
141. Parker, L. 2015. Eight million tons of plastic dumped in ocean every year. *National Geographic*, February 13.
142. Ocean Conservancy. 2019. To the beach and beyond: Breaking down the 2018 international coastal cleanup results. http://oceanconservacy.org/.

143. Toussaint, K. 2020. Food wrappers just passed cigarette butts as the most common beach trash. *Fast Company*, September 8.
144. Lohan, T. 2019. Cigarette waste: New solutions for the world's most littered trash. *The Revelator*, June 24.
145. EPA (Environmental Protection Agency). 2017. *Learn about ocean dumping.* Washington, D.C.: Office of Water.
146. EPA (Environmental Protection Agency). 2018. *EPA history: Marine Protection, Research and Sanctuaries Act (Ocean Dumping Act).* Washington, D.C.: Office of Water.
147. Copeland, C. 1999. Ocean Dumping Act: A summary of the law. Report RS 20028. Washington, D.C.: Congressional Research Service.
148. Anonymous. 1992. World's oceans are sending an S.O.S. *The New York Times Magazine*, May 3.
149. IMO (International Maritime Organization). 2005. Introduction to IMO. http://www.imo.org/About/mainframe.asp?topic_id=3.
150. White House. 2010. Executive order 13547--Stewardship of the ocean, our coasts, and the Great Lakes. Washington, D.C.: Office of the Press Secretary, July 19.
151. White House. 2013. Obama administration releases plan to promote ocean economy and resilience. Washington, D.C.: Office of the Press Secretary, April 16.
152. Malakoff, D. 2018. Trump's new oceans policy washes away Obama's emphasis on conservation and climate. *Science Magazine*, July 19.
153. EPA (Environmental Protection Agency). 2004. London convention. http://www.epa.gov/owow/ocpd/icnetbri.html.
154. IMO (International Marine Organization). 2016. *London Convention and Protocol. The Role and contribution to protection of the marine environment.* London: Office of the London Convention and Protocol.
155. NOAA (National Oceanic and Atmospheric Administration). 2005. Act to Prevent Pollution from Ships. http://www.csc.noaa.gov/opis/html/summary/apps.htm.
156. Saul, J. 2016. U.N. sets rules to cut sulfur emissions by ships from 2020. *Reuters*, October 27.
157. EC (European Commission). 2016. Legislation: The marine directive. http://ec.europa.eu/environment/marine/eu-coast-and-marine-policy/marine-strategy-framework-directive/index_en.htm.
158. EC (European Commission). 2016. Our oceans, seas and coasts: Descriptor 10: Marine litter. http://ec.europa.eu/environment/marine/good-environmental-status/descriptor-10/index_en.htm.
159. Galgani, F., et al. 2013. Marine litter within the European Marine Strategy Framework Directive. *ICES J. Marine Sci.* 70(6):1055–64.
160. Hoornweg, D., P. Bhada-Tata, and C. Kennedy. 2013. Environment: Waste production must peak this century. *Nature* 502:615–7.
161. EPA (Environmental Protection Agency). 2016. *Recycling basics.* Washington, D.C.: Office of Land and Emergency Management.
162. EPA (Environmental Protection Agency). 2016. *Advancing sustainable materials management: Facts and figures.* Washington, D.C.: Office of Land and Emergency Management.
163. Anonymous. 2015. Recycling rates around the world. *Planet Aid*, September 2.
164. The Aluminum Association. 2016. Recycling. http://www.aluminum.org/industries/production/recycling.
165. Gelles, D. 2016. Skid in oil prices pulls the recycling industry down with it. *The New York Times*, February 18.
166. Davis, A. C. 2016. American recycling is stalling, and the big blue bin is one reason why. *The Washington Post*, June 20.

167. Howard, B. C. 2015. Watch: Cigarette butts, world's #1 litter, recycled as park benches. *National Geographic*, May 5.
168. Rogers, P. 2019. Smoking banned on all California state beaches and state parks. *The Mercury News*, October 12.
169. Ringstrom, A. 2015. Recycling -- fashion world's antidote to environmental concerns. *Reuters*, August 24.
170. Mathiesen, K. 2016. Could a new plastic-eating bacteria help combat this pollution scourge? *The Guardian*, March 10.
171. Espinosa, M. J. C., A. C. Blanco, and T. Schmidgall, et al. 2020. Toward biorecycling: Isolation of a soil bacterium that grows on a polyurethane oligomer and monomer. *Front. Microbiol.* 11:404.
172. EMF (Ellen Macarthur Foundation). 2016. Circular economy. https://www.ellenmacarthurfoundation.org/circular-economy.
173. EC (European Commission). 2016. Circular economy strategy. http://ec.europa.eu/environment/circular-economy/index_en.htm.
174. Moodie, A. 2015. A zero waste business policy is now easier to implement than you think. *The Guardian*, October 1.
175. Kwai, I. 2020. Super-enzyme speeds up breakdown of plastic, researchers say. *The New York Times*, September 29.

3 Drugs and Alcohol

3.1 INTRODUCTION

Drugs and alcohol are the subjects of this chapter. The subject differs from most of the book's other chapters in the respect that the health and policies issues result from essentially a self-assumed hazard. A person may choose to use addictive drugs or consume alcohol in excessive qualities. In contrast, few people choose to breath polluted air, drink contaminated water, or ingest unsafe food. This matter of self-acceptance of health risk is important for public health issues, in particular, in the area of risk communication.

Rather than proceeding, it is important to define the terms that comprise the subject of this chapter. A *drug* is any chemical that is used as a medicine. A drug is also defined as a chemical or other substance that is illegally used, sometimes to improve performance in an activity or because a person cannot stop using it. *Alcohol* is defined in chemistry as any organic compound whose molecule contains one or more hydroxyl (–OH) groups attached to a carbon atom. For the purposes of this chapter, the definition of *alcohol* is a liquid that contains ethanol and has the potential to intoxicate drinkers. In this chapter, ethanol and alcohol are considered synonyms and used interchangeably.

> A drug is any chemical that is used as a medicine. A drug is also defined as a chemical or other substance that is illegally used.

While the subject of drugs and alcohol might seem to some a bit astray from the book's central theme – environmental public health – both drugs and alcohol can have adverse health effects under certain circumstances, thereby constituting a matter of public health. And it must also be acknowledged that both drugs and alcohol have beneficial health effects, but under controlled situations. Drugs are beneficial to health as medical prescriptions or taken under conditions of non-prescribed social acceptance, e.g., an efficacious folk remedy. Similarly, alcohol when consumed in moderation can be helpful for stress reduction or enhanced social communication.

But misuse of drugs and alcohol can obviate their positive health qualities. The misuse of therapeutic drugs can ensue from bogus prescriptions, theft, and commercial fraud by pharmaceutical sources. And there also exists the whole social agenda of the criminal activity involving distribution of addictive drugs such as heroin. Similarly, alcohol misuse can result in health and economic detriment to individuals, families, and social structures. This chapter focuses on the presence of misused legal drugs, illegal drugs, and intemperate consumption of alcohol, i.e., ethanol, in the physical environment, thereby presenting a hazard to human and

> This chapter focuses on the presence of misused legal drugs, illegal drugs, and intemperate consumption of alcohol.

ecological health. Policies to prevent or reduce the adverse health and social effects of this pair of subjects are presented for both domestic and international sources.

To be described are a précis of drug and alcohol developments across spans of time, the health consequences of their ill-advised uses, U.S. and global policies on control of drug misuse, and some hazard interventions postulated to reduce their hazard to health. The purpose of this chapter is not to question the beneficial uses of legal drugs and alcohol, but rather the public and ecological health adverse effects of drug and alcohol misuse.

3.2 DRUGS AND PUBLIC HEALTH

Modern medicine relies on drug therapy as a cornerstone of the medical profession. From a simple analgesic such as aspirin for relief of a headache to a complex immunotherapy drug for treatment of kidney cancer, drugs are indispensable for patient and physician alike. Humankind and other living creatures such as domestic companions like canines and farm animals such as dairy cows now rely on various forms of drugs for prevention of illness and treatment of diseases. Public health has improved in both practice and outcome as a consequence of drug therapy.

And yet there is an unwelcome aspect of the development and use of drugs in modern global society. One has only to note frequent, sometimes daily, news accounts of apprehension of illegal drug purveyors, persons who have chosen to sell or otherwise distribute addictive drugs such as heroin to persons who become victims of the drug culture in the U.S. and other countries. These kinds of drugs are called nontherapeutic for the purposes of this chapter, but "illegal" would be an appropriate synonym. A brief history of therapeutic drugs follows.

> Modern medicine relies on drug therapy as a cornerstone of the medical profession.

3.2.1 Précis History of Therapeutic Drugs

Drug discovery and development have been with us since the early days of human civilization. Folk medicines were primarily of plant origin and were complemented with minerals and animal substances. Although early folk remedies were discovered separately in different civilizations, they often used the same plants and herbs for similar diseases. Many early medicinal treatments originated in China and are referred to as traditional Chinese medicines (TCMs). These were developed as early as 3500 BCE, and many of these early TCM recipes were recorded and are still available today. The Chinese pharmacopoeia, or list of medicinal drugs with their effects and directions for their use, is quite large. Some of these herbal medicines contain active ingredients used in modern medicine, including reserpine, originally derived from Rauwolfia evergreen trees and shrubs. Reserpine is used as an antihypertensive, which lowers blood pressure [1].

Development of medicines was also well underway in Egypt by 3000 BCE, at which time a medicinal papyrus contained 877 prescriptions for various maladies, including gynecology and eye and skin problems. Indian folk remedies also existed

3,000–5,000 years ago, and are referred to as Ayurvedic medicine. They were based primarily on herbal preparations, including cardamom and cinnamon. Some Ayurvedic ingredients are still used in Western medicines today.

> Many early medicinal treatments originated in China, and are referred to as TCMs. These were developed as early as 3500 BCE.

The Greeks drew from Egyptian, Chinese, and Indian medical ideas to develop their own medicinal treatments. Importantly, Greek society advanced medicine and drug development by rejecting the notion that diseases are caused by supernatural causes or spells. In fact, they concluded that diseases arise from natural causes that could be treated with their medicinal formulations. The Romans also developed their own medicinal treatments, led by Claudius Galen who wrote on all aspects of medicine in 400 different manuscripts. With the decline of the Roman Empire, medicinal expertise moved to the Arab world. Starting in the fourth century, Greek and Roman medicinal development was completely neglected in Europe, as the use of folk remedies in the west became branded as witchcraft and sorcery [1].

The development of medicinal remedies in Europe restarted during the Renaissance (14th–17th centuries). The physician Paracelsus brought opium into the European pharmacopoeia in the early 16th century, and he also developed the modern concept of dose dependency for drug action and drug toxicity. He used metal salts for treatment, especially those of mercury for treatment of syphilis, developing the predecessor of current chemical therapeutics. Although advances were made in the 19th century, only a few true drugs were available at the beginning of the 20th century [1]. Scientific method then began to drive drug development, with systematic research leading to the synthesis of drugs via chemical methods, heralding the beginning of the modern pharmaceutical industry.

> The development of medicinal remedies in Europe restarted during the Renaissance (14th–17th centuries).

3.2.2 Précis History of Drug Abuse in the U.S.

A history of drug abuse was compiled by a drug affliction center [2]. A synopsis of the center's report follows. Drug abuse has been reported in many countries for hundreds of years, including in the U.S. Drug abuse has been a problem from the foundation of the country. The U.S. is a relatively young country and drugs from other areas of the world were already used for medicinal purposes in the early history of the country. Before the 19th century, substance abuse was primarily related to plant products and alcohol. The drugs that were commonly used in the early history of the country were obtained from poppy plants or similar plants that helped reduce feelings of pain. Opium and alcohol were common substances that were used in early American medical history.

> Opium and alcohol were common substances used in early American medical history.

Doctors would use opium to dull pain during surgical or medical procedures. When opium was not available, the doctors would turn to alcohol to dull or mask the symptoms of the health problem. Although the drugs were used for medical purposes, the addictive qualities of the drugs were not fully understood before the American Civil War. Many soldiers were given opium to dull pain during surgery and became addicted to the drug. During the 1800s, developments in medicine led to the creation of morphine, codeine, and cocaine. Initially, the drugs were unregulated and readily available. When it became clear that the drugs were a serious problem, regulations were developed and laws were made to help contain the problem.

In 1906, Congress enacted the Pure Food and Drug Act, leading to regulations that made it harder to obtain certain substances. By 1918, clinics were established for drug maintenance, and addiction to narcotics became illegal [2]. Drugs were no longer readily available and easy to obtain. Certain substances were banned due to the dangers associated with addiction, and medicinal medications were only available through a doctor's prescription. As it became obvious that substance abuse was still a problem, addiction became a matter of public health and safety. By 1939, the Federal Bureau of Narcotics began taking a harsh stand against illegal drug abuse and started prosecuting medical doctors who enabled addiction by giving prescriptions that violated laws.

Despite efforts to prevent substance abuse and the laws that regulated drugs on a federal and U.S. state and territorial levels, substance abuse did not stop with the legal changes. Many men and women continued to abuse drugs and alcohol. During the 1960s, consumers began using drugs for recreational purposes. The particular drugs that were abused during the 1960s and 1970s varied. The drugs that were commonly abused included: marijuana, cocaine, heroin,

LSD, and hallucinogenic mushrooms. Although the hippie culture was known for abusing hallucinogenic substances and marijuana, other drugs were still a problem. Opiates remained a challenge due to the addictive nature of the drugs and the medical use of the drugs. Although cocaine was still abused during the 1960s and 1970s, it was a secondary drug and did not gain the popularity of hallucinogenic substances.

The 1980s saw a shift in the type of drugs that were abused in the U.S. The drug of choice shifted from hallucinogens to cocaine, which resulted in a wave of violence and violent crimes. Throughout the 1980s and 1990s, drug abuse began to increase. By the year 2000, the number of arrests related to illicit substance abuse increased dramatically. More than 1.5 million arrests were related to drug abuse. Even in crimes that were not directly associated with buying or selling drugs, many criminals were also abusing drugs at the time of the crime, which resulted in a secondary charge [2].

> The 1980s saw a shift in the type of drugs that were abused in the U.S. The drug of choice shifted from hallucinogens to cocaine.

According to the FBI's data, law enforcement agencies in the U.S. made more than 1.57 million arrests for drug law violations in 2016, a 5.63% increase over the previous year. The number of arrests represents one drug arrest every 20 seconds – and more than three times more arrests than for all violent crimes combined [3]. In 2015, the most recent year for which data are available, 84% of all drug arrests were for simple low-level drug possession – while 43% of all drug arrests (643,121) were

Drugs and Alcohol

for marijuana law violations, and 39% of all drug arrests (574,641) were simply for marijuana possession. And as mentioned in Volume 1, Chapter 5, several U.S. states have approved the sale and use of recreational marijuana, a policy that will surely decrease the number of arrests for marijuana law violations. The smuggling of marijuana, in particular, and other illegal drugs into the U.S. is both a major legal issue as well as a public health problem. Illustrated in Figure 3.1 are bags of marijuana confiscated in 2019 by U.S. Border agents at the port of San Diego. Persons caught smuggling illegal drugs face criminal charges and penalties.

3.2.3 Impact of Drug Abuse on Human Health

Commonly used illegal drugs can cause serious adverse health effects in users. Listed in Table 3.1 are commonly used illegal drugs and their long-term health risks. Each drug also causes short-term health effects that are not listed in the table, since the chronic effects are more relevant for public health purposes. Particularly important are the opioids, heroin and opium, since both can be highly addictive and linked to fatal overdose.

According to the WHO, about 275 million people worldwide, which is about 5.6% of the global population aged 15–64 years, used illegal drugs at least once during 2016 [4]. Some 31 million of people who use illegal drugs suffer from drug use disorders, meaning that their drug use is harmful to the point where they might need medical treatment. Initial estimations suggest that, globally, 13.8 million young people aged 15–16 years used cannabis (aka marijuana) in the past year, equivalent to a rate of 5.6%. About 450,000 people died as a result of drug use in 2015, according to the

> According to the WHO, about 275 million people worldwide, which is about 5.6% of the global population aged 15–64 years, used drugs at least once during 2016.

FIGURE 3.1 Packages of marijuana seized by U.S. Border Patrol. (U.S. Border Patrol [23].)

TABLE 3.1
Commonly Used Illegal Drugs and Long-Term Health Risks [48]

Drug	Examples	Street Names	Health Risks
Cannabinoids	Marijuana	Pot, weed, Mary Jane, others	Cough, frequent respiratory infections, possible mental health decline, addiction
	Hashish	Hash, hemp, boom, others	Same
Opioids	Heroin	Horse, H, smack, others	Constipation, endocarditis, hepatitis, HIV, addiction, fatal overdose
	Opium	Big O, gum, hop, others	Same
Stimulants	Cocaine	Blow, coke, crack, others	Weight loss, insomnia, cardiac or cardiovascular complications, stroke, seizures, addiction, nasal damage (snorting)
	Amphetamine	Bennies, speed, uppers, others	Same
	Methamphetamine	Meth, crank, crystal, others	Same, with severe dental problems
Club Drugs	MDMA	Ecstasy, Adam, Molly, others	Sleep disturbances, depression, impaired memory, hyperthermia, addiction
	Flunitrazepam	R2, roofies, roach, others	Addiction

WHO. Of those deaths, 167,750 were directly associated with drug use disorders (mainly overdoses). The rest were indirectly attributable to drug use and included deaths related to HIV and hepatitis C acquired through unsafe injecting practices.

3.2.3.1 U.S. Opioids Epidemic

According to the WHO data, opioids continued to cause the most harm, accounting for 76% of deaths where drug use disorders were implicated. People who inject drugs – some 10.6 million worldwide in 2016 – endure the greatest health risks. More than half of them live with hepatitis C, and one in eight live with HIV [4].

In the U.S., according to CDC data, "70,237 drug overdose deaths occurred in the U.S. in 2017 [6]. The age-adjusted rate of overdose deaths increased significantly by 9.6% from 2016 (19.8 per 100,000) to

---

PURDUE PHARMA GUILTY PLEA

Washington, D.C. – November 24, 2020. Purdue Pharma pled guilty to criminal charges for opioid sales. The U.S. Department of Justice announced Pharma faced penalties of about $8.3 billion for its role in the opioids epidemic. Additionally, the company's family owners agreed in October 2020 to pay $225 million in civil penalties [5].

Drugs and Alcohol

2017 (21.7 per 100,000)". Opioids – mainly synthetic opioids (other than methadone) – are currently the main driver of drug overdose deaths. Opioids were involved in 47,600 overdose deaths in 2017 (67.8% of all drug overdose deaths). According to CDC, as of 2017 in the U.S., more than 130 people die every day from opioid overdose [6].

Since 2000, more than 300,000 Americans have lost their lives to an opioid overdose. The misuse of and addiction to opioids, including prescription pain relievers, heroin, and synthetic opioids such as fentanyl, has led to a national crisis that affects not only public health, but also the social and economic welfare of the country [7]. In 2017, the U.S. states and one district with the highest rates of death due to drug overdose were West Virginia (57.8 per 100,000), Ohio (46.3 per 100,000), Pennsylvania (44.3 per 100,000), the District of Columbia (44.0 per 100,000), and Kentucky (37.2 per 100,000) [6].

In 2017, the latest year for which complete data are available from the CDC, more than 70,000 people died from drug overdoses. Opioids such as fentanyl and heroin represented about 68% of those deaths [7]. CDC's National Center for Health Statistics found that fentanyl is the drug most commonly identified in fatal overdoses. In 2017, fentanyl was associated with 38.9% of all drug overdose deaths, an increase from 2016, when it was associated with 29% of all fatal overdoses. In 2017, heroin was associated with 22.8% of all fatal overdoses. Cocaine, a stimulant, was involved in 21.3% and methamphetamine, also a stimulant, was involved in 13.3% [7].

Related to prevention policies concerning the U.S. opioid overdose epidemic is a public health action taken by the state of Pennsylvania, which in 2018 approved a nonprofit organization's proposed creation of a safe injection program [8]. The "safehouse" would provide "consumption rooms" for drug users to inject using sterile equipment under the supervision of medically trained staff. Supervised injection sites are intended to prevent opioid users from sharing contaminated needles. The U.S. Department of Justice litigated the state in an effort to prevent the safehouse's establishment. Pennsylvania federal judge ruled in October 2019 that the nonprofit's plan to open the nation's first safe injection site does not violate federal law. Subsequent to the court's decision, the Trump administration's Department of Justice announced its decision to appeal the decision.

> Since 2000, more than 300,000 Americans have lost their lives to an opioid overdose. The misuse of and addiction to opioids has led to a national public health crisis.

3.2.3.2 Effects of Opioid Epidemic on Children

The adverse consequences of the U.S. opioid epidemic on children were extensively studied by the United Hospital Fund [9]. "In 2017, an estimated 2.2 million children – approximately 2.8% of the 74.3 million children in the U.S. were directly affected by parental opioid use or their own use. Approximately 2 million young people were affected primarily by parental use: they were either living with a parent with opioid use disorder, had lost apparent to an opioid-related death (at any time in their life), had a parent in prison or jail because of opioids, or had been removed from their home due to an opioid-related issue. An additional 170,000 children had opioid use disorder (OUD) themselves or had accidentally ingested opioids. Most young people (1.4 million) affected by the epidemic are primarily influenced by living in a home with a parent with OUD" [9].

- In 2017, an estimated 2.2 million children and adolescents had a parent with opioid use disorder (OUD) or had OUD themselves.
- If current trends continue, an estimated 4.3 million children will have had OUD or a parent with OUD by 2030.
- By 2030, the cumulative, lifetime cost of the "ripple effect" will be $400 billion (this includes additional spending in health care, special education, child welfare, and criminal justice stemming from the multiple impacts of parental opioid use disorder on a child's physical, mental, and social emotional health; it does not include productivity losses or missed opportunities).
- The rate of children affected by the opioid epidemic in 2017 varied significantly from state to state.

In 2017, 28 out of every 1,000 children in the U.S. were affected by opioids. West Virginia had the highest rate of children affected, with 54 out of 1,000 – at least twice the rate of 17 other states. New Hampshire (51 out of 1,000) and Vermont (46 out of 1,000) had the second and third highest rates, respectively. In contrast, California had the lowest rate, with 20 children per 1,000. Children affected by the opioid epidemic will likely incur higher expenses during childhood, a trend that persists into adulthood. "This lifetime societal cost is estimated to be $180 billion for the 2.2 million children affected in 2017. This cost has two components: $117.5 billion incurred during the years of childhood that stems from higher expenses on health care, child welfare, and special education; and $62.1 billion in long-term expenses that accrue during adulthood. The $180 billion estimate does not include missed opportunities or productivity losses, which could be significant. The bulk of the costs that accrue during childhood are for additional general health care and special education services" [9].

ENFORCEMENT EXAMPLE

Boston, January 23, 2020: The founder of an Arizona pharmaceutical company was ordered Thursday in federal court to spend 5½ years in prison for orchestrating a bribery and kickback scheme prosecutors said helped fuel the U.S. opioid crisis [10].

Perspective: The public health and societal damage done to the U.S. population attributable to the opioid epidemic has appropriately received considerable response in news and social media outlets. However, this response has not been as effective in communicating the cost in its fullest context of the epidemic on the well-being of children. A society that fails to protect its children will not last.

3.2.4 Impact of Drug Abuse on Ecosystem Health

Drugs can become a hazard to ecosystem health should they be inappropriately deposited in environmental media. A review of the various ways drugs can enter environmental media identified the following main routes [11]:

- Drug residues can enter surface waters in the manufacturing process.
- Humans metabolize drugs and then excrete trace amounts of them into sewer systems, eventually exiting waste water treatment systems into water supplies.
- Veterinary pharmaceuticals used on pasture animals are excreted into soils and surface waters.
- Using manure as a fertilizer can spread the medications used in livestock.
- Unused medications are often disposed into the water supply through sinks, toilets, and landfills.
- Direct emissions are released into the air during manufacturing.

> **OPIOID DRUG FELONY**
>
> Abingdon, VA: October 2, 2019
>
> A Virginia doctor was sentenced on October 2, 2019, to 40 years in prison for illegally prescribing opioids. He was convicted in May of more than 800 counts of illegally distributing opioids, including oxycodone and oxymorphone. Authorities said the doctor prescribed more than 500,000 doses of opioids to patients from Virginia, Kentucky, West Virginia, Ohio and Tennessee [12].

The degree of hazard will depend on the pharmacologic properties of the drugs, the concentration of the drugs in environmental media, and the susceptibility of ecosystem components to the drugs. Perhaps water supplies are the environmental medium most at risk of drug contamination. In a major study of this issue, U.S. Geological Survey (USGS) researchers in 2018 analyzed samples for 21 different hormones and 103 pharmaceuticals at 1,091 sites around the U.S. in untreated groundwater aquifers [13]. They found at least one hormone or pharmaceutical in 7% of the 844 aquifers at depths used for public water supply and 14% of 247 sites at aquifers used for domestic supply. The sites tested represent about 60% of all the water pumped for drinking water supply across the country, providing water for about 80 million people. Although the drug levels were below contemporary levels of health concern, nevertheless, the actual effects of chronic low-dose exposure to these drugs is unknown. The effects of these drugs on marine life is unknown, but caution should be exercised, given the possibilities of their causing endocrine disruption, as related in Volume 1, Chapter 6 (Toxic Substances in the Environment).

Another consideration regarding the impact of illegal drugs on ecosystems is the environmental damage that results from the deforestation in rainforests and mountain areas such as Columbia's Sierra Nevada de Santa Marta [14]. Deforestation occurs there in order to make room for growing drug-producing plants and their processing into illegal drug products. Drug production in places like this can lead to the loss of complex ecosystems and watercourses before the crops are even planted, followed by the application of pesticides, fungicides, and various chemicals that are part of the agricultural activity to suppress the natural plant growth around and to enhance the monoculture of coca plants.

"Once the coca crop is cultivated, more damage is done in conversion from coca leaves into coca paste and

> Deforestation occurs in order to make room for growing drug-producing plants and their processing into illegal drug products.

eventually high-quality, concentrated cocaine through an even greater application of chemicals including "gasoline, potassium permanganate and a variety of solvents. During production, there is a large amount of wastage and spillage, and that remaining chemicals are often just dumped into watercourses. It's a similar story for heroin, most commonly produced from opium poppies in Afghanistan" [14].

3.2.5 U.S. Policies on Drug Abuse

U.S. federal government legislative concerns about drugs date from the early 20th century. In 1906, Congress enacted the Pure Food and Drug Act, as discussed in Volume 1, Chapter 4, in regard to food safety. The initial concerns circa 1906 were food contamination and bogus medicines, but as time passed into the 1960s, concern in the U.S. grew over the wider presence of addictive drugs. This concern stimulated federal policies and statutes on drug abuse prevention and control, as will be subsequently described.

3.2.5.1 Pure Food and Drug Act, 1906

The first Pure Food and Drug Act was passed in 1906. The purpose was to protect the public against adulteration of food and from products identified as healthful but without scientific support [15]. The original Pure Food and Drug Act was amended in 1912, 1913, and 1923. A greater extension of its scope took place in 1933. Congressional action was fueled by heightened public awareness of safety issues stemming from careless food preparation procedures and the increasing incidence of drug addiction from patent medicines, both accidental and deliberate. The Act was to be applied to goods shipped in foreign or interstate commerce.

The Act's purpose was to prevent adulteration or misbranding of products covered under the Act. Adulteration was defined in various ways. For a confectionary, adulteration would be the result of any poisonous color or flavor, or of any other ingredients harmful to human health. Food was adulterated if it contained filthy or decomposed animal matter, poisonous or deleterious ingredients, or anything that attempted to conceal inferior components. Provisions in Act included creation of the Food and Drug Administration, which was entrusted with the responsibility of testing all foods and drugs destined for human consumption; the requirement for prescriptions from licensed physicians before a patient could purchase certain drugs; and the requirement of label warnings on habit-forming drugs.

From this somewhat humble beginning has come the present day Food and Drug Administration (FDA). The agency states its current mission as: "The Food and Drug Administration is responsible for protecting the public health by ensuring the safety, efficacy, and security of human and veterinary drugs, biological products, and medical devices; and by ensuring the safety of our nation's food supply, cosmetics, and products that emit radiation" [15].

Pertaining to drugs, FDA must review and approve drugs that may be prescribed for medical treatment of humans or for veterinary purposes. The review process involves the following steps, as described by a pharmaceutical support group, noting that it takes on average 12 years and more than $350 million to get a new drug from

Drugs and Alcohol

the laboratory onto the pharmacy shelf [16]. Once a company develops a drug, it undergoes around 3.5 years of laboratory testing before an application is made to the FDA to begin testing the drug in humans. Only one in 1,000 of the compounds that enter laboratory testing will ever make it to human testing.

If the FDA gives it approval, the "investigative" drug will then enter three phases of clinical trials: Phase 1 uses 20–80 healthy volunteers to establish a drug's safety and profile (about 1 year). Phase 2 employs 100–300 patient volunteers to assess the drug's effectiveness (about 2 years). Phase 3 involves 1,000–3,000 patients in clinics and hospitals who are monitored carefully to determine effectiveness and identify any adverse reactions (about 3 years). The company then submits an application (usually about 100,000 pages) to the FDA for approval, a process that can take up to 2.5 years. After review and final approval by FDA, the drug becomes available for physicians to prescribe. At this stage, the drug company must continue to report cases of adverse reactions and other clinical data to the FDA [16].

> Pertaining to drugs, FDA must review and approve drugs that may be prescribed for medical treatment of humans and for veterinary purposes.

The FDA's Center for Drug Evaluation and Research approved 59 novel drugs in 2018, breaking its record of 53 drugs in 1996. The FDA's 5-year annual average is now 43 drugs per year [17].

3.2.5.2 Food, Drug, Cosmetic Act, 1938

The bill that would replace the 1906 Pure Food and Drug Act was ultimately enhanced and passed in the wake of a therapeutic disaster in 1937. A Tennessee drug company marketed a form of the new sulfa wonder drug that would appeal to pediatric patients, Elixir Sulfanilamide. However, the solvent in this untested product was a highly toxic chemical analogue of antifreeze; more than 100 people died, many of whom were children, producing a public outcry for action [15]. The result was the Food, Drug, and Cosmetic Act of 1938, signed into law by President Franklin Roosevelt (D-NY) on June 25, 1938.

The new law brought cosmetics and medical devices under FDA's control, and it required that drugs be labeled with adequate directions for safe use. Moreover, it mandated premarket approval by FDA of all new drugs, such that a manufacturer would have to prove to FDA that a drug was safe before it could be sold. It irrefutably prohibited false therapeutic claims for drugs, although a separate law granted the Federal Trade Commission jurisdiction over drug advertising. The act also corrected abuses in food packaging and quality, and it mandated legally enforceable food standards. Tolerances for certain poisonous substances were addressed. The law formally authorized factory inspections, and it added injunctions to the enforcement tools at the agency's disposal [15]. This 1938 law and its subsequent amendments gave FDA powerful authorities to regulate drugs used in the U.S. for medical purposes.

> The new law brought cosmetics and medical devices under FDA's control, and it required that drugs be labeled with adequate directions for safe use.

3.2.5.3 Comprehensive Drug Abuse Prevention and Control Act, 1970

Recreational drug use became relatively common in the U.S. in the late 1960s, although people have been altering their consciousness with substances since the beginning of civilization. In an effort to combat illegal drug use, Congress passed the Comprehensive Drug Abuse Prevention and Control Act in 1970 [18]. The drug laws prior to this Act were not adequate to address, for example, the illegal use of legally manufactured drugs, such as amphetamines and barbiturates. The Act was signed into law by President Richard M. Nixon (R-CA) on October 27, 1970.

The Act consists of two main parts, Title II and Title III. In addition to classifying and outlawing certain drugs, the Act was also created to research drug abuse and to provide treatment. Changes in drug use in the U.S. after 1970 have at times resulted in amendments to the Act.

This Act's stated purpose is " to amend the Public Health Service Act and other laws to provide increased research, into, and prevention of, drug abuse and drug dependence; to provide for treatment and rehabilitation of drug abusers and drug dependent persons; and to strengthen existing law enforcement authority in the field of drug abuse" [18].

3.2.5.4 The Controlled Substances Act, 1970

The Controlled Substances Act (CSA), Title II of the Comprehensive Drug Abuse Prevention and Control Act of 1970, is the legal foundation of the federal government's fight against the abuse of drugs and other substances [18]. This law is a consolidation of numerous laws regulating the manufacture and distribution of narcotics, stimulants, depressants, hallucinogens, anabolic steroids, and chemicals used in the illicit production of controlled substances. The act also provides a mechanism for substances to be controlled, added to a schedule, decontrolled, removed from control, rescheduled, or transferred from one schedule to another.

The Controlled Substances Act (CSA) places all substances that were in some manner regulated under existing federal law into one of five schedules. This placement is based upon the substance's medical use, potential for abuse, and safety or dependence liability. The CSA also provides a mechanism for substances to be controlled (added to or transferred between schedules) or decontrolled (removed from control). Proceedings to add, delete, or change the schedule of a drug or other substance may be initiated by the Drug Enforcement Administration (DEA), the Department of Health and Human Services (HHS), or by petition from any interested party. Excerpts of the findings required for each of the schedules follow [18]:

> The Controlled Substances Act is the legal foundation of the federal government's fight against the abuse of drugs and other substances.

1. Schedule I. – (A) The drug or other substance has a high potential for abuse. (B) The drug or other substance has no currently accepted medical use in treatment in the U.S. [..].
2. Schedule II. – (A) The drug or other substance has a high potential for abuse. (B) The drug or other substance has a currently accepted medical

use in treatment in the U.S. or a currently accepted medical use with severe restrictions [..].
3. Schedule III. – (A) The drug or other substance has a potential for abuse less than the drugs or other substances in schedules I and II. (B) The drug or other substance has a currently accepted medical use in treatment in the U.S. [..].
4. Schedule IV. – (A) The drug or other substance has a low potential for abuse relative to the drugs or other substances in schedule III. (B) The drug or other substance has a currently accepted medical use in treatment in the U.S.
5. Schedule V. – (A) The drug or other substance has a low potential for abuse relative to the drugs or other substances in schedule IV. (B) The drug or other substance has a currently accepted medical use in treatment in the U.S.

3.2.5.5 U.S. Federal Agencies with Illegal Drug Authorities

Drugs are an integral component of modern life. Used properly, they are essential for medical treatment of many illnesses and diseases for humans, domestic animals, and livestock. Drugs are a substantial force in U.S. commerce and elsewhere. But it is vital that the drugs used for medical care are efficacious, which is why there exist federal statutes and other means to control prescription drugs. On the other hand, some drugs, e.g., opioids, can become abused, causing great adverse social, personal, and economic consequences. As a product of the essential role of therapeutic drugs and the widespread awareness of drug abuse, the U.S. federal government has assumed a major role in drug matters, as subsequently described by the work of several federal agencies.

3.2.5.5.1 Food and Drug Administration

The Food and Drug Administration (FDA) is responsible for protecting the public health by ensuring the safety, efficacy, and security of human and veterinary drugs, biological products, and medical devices; and by ensuring the safety of our nation's food supply, cosmetics, and products that emit radiation [15]. The FDA also has responsibility for regulating the manufacturing, marketing, and distribution of tobacco products to protect the public health and to reduce tobacco use by minors. Concerning drugs, the FDA has regulatory authority over both prescription drugs (both brand-name and generic) and non-prescription (over-the-counter) drugs.

3.2.5.5.2 Drug Enforcement Administration

The Drug Enforcement Administration (DEA) is a U.S. federal law enforcement agency under the U.S. Department of Justice, tasked with combating drug trafficking and distribution within the U.S. The DEA was founded July 1, 1973, and signed into law by President Richard Nixon (R-CA). The agency is headquartered in Springfield, VA. The agency states its mission "is to enforce the controlled substances laws and regulations of the U.S. and bring to the criminal and civil justice system of the U.S., or any other competent jurisdiction, those organizations and principal members of organizations, involved in the growing, manufacture, or distribution of controlled substances appearing in or destined for illicit traffic in the U.S.; and to recommend and support non-enforcement programs aimed at reducing the availability of illicit controlled substances on the domestic and international markets" [19]. The DEA is the U.S. federal agency tasked solely with enforcement of illegal drug laws and regulations in the U.S. The agency works closely with other federal agencies and U.S. state and local law enforcement departments in the pursuit of illegal drug rings, their individuals, and their illegal drug sources.

> The DEA is the U.S. federal agency tasked solely with enforcement of illegal drug laws and regulations in the U.S.

3.2.5.5.3 Federal Bureau of Investigation

The Bureau of Investigation was created on July 26, 1908, after the Congress had adjourned for the summer [20]. Attorney General Bonaparte, using Department of Justice expense funds, hired 34 people, including some veterans of the Secret Service, to work for a new investigative agency. Bonaparte notified the Congress of these actions in December 1908.

The bureau's first official task was visiting and making surveys of the houses of prostitution in preparation for enforcing the "White Slave Traffic Act," or Mann Act, passed on June 25, 1910. In 1932, the bureau was renamed the United States Bureau of Investigation. The following year it was linked to the Bureau of Prohibition and rechristened the Division of Investigation (DOI) before finally becoming an independent service within the Department of Justice in 1935. In the same year, its name was officially changed from the DOI to the present-day Federal Bureau of Investigation, or FBI [20]. The agency cites its mission as "To protect the American people and uphold the Constitution of the United States." The FBI is headquartered in Washington, D.C., with 56 field offices located in major cities throughout the U.S., and more than 350 satellite offices called resident agencies in cities and towns across the nation [20]. The FBI is the primary federal agency tasked with investigation of criminal activities across the nation. The FBI works closely with the DEA in the investigation of criminal activities involving illegal drugs and alcohol.

3.2.5.5.4 U.S. Coast Guard

The Coast Guard is both a federal law enforcement agency and a military force. It was established in August 1790 when President George Washington signed the Tariff Act that authorized the construction of 10 vessels, referred to as The Revenue Cutter Service to enforce federal tariff and trade laws and to prevent smuggling. In 1915, The Revenue Cutter Service was merged with the U.S. Life-Saving Service, and was officially renamed the Coast Guard. In 1967, the Coast Guard was transferred to the Department of Transportation. Then in 2003 the Coast Guard was again transferred, this time to the Department of Homeland Security [21].

> Coast Guard drug interdiction accounts for more than half of all U.S. government seizures of cocaine each year.

The Coast Guard is the nation's first line of defense against drug smugglers seeking to bring illegal substances into the U.S. The Coast Guard coordinates closely with other federal agencies and countries within a vast 6 million square-mile region to disrupt and deter the flow of illegal drugs. Coast Guard drug interdiction accounts for

Drugs and Alcohol

FIGURE 3.2 U.S. Coast Guard boarding a "narco sub." (U.S. Coast Guard [21].)

more than half of all U.S. government seizures of cocaine each year [21]. Illustrated in Figure 3.2 are U.S. Coast Guard officers in September 2019 boarding a "narco sub" off the coast of California. The sub contained 12,000 pounds of cocaine with a street value of more than $165 million [22].

3.2.5.5.5 U.S. Border Patrol

A major federal agency with enforcement of illegal drug laws is the U.S. Border Patrol. Its history dates to the early 20th century, having been preceded by immigration officers on horseback along the Canada and Mexico borders. On May 28, 1924, Congress passed the Labor Appropriation Act of 1924, officially establishing the U.S. Border Patrol for the purpose of securing the borders between inspection stations. In 1925, its duties were expanded to patrol the seacoast. In 1932, the Border Patrol was placed under the authority of two directors, one in charge of the Mexican border office in El Paso, the other in charge of the Canadian border office in Detroit. Liquor smuggling was a major concern because it too often accompanied alien smuggling [23]. Legislation in 1952 codified and carried forward the essential elements of the 1917 and 1924 acts. The same year, Border Patrol agents were first permitted to board and search a conveyance for illegal immigrants anywhere in the U.S. For the first time, illegal entrants traveling within the country were subject to arrest. The U.S. Boarder Patrol is headquartered in Washington, D.C.

At the nation's more than 300 ports of entry, customs and border patrol officers screen all foreign visitors, returning U.S. citizens, and imported cargo that enters the U.S. Border Patrol officers also inspect visitors and cargo for the presence of illegal drugs. The agency reported the following illegal drug seizures nationwide in fiscal years 2017 and 2018, respectively: cocaine (9,346; 6,550), heroin (953; 568), marijuana (861,231; 461,030), methamphetamine (10,328; 11,314), fentanyl (181; 388) [24].

3.2.5.5.6 National Institute on Drug Abuse

The mission of the National Institute on Drug Abuse (NIDA) is to advance science on the causes and consequences of drug use and addiction and to apply that knowledge to improve individual and public health. The NIDA is a component of the National Institutes of Health (NIH), based in Rockville, Maryland. The Institute " addresses the most fundamental and essential questions about drug abuse – from detecting and responding to emerging drug abuse trends and understanding how drugs work in the brain and body, to developing and testing new approaches to treatment and prevention. NIDA also supports research training, career development, public education, public-private partnerships, and research dissemination efforts. Through its Intramural Research Program, as well as grants and contracts to investigators at research institutions around the country and overseas" [25].

3.2.6 OREGON'S DECRIMINALIZATION OF HARD DRUGS

In the November 2020 election, Oregon became the first U.S. state to decriminalize the possession of small amounts of hard drugs [26]. The Oregon drug initiative will allow people arrested with small amounts of hard drugs, including heroin, cocaine, and meth, to avoid going to trial, and possible jail time. Persons arrested would pay a $100 fine and attend an addiction recovery program. Treatment centers will be funded by revenues from legalized marijuana, which was approved in Oregon several years ago.

> In the November 2020 election, Oregon became the first U.S. state to decriminalize the possession of small amounts of hard drugs.

Supporters of the initiative assert the "Drug Addiction Treatment and Recovery Act" will transition Oregon's drug policy from a punitive, criminal approach to "a humane, cost-effective, health approach." Only small amounts of drugs are decriminalized, such as less than 1 g of heroin or MDMA; 2 g of cocaine or methamphetamine; 12 g of psilocybin mushrooms; and 40 doses of LSD, oxycodone or methadone. Crimes that are associated with drug use, such as manufacturing drugs, selling drugs and driving under the influence, are still criminal offenses.

3.2.7 UN POLICIES ON ILLEGAL DRUGS

The production, transportation, sale, and use of illegal drugs are a global enterprise. It is not hyperbole to state that few if any countries are spared the ignominy of the presence of illegal drugs. This kind of commerce is not new, probably dating to the time that alcohol and opium were found to have pain relief properties. What is relatively new is the geographic extent

> The UN's work in countering the world drug problem is based on three major international drug control treaties.

and commercial worth of the illegal drug trade. Because illegal drugs are a problem for the maintenance of stable societies, individual countries, regional associations, and international bodies such as the UN have developed policies to prevent illegal drug commerce and to provide public health and other social benefits to persons caught up in the drug traffic.

The UN's work in countering the world drug problem is based on three major international drug control treaties: the Single Convention on Narcotic Drugs of 1961 (as amended in 1972), the Convention on Psychotropic Substances of 1971, and the United Nations Convention against Illicit Traffic in Narcotic Drugs and Psychotropic Substances of 1988. These three conventions impart important functions to the Commission on Narcotic Drugs and to the International Narcotics Control Board [27].

The Single Convention on Narcotic Drugs of 1961 aimed to combat drug abuse by coordinated international action [28]. First, the Convention sought to limit the possession, use, trade in, distribution, import, export, manufacture and production of drugs exclusively to medical and scientific purposes, second, to combat drug trafficking through international cooperation to deter and discourage drug traffickers. This Convention functioned by treaties between the UN and its Member Countries, with a focus on international cooperation on prevention of drug trafficking.

The Convention on Psychotropic Substances of 1971 is a UN treaty designed to control psychoactive drugs such as amphetamine-type stimulants, barbiturates, benzodiazepines, and psychedelics. The Convention was signed in Vienna, Austria, on February 21, 1971 [29]. In retrospect, the Single Convention on Narcotic Drugs of 1961's scope was limited to drugs with cannabis, coca, and opium-like effects. The Convention, which contains import and export restrictions and other rules aimed at limiting drug use to scientific and medical purposes, came into force on August 16, 1976. As of 2013, 183 member states are Parties to the treaty. Many laws have been passed to implement the Convention, including the U.S. Psychotropic Substances Act. In 1991, the UN created the United Nations International Drug Control Program, directed by a senior UN officer.

The United Nations Convention against Illicit Traffic in Narcotic Drugs and Psychotropic Substances of 1988 was introduced following the political and sociological developments in the 1970s and 1980s [30]. The growing demand for cannabis, cocaine, and heroin for recreational purposes, mostly in the developed world, had triggered an increase of illicit production in geographical areas where cannabis, coca, and opium had been traditionally cultivated. With the rising size of the illicit drug trade, international drug trafficking became a multibillion-dollar business dominated by criminal groups, providing grounds for the creation of the 1988 Convention and the consequential escalation of the war on drugs.

The Preamble notes that previous enforcement efforts had not stopped drug use, warning of "steadily increasing inroads into various social groups made by illicit traffic in narcotic drugs and psychotropic substances." Much of the treaty is devoted to fighting organized crime by mandating cooperation in tracing and seizing drug-related assets [30]. Article 5 of the Convention requires its parties to confiscate proceeds from drug offenses. It also requires parties to empower its courts or other competent authorities to order that bank, financial, or commercial records be made available or seized. The Convention further states that a party may not decline to act on this provision on the ground of bank secrecy. Article 6 of the Convention provides a legal basis for extradition in drug-related cases among countries having no other extradition treaties. In addition, the Convention requires the parties to provide mutual legal assistance to one another upon request, for purposes of searches, seizures, service of judicial documents, and so on.

3.3 ALCOHOL AND PUBLIC HEALTH

Setting aside the **putative** social benefits of alcohol consumption in moderation, excess (emphasis added) consumption of alcoholic beverages is a well-known, well-documented hazard to an individual's health and well-being, as will be summarized in the following sections. The old dictum in toxicology that "the dose makes the poison" applies equally well to alcohol consumption and arsenic ingestion. In particular, it is important in a public health sense to know that alcohol abuse afflicts more than just the imbiber. There are social impacts on familiar relations, community well-being, and economic systems such as increased healthcare costs.

3.3.1 Précis History of Alcohol Discovery and Use

It's likely that alcohol production started when early farmers noted the fermentation that took place in fallen fruit. They may have found the fizzy flavor and sharp aroma pleasing. Trial and error using different fruits and grains finally resulted in formulas that could be refined and repeated for a palatable alcoholic beverage. Alcohol manufacture started in an organized fashion about 10,000 years ago, when a fermented drink was produced from honey and wild yeasts [31]. Interestingly, this same beverage, mead, is still produced and consumed today after 100 decades.

It is possible that alcoholic beverages were used in China well before they were used in the West. It is thought that alcoholic drinks were used as part of celebrations, when taking an oath of office or going into battle, as well as occasions such as births, deaths, and marriages. Elsewhere, by 6000 BCE, grapevines were being cultivated in the mountains between the Black and Caspian Seas for the purpose of making wine. In another 2,000 years, Mesopotamia (present-day Iraq) had a thriving winemaking enterprise. Around 3000 BCE, wine production and shipping throughout the Mediterranean were important businesses for Egypt's commerce and its wine merchants.

> It is possible that alcoholic beverages were used in China well before they were used in the West.

Romans made wine from the wild grapes that grew in the countryside – grapes with the yeasts necessary for fermentation already growing on their skin. The Romans developed a way of letting a fine vintage age, using an amphora, a large, tapered two-handled jar. It was filled with nearly seven gallons of wine and then sealed, protected from the air while it matured. Around 1500 BCE, the Roman god Dionysus began to appear in literature. Dionysus (Bacchus in Greek myth) was the god of the grape harvest and winemaking. A cult grew around the belief that wine could be used in rituals to return to a more innocent, aware state. Increasing drunkenness began to accompany a Roman decline in simplicity and honesty and a rise in raw ambition, corruption, and regular, heavy drinking [31].

By 800 BCE, barley and rice beer began to be produced in India. Elsewhere, in Rome, one emperor after another became known for abusive drinking. After 69 CE, these reports dropped off and it is thought that drinking might have declined substantially over the whole Roman Empire. In 600 CE, the Prophet Muhammed ordered his adherents to refrain from drinking alcohol. Buddhists and Hindu Brahmins also abstained.

The Middle Ages in Europe saw extensive development of choices of wines, beer, and mead (alcoholic beverage made from honey). Wines stayed the most popular choice in the regions that became Italy, Spain, and France. Monks began to brew nearly all the beer of good quality, which by this time contained hops, plus wine for celebrating mass. They eventually added brandy to their list of wares. Beer manufacturing began to grow in Germany, with cities competing for the best products. By the end of the Middle Ages, beer and wine production made its way to Scotland and England and quickly became important industries. In the 1600s, drunkenness became a widespread problem in England, with both beer and wine commonly abused. When religious groups fled to America in the next century, many formed temperance societies in the new country.

> The Middle Ages in Europe saw extensive development of choices of wines, beer, and mead (alcoholic beverage made from honey).

The first distillery in America was established on Staten Island and hops were grown in Massachusetts to supply the breweries. Massachusetts also had a rum distillery, started in 1657 in Boston. This would soon become New England's most prosperous industry and gave rise to smuggling activities along the coast, as alcohol production was taxed in the colonies.

In the sixteenth century, alcohol (called "spirits") was used largely for medicinal purposes. The early 1700s in England saw the production of millions of gallons of gin, which is alcohol flavored with juniper berries. By 1733, the London area alone produced 11 million gallons of gin. The poor in London found relief from the difficulties of urban poverty in cheap liquor. Taxes on gin were soon increased to try to reduce the epidemic of drunkenness that followed [36].

At the beginning of the eighteenth century, the British parliament passed a law encouraging the use of grain for distilling spirits. Cheap spirits flooded the market and reached a peak in the mid-eighteenth century. In Britain, gin consumption reached 18 million gallons and alcoholism became widespread [32]. The nineteenth century brought a change in attitudes and the temperance movement began promoting the moderate use of alcohol—which ultimately became a push for total prohibition. In 1920, the U.S. passed the 18th Amendment to the U.S. Constitution, prohibiting the manufacture, sale, import, and export of intoxicating liquors. As a consequence, an illegal alcohol trade boomed and by 1933, the prohibition of alcohol was repealed via the 21st Amendment to the U.S. Constitution. In 2019, an estimated 15 million Americans suffer from alcoholism and 40% of all car accident deaths in the U.S. involve alcohol [32].

3.3.2 Forms and Prevalence of Alcohol Consumption

The three types of alcohol that humans use every day are methanol, isopropanol, and ethanol. Only ethanol is safe to drink; the other two forms of alcohol are poisonous if consumed in health excess by humans, but serve useful industrial and commercial purposes such as solvents and for medicinal purposes. Ethanol (or ethyl alcohol) is the type of alcohol that more than 2 billion people drink every day. This type of alcohol is produced by the fermentation of yeast, sugars, and starches [33]. There are two categories of alcoholic beverages: distilled and undistilled.

Undistilled drinks are also called fermented drinks. Fermentation is the process of using bacteria or yeast to chemically convert sugar into ethanol. Wine and beer are both fermented, undistilled alcoholic beverages. Wineries ferment grapes to make wine and breweries ferment barley, wheat, and other grains to make beer. Distillation is a process that follows fermentation. It converts a fermented substance into one with an even higher concentration of alcohol, e.g., whiskey. Distillation concentrates alcohol by separating it from the water and other components of a fermented substance. Liquors and spirits are distilled alcoholic beverages. The principal types of alcoholic beverage and their ethanol volume are shown in Table 3.2. Beer is the most often consumed alcoholic beverage worldwide. In fact, after water and tea, beer is the most commonly-consumed drink in the world. Beer is also most likely the oldest alcoholic drink in history [33]. Illustrated in Figure 3.3 are some varieties of alcoholic beverages commonly available in U.S. stores. Noteworthy is the wide variety of alcoholic beverages, including beer, wine, and various spirits. As the figure suggests, alcohol sales are big business in the U.S. and elsewhere. According to an industry source,

TABLE 3.2
Types of Alcoholic Drinks by Alcohol Content (ABV) [33]

Alcohol	Form	Fermentation Base	ABV (%)
Beer	Standard	Grain	4–6
	Lite		2–4
	Malt liquors		6–8
Wine	Standard	Fruit	<14
	Champagne		10–12
	Fortified (e.g., sherry)		20
Cider	Hard	Apples	5
Mead		Honey	10–14
Saké		Rice	16
Gin		Juniper berries	35–55
Whiskey		Grain	40–50
Rum	Standard	Sugarcane/Molasses	40
	Overproof		>57.5
Tequila		Mexican agave	40
Vodka		Grains/Potatoes	40

ABV, alcohol by volume

Drugs and Alcohol

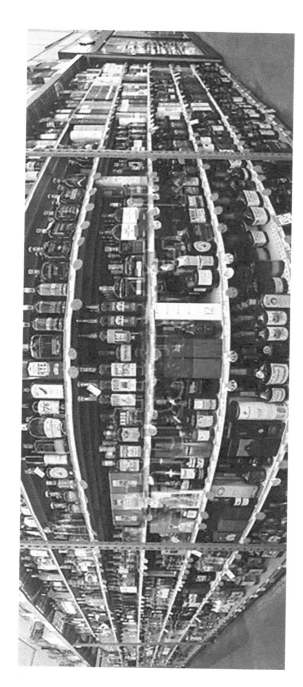

FIGURE 3.3 Typical display of alcoholic beverage bottles in a store. (Shuttercock. Editorial Content. https://www.shutterstock.com/image-photo/chatham-nj-march-11-2017-spirits-603314933.)

U.S. alcohol sales in 2018 reached $253.8 billion, which was an increase of 5.1% over 2017 sales [34]. For sake of comparison, the 2018 budget for the U.S. National Institutes of Health was $31.7 billion.

> There are two categories of alcoholic beverages: distilled and undistilled.

3.3.3 COVID-19 AND ALCOHOL CONSUMPTION

As discussed in Chapter 1, the global pandemic of COVID-19 had an unprecedented effect on global health, commerce, and policies. Social distancing was among the public health measures advocated to prevent the pandemic's spread and consequences. Effecting this strategy led to the closure or limited use of bars, restaurants, conventions, sporting events, and other assemblies where people would normally congregate in close proximity. Persons at potential risk of exposure to the virus were advised to remain at home, thereby minimizing contact with other persons who might be infected with the virus. Needless to say, these social changes had an effect on alcohol consumption.

Research by public health investigators found evidence of changes in alcohol use and associated consequences during the COVID-19 pandemic [35]. On average, alcohol was consumed 1 day more per month by 3 of 4 adults. For women, there was also a significant increase of 0.18 days of heavy drinking, from a 2019 baseline of 0.44 days, which represents an increase of 41% over baseline. This equates to an increase of 1 day for 1 in 5 women. Further, for women, there was an indication of increased alcohol-related problems independent of consumption level for nearly 1 in 10 women. One can speculate that stress caused by the pandemic was associated with increased alcohol consumption.

3.3.4 IMPACT OF ALCOHOL ON HUMAN HEALTH

The WHO has framed the global health toll caused by excess alcohol consumption, commenting "Alcohol consumption is a causal factor in more than 200 disease and injury conditions. Drinking alcohol is associated with a risk of developing health problems such as mental and behavioral disorders, including alcohol dependence, major noncommunicable diseases such as liver cirrhosis, some cancers and cardiovascular diseases, as well as injuries resulting from violence and road clashes and collisions."

A significant proportion of the disease burden attributable to alcohol consumption arises from unintentional and intentional injuries, including those due to road traffic crashes, violence, and suicides, and fatal alcohol-related injuries tend to occur in relatively younger age groups.

The latest causal relationships are those between harmful drinking and incidence of infectious diseases such as tuberculosis as well as the incidence and course of HIV/AIDS. Alcohol consumption by an expectant mother may cause fetal alcohol syndrome and preterm birth complications. The WHO has also provided in 2018 the following statements of fact [36,37]:

Drugs and Alcohol

- Worldwide, 3 million deaths every year result from harmful use of alcohol; this represents 5.3% of all deaths. More than three quarters of these deaths were among men.
- The harmful use of alcohol is a causal factor in more than 200 disease and injury conditions.
- Of all deaths attributable to alcohol, 28% were due to injuries, such as those from traffic crashes, self-harm, and interpersonal violence; 21% due to digestive disorders; 19% due to cardiovascular diseases, and the remainder due to infectious diseases, cancers, mental disorders and other health conditions.
- Globally an estimated 237 million men and 46 million women suffer from alcohol-use disorders with the highest prevalence among men and women in the European region (14.8% and 3.5%) and the Region of Americas (11.5% and 5.1%). Alcohol-use disorders are more common in high-income countries.
- Alcohol consumption causes death and disability relatively early in life. In the age group 20–39 years approximately 13.5% of the total deaths are alcohol-attributable.
- There is a causal relationship between harmful use of alcohol and a range of mental and behavioral disorders, other noncommunicable conditions as well as injuries.
- The latest causal relationships have been established between harmful drinking and incidence of infectious diseases such as tuberculosis as well as the course of HIV/AIDS.
- Beyond health consequences, the harmful use of alcohol brings significant social and economic losses to individuals and society at large [34].

In the U.S., as characterized by the CDC, excessive alcohol use led to approximately 88,000 deaths and 2.5 million years of potential life lost (YPLL) each year in the U.S. from 2006 to 2010, shortening the lives of those who died by an average of 30 years [38]. Further, excessive drinking was responsible for 1 in 10 deaths among working-age adults aged 20–64 years. The economic costs of excessive alcohol consumption in 2010 were estimated at $249 billion. In the U.S., a standard drink is defined as 0.6 ounces (14.0 grams) of pure alcohol. Generally, this amount of pure alcohol is found in 12-ounces of beer (5% alcohol content), 8-ounces of malt liquor (7% alcohol content), or 5-ounces of wine (12% alcohol content).

> WHO: Worldwide, 3 million deaths every year result from the harmful use of alcohol, which is a causal factor in more than 200 disease and injury conditions.

A study of alcohol-related mortality in the U.S., 1999–2017 was performed by researchers with the National Institute on Alcohol Abuse and Alcoholism [39]. The study was prompted by data showing that alcohol consumption, alcohol-related emergency department visits, and hospitalizations had all increased in the last two decades in the U.S., particularly among women and people middle-aged and older.

Mortality data from the National Center for Health Statistics were analyzed to estimate the annual number and rate of alcohol-related deaths by age, sex, race, and ethnicity between 1999 and 2017 among people aged 16+. Mortality data contained details from all death certificates filed nationally.

Results found that the number of alcohol-related deaths per year among people aged 16+ doubled from 35,914 to 72,558, and the rate increased 50.9% from 16.9 to 25.5 per 100,000. Nearly 1 million alcohol-related deaths (944,880) were recorded between 1999 and 2017. In 2017, 2.6% of roughly 2.8 million deaths in the United States involved alcohol. Nearly half of alcohol-related deaths resulted from liver disease (30.7%; 22,245) or overdoses on alcohol alone or with other drugs (17.9%; 12,954). Rates of alcohol-related deaths were highest among males, people in age-groups spanning 45–74 years, and among non-Hispanic (NH) American Indians or Alaska Natives [39]. These results depict a troubling public health pattern of alcohol association with premature death in the U.S.

The CDC considers excessive drinking as including binge drinking, heavy drinking, and any drinking by pregnant women or persons younger than age 21 years. Binge drinking, the most common form of excessive drinking, is defined as consuming for women, 4 or more drinks during a single occasion. For men, binge drinking is 5 or more drinks during a single occasion. Heavy drinking is defined as consuming for women, 8 or more drinks per week and for men, 15 or more drinks per week. Over time, CDC notes that excessive alcohol use can lead to the development of chronic diseases and other serious problems including [38]:

> CDC: Excessive alcohol use led to approximately 88,000 deaths and 2.5 million YPLL annually in the U.S.

- High blood pressure, heart disease, stroke, liver disease, and digestive problems.
- Cancer of the breast, mouth, throat, esophagus, liver, and colon.
- Learning and memory problems, including dementia and poor school performance.
- Mental health problems, including depression and anxiety.
- Social problems, including lost productivity, family problems, and unemployment.
- Alcohol dependence or alcoholism.

Perspective: Reducing the risk of the long-term health risks of excessive alcohol consumption can occur by decreasing alcohol consumption. The facts and figures of alcohol abuse, as cited by the WHO and the CDC, are alarming in a public health and community well-being sense. Public health agencies and environmental health policymakers bear a shared responsibility to make alcohol abuse a subject of education and awareness.

3.3.5 Impact of Alcohol on Ecosystem Health

There are a number of ways in which alcohol production negatively impacts the planet, starting with the process of growing the ingredients necessary to produce alcohol. Grains, potatoes, rice, botanicals, sugar cane, and agave are all significant ingredients in the alcohol industry, each of which requires a significant amount of water, fertilizer, land, and use of machinery. In essence, these resources are being used to produce beverages that are not necessary for human survival, and which could be diverted to providing food aid for those in need [40].

> Disposal of alcohol waste must comply with EPA regulations.

There are also issues with packaging and distributing alcoholic beverages. An enormous number of bottles, cans, kegs, plastic wrappers, and cardboard boxes are necessary for shipping and distributing alcoholic beverages all over the world. While some of these resources are reusable and recyclable (e.g., paper and aluminum), many are just as easily broken and damaged, creating a need for more to be made. It is also important to look at the physical transportation costs that come with distributing alcohol. Tequila is a product that can only be produced in Mexico, and Scotch whisky can only be made in Scotland, yet both are readily and widely available all around the world.

The containers, bottles, cans, paper boxes all constitute potential solid waste environmental issues. Disposal of alcohol waste must comply with EPA regulations. Specifically, if the waste is contaminated with a solution that is at least 50% water and less than 24% alcohol by volume, it can be disposed of in accordance with the applicable rules for solid waste. If the waste exceeds these specifications, it might qualify as hazardous waste, requiring local waste managers to scrutinize and act accordingly [40].

3.3.6 U.S. Policies on Alcohol Production

Alcohol was an essential companion of the European colonists as they gradually emigrated to what became the U.S. Alcohol was thought to be essential due to distrust and uncertainty of drinking water supplies. But it was also because alcohol had become a steadfast component of social and family life in the countries of the emigrants. Beer, cider, and mead were favorite alcoholic beverages in the lives of the colonists. In time, as noted in a subsequent section on the history of alcohol, rum production became a major commercial enterprise in the emerging colonial commercial strata. Alcohol production and the use of its products were relatively free of government oversight and control, a situation that remains today. To be discussed in this section are the federal government policies pertaining to alcohol production and distribution. Perhaps the most interesting federal alcohol policy came early in the 20th century, as described in the following section.

3.3.6.1 18th Amendment to the U.S. Constitution

"After one year from the ratification of this article the manufacture, sale, or transportation of intoxicating liquors within, the importation thereof into, or the exportation thereof from the U.S. and all territory subject to the jurisdiction thereof for beverage purposes is hereby prohibited." With these words began a national experiment in the U.S. to control the perceived adverse consequences of alcohol consumption. The experiment was embodied in the 18th amendment to the U.S. Constitution [41]. (The first ten amendments to the US Constitution are called the Bill of Rights.)

This amendment was proposed in Congress on December 18, 1917, and ratified on January 16, 1919. Thus began the period of what became known as Prohibition. Though Congress provided states with a period of 7 years in which to ratify the amendment, approval took just over a year, such was the prevailing spirit among lawmakers. (To become part of the Constitution, an amendment must be ratified by either – as determined by Congress – the legislatures of three-quarters of the states or state-ratifying conventions in three-quarters of the states.)

> The 18th Amendment prohibited the manufacture, sale, or transportation of intoxicating liquors in the U.S.

One consequence of Prohibition was the almost instant presence of criminal activities that supplied alcohol to the U.S. public. For instance, alcohol produced in Canada was transported illegally into the U.S. by organized gangs of criminals, some of whom like Chicago's Al Capone became notorious characters in U.S. history. Gang warfare became a major urban problem when gangs competed for shares of the illegal alcohol trade. Prohibition also created novel methods to entertain U.S. partygoers. One method was to convert existing night clubs into what became known as "speak-easy" clubs, where illegally purchased alcohol was available, and entrance into the club was controlled by guards who validated identity cards.

In the early decades of the twentieth century, the Temperance Movement (which advocated abstinence from alcohol) was steadily growing: Thirteen of 31 states had outlawed the manufacture and sale of alcohol by 1855. During the 1870s, temperance also became one of the cornerstones of the growing women's movement in the U.S. As the nation's women, joined by other activists, mobilized to gain suffrage (the right to vote), they also espoused sweeping cultural changes. Outlawing the manufacture and consumption of alcoholic beverages, which were viewed by many women to be a corrupt influence on American family life, was one such initiative. Even President Woodrow Wilson (D-NJ) supported Prohibition, as one of the domestic policies of his New Freedom program [41].

The U.S. Constitution's 21st Amendment repealed the 18th Amendment, which prohibited the manufacture, transportation, and sale of alcoholic beverages. It was passed in 1933, thus ending 13 years of Prohibition. The repeal generated a number of positive effects, such as giving adults the personal freedom to drink again and weakening organized crime's grip on power.

3.3.6.2 Federal Alcohol Administration Act, 1935

The U.S. federal government levied a tax on alcoholic beverages almost from the beginning of the nation. This continued until the early 20th century when the prohibition of the sale of intoxicating liquors in the U.S., by the 18th amendment to the U.S. Constitution, came to an end with the ratification of the twenty-first amendment by the states on December 5, 1933. Quickly, laws had to be repealed, amended or drawn up to control production and distribution of alcoholic beverages and to assure the collection of taxes

> The Federal Alcohol Administration Act of 1935 gave the federal government the authority and resources to levy an excise tax on alcoholic beverages.

on such products. Thus, on August 29, 1935, the Congress passed the Federal Alcohol Administration Act "To further protect the revenue derived from distilled spirits, wine, and malt beverages, to regulate interstate and foreign commerce, and enforce the postal laws with respect thereto, to enforce the twenty-first amendment, and for other purposes" [42]. This Act gives the federal government the authority and resources to levy an excise tax on alcoholic beverages. The U.S. states have their own taxing authorities concerning alcoholic beverages.

Historically, the Federal Alcohol Administration was a U.S. government agency created in 1935 (as part of the Department of the Treasury) by the Federal Alcohol Administration Act. It was created to regulate the alcohol industry after the repeal of Prohibition, replacing a previous body (the Federal Alcohol Control Administration), which did not have statutory powers. The Act still partly continues in force, underpinning the powers of the Alcohol and Tobacco Tax and Trade Bureau (TTB) [42]. The federal government, U.S. states, and local governments all collect excise taxes[1] on alcohol beverages. The federal government collects about $1 billion per month from excise taxes on spirits, beer, and wine. Additional taxes are imposed by state governments. For example, in Georgia the state and its counties add an aggregate excise tax of $4.62 per gallon of spirits and 85 cents to beer and $2.34 to wine [43].

3.3.6.3 U.S. Bureau of Alcohol, Tobacco, Firearms, and Explosives

The only federal government organization bearing the word "alcohol" in its title is the Bureau of Alcohol, Tobacco, Firearms, and Explosives (ATF), which is a federal law enforcement organization within the U.S. Department of Justice. Its responsibilities include the investigation and prevention of federal offenses involving the unlawful use, manufacture, and possession of firearms and explosives; acts of arson and bombings; and illegal trafficking and tax evasion of alcohol and tobacco products. The ATF also regulates via licensing the sale, possession, and transportation of firearms, ammunition, and explosives in interstate commerce.

As background, in the early 1950s, the Bureau of Internal Revenue was renamed "Internal Revenue Service" (IRS), and its Alcohol Tax Unit (ATU) was given the additional responsibility of enforcing federal tobacco tax laws. At this time, the name of the ATU was changed to the Alcohol and Tobacco Tax Division (ATTD). In 1968, with the passage of the Gun Control Act, the agency changed its name again, this time to the Alcohol, Tobacco, and Firearms Division of the IRS and first began to be referred to by the initials "ATF." In Title XI of the Organized Crime Control Act of 1970, Congress enacted the Explosives Control Act, which provided for close regulation of the explosives industry and designated certain arsons and bombings as federal crimes. The Secretary of the Treasury was made responsible for administering the regulatory aspects of the new law and was given jurisdiction over criminal violations relating to the regulatory controls. These responsibilities were delegated to the ATF division of the IRS. The Secretary of the Treasury and the U.S. Attorney General were given concurrent jurisdiction over arson and bombing offenses [44].

---

[1] Excise tax is a flat-rate tax that applies to specific goods, services, and activities. In the U.S., goods like alcohol and services like indoor tanning are assessed an excise tax, which applies to each unit or occurrence regardless of its cost.

On July 1, 1972, the ATF was officially established as an independent bureau within the U.S. Treasury Department. This transferred the responsibilities of the ATF division of the IRS to the new Bureau of Alcohol, Tobacco, and Firearms, which brought the organization into a new era where federal firearms and explosives laws addressing violent crime became the primary mission of the agency. However, taxation and other alcohol issues remain principal duties because the ATF is responsible for collecting federal taxes on alcohol and tobacco sales [44].

3.3.7 Global Policies on Alcohol Production

Unlike global policies between nations for combating illegal drugs, no similar activity exists for alcohol commerce. Rather, alcoholic beverages are a matter of individual national policies. For some countries, e.g., Islamic republics, alcohol is banned in accord with religious education and practice. In other countries, e.g., European and North American, alcohol production, distribution, and consumption are matters of tax revenue and social systems that enforce policies against alcohol-fueled crimes such as operating a vehicle while under the influence of alcohol.

One global policy that is important to relate for public health purposes is a WHO program that commenced in 2010. In that year, delegations from all 193 Member States of WHO reached consensus at the World Health Assembly on a global strategy to confront the harmful use of alcohol. In 2008, WHO had begun the process of drafting a global strategy to reduce the harmful use of alcohol. On May 21, 2010, the Sixty-third session of the World Health Assembly adopted by consensus resolution WHA63.13, which endorses the global strategy. The WHO noted that implementation of the global strategy would require active collaboration with Member States, with appropriate engagement of international development partners, civil society, the private sector, as well as public health and research institutions [45]. The four priority areas for global action were stated as:

- public health advocacy and partnership;
- technical support and capacity building;
- production and dissemination of knowledge;
- resource mobilization.

A primary effort by WHO in implementing its global strategy on alcohol is the establishment of research data on the adverse health effects of alcohol and to distribute the information to its Member States, with advice on how to use the health information for purposes of health policy.

The WHO's efforts on pursuing a global strategy on alcohol were illustrated in 2019 when a WHO report examined the effects of alcohol control measures on mortality and life expectancy in Russia [46]. The WHO observed that Russians are officially drinking less alcohol and as a consequence are living longer: life expectancies reached an historic peak in 2018—almost 68 years for men and 78 years for women. Alcohol control measures in Russia began in 1995 when the government slowly introduced alcohol production control measures, including Federal Law number 171, and restrictions on licensing and advertising, but with limited success. In 2003, Russia reached its all-time high: the total adult per capita alcohol consumption was 20.4 liters. This peak coincided with almost half of all deaths in working-age men, in a typical Russian city, being attributed to hazardous drinking. Since 2011, Russia has taken an active role in implementing policies to reduce the harmful use of alcohol by further increasing excise taxes, raising the minimum unit price of alcohol, and substantially reducing the availability of retail alcohol. As a result of government actions, the period from 2003 to 2017 saw the prevalence of alcohol dependence in patients registered in state-run treatment services fall by 38%, the prevalence of harmful use of alcohol drop by 54%, and the prevalence of alcoholic psychosis reduce by 64%.

3.4 LOW-RISK DRINKING GUIDELINES

Many countries propose low-risk drinking guidelines (LRDGs) to mitigate alcohol-related harms. In an attempt to mitigate these alcohol-caused harms, at least 50 countries worldwide have established LRDGs to provide information to drinkers wishing to reduce the risk of both acute harm related to intoxication and chronic harm related to long-term alcohol use. LRDGs vary substantially by jurisdiction and usually provide both daily and weekly thresholds that should not be exceeded. In fact, a review of 37 international LRDGs found that daily alcohol limits differed by a factor of almost six [47].

The 2015–2020 *Dietary Guidelines for Americans* define moderate drinking as up to 1 drink per day for women of legal drinking age and up to 2 drinks per day for men of legal drinking age [48]. Avoidance of alcohol altogether includes, among other reasons: taking medications that interact with alcohol, managing a medical condition that can be made worse by drinking, underage, planning to drive a vehicle or operate machinery.

Canada's Low-Risk Alcohol Drinking Guidelines recommend: No more than 10 drinks a week for women, with no more than 2 drinks a day most days. No more than 15 drinks a week for men, with no more than 3 drinks a day most days. Not drinking on some days each week. It should be noted that these are guidelines. They are intended to provide useful advice for healthcare providers and public health specialists.

3.5 HAZARD INTERVENTIONS

History has shown that the presence of illegal drugs in a society will result in adverse consequences that will in the extreme destroy the society's social structure, commencing with family units. Similarly, individuals' intemperate consumption of alcohol can yield deleterious social impacts, including adverse health effects, job loss, and family dissolution. For these reasons, interventions to prevent these kinds of social consequences are suggested.

1. Support for programs of prevention of illegal drugs is vital, particularly public health programs of education.
2. Family members should be knowledgeable about the hazards of inappropriate use of prescribed drugs as well as the danger of illegal addictive drugs.
3. Prescription drugs should be maintained in a safe, secure location within a family residence.

4. School health programs should educate youth about the hazards of illegal drugs.
5. National and state government programs of illegal drug control should be supported by policymakers and individuals who elect them.
6. Families should secure alcohol products in safe, secure locations in order to prevent young persons from binge drinking.
7. School health programs should educate youth about the hazards of alcoholic beverages.
8. State and local governments should support the availability of alcoholism treatment programs since they derive revenue from alcoholic beverage excise taxes.

> Families should secure alcohol products in safe, secure locations in order to prevent young persons from binge drinking.

3.6 SUMMARY

This book is about hazards to health that we experience in our physical environment. Hazards to health such as air pollution and contaminated water sources are presented as hazards unwelcome to those exposed to them. In contrast, this chapter is about two hazards to health, drugs and alcohol, that humans choose to impose upon themselves. In that regard, this chapter is in accord with Volume 1, Chapter 5, where the self-imposed health hazards of tobacco, vaping, and marijuana smoking were discussed.

Regarding drugs, modern medicine would be far less effective sans modern drug therapy. Whether a simple drug such as aspirin for minor pain relief or a complex drug for immunotherapy of cancer, medical practice relies on their pharmacological partner in the treatment of human maladies. And yet, humans have unfortunately found ways to abuse drugs, as discussed in this chapter. In response, societies have developed policies to combat the presence and use of drugs termed illegal. The problem of drug abuse is global, as noted by the WHO. How to effectively prevent the presence of illegal drugs is a challenge to policymakers, as noted in this chapter. The importance of education about the adverse health and societal impacts of illegal drugs cannot be overstated, commencing with education of young children at home and in school.

Alcohol as a public health issue is the second subject of this chapter. Humankind has long favored alcoholic beverages for social communication purposes and personal enjoyment. While temperate use of alcohol seems beneficial, intemperate, overindulgence of alcohol can produce adverse health and economic impacts. The problems caused by alcohol intoxication are myriad, especially under chronic conditions. As with the problem of dealing with illegal drugs, education about the problems caused by unwise use of alcohol is basic to prevention of health and socioeconomic problems associated with alcohol.

REFERENCES

1. Staff. 2017. The history of therapeutic drug development. *Twist Bioscience*. https://twistbioscience.com/company/blog/the-history-of-therapeutic-drug-development.
2. Staff. 2019. *The history of drug abuse in the United States*. Battle Creek, MI: A Forever Recovery.

3. Newman, T. and T. McDonald. 2017. New FBI report shows drug arrests increased in 2016, as drug war rages on. Press Release. Drug Policy Alliance, September 25.
4. WHO (World Health Organization). 2019. *World Drug Report 2019: 35 million people worldwide suffer from drug use disorders while only 1 in 7 people receive treatment.* Geneva: Office of Director General, Public Relations.
5. Hoffman, J. and K. Benner. 2020. Purdue Pharma pleads guilty to criminal charges for opioid sales. *The New York Times*, October 21.
6. CDC (Centers for Disease Control and Prevention). 2019. Drug overdose deaths. https://www.cdc.gov/drugoverdose/data/statedeaths.html.
7. Kounang, N. 2019. Fentanyl is the deadliest drug in the US, but in some places, meth kills more. *CNN Health*, October 25.
8. Weixel, N. 2019. Federal judge rules first U.S. supervised opioid injection site legal. *The Hill*, October 2.
9. Brundage, S. C., A. Fifield, and L. Partridge. 2019. *The ripple effect*. New York: United Hospital Fund.
10. Krupa, C. 2020. Ex-pharmaceutical exec gets 5½ years for pushing opioid spray. *NBC News*, January 23.
11. Foundation Recovery Network. The environmental impact of growing drugs. https://www.dualdiagnosis.org/the-environmental-impact-of-growing-drugs/.
12. Hassan, A. 2019. A doctor who prescribed 500,000 doses of opioids is sent to prison for 40 years. *The New York Times*, October 2.
13. Mammoser, G. 2018. Experts find drug contaminants in groundwater: What you need to know. *Healthline*, December 4.
14. Dimech, T. 2015. What is waste from the illegal drug trade doing to the environment? *Resource Magazine*, November 16.
15. Staff. 2019. Pure Food and Drug Act: A muckraking triumph. *United States History*. https://www.u-s-history.com/pages/h917.html.
16. Staff. 2019. New drug approval process. *Drugs.com*. https://www.drugs.com/fda-approval-process.html.
17. FDA (Food and Drug Administration). 2019. 2018 FDA drug approvals. *FDA News*, January 15.
18. Staff. 2019. *Comprehensive Drug Abuse Prevention and Control Act of 1970*. Atlanta, GA: Abt Law Firm, LLC.
19. DEA (drug enforcement Administration). 2016. DEA mission statement. https://www.dea.gov/mission.
20. DoJ (U.S. Department of Justice). 2019. *Organization, mission and functions manual: Federal bureau of investigation*. Washington, D.C.: Office of Director.
21. Coast Guard. 2019. Drug interdiction. https://www.gocoastguard.com/about-the-coast-guard/discover-our-roles-missions/drug-interdiction.
22. Johnson, L. M. 2019. Million in cocaine. *CNN*, September 24.
23. U.S. Customs and Border Protection. 2019. Border patrol history. https://www.cbp.gov/border-security/along-us-borders/history.
24. U.S. Customs and Border Protection. 2019. CBP enforcement statistics FY 2019. https://www.cbp.gov/newsroom/stats/cbp-enforcement-statistics.
25. NIDA (national institute on Drug Abuse). 2019. 2016–2020 NIDA strategic plan. https://www.drugabuse.gov/about-nida/strategic-plan/nidas-mission.
26. Levine, S. 2020. Oregon becomes first US state to decriminalize possession of hard drugs. *The Guardian*, November 4.
27. UN (United Nations). 2019. United Nations and the rule of law: Drug trafficking. https://www.unodc.org/unodc/en/drug-trafficking/index.html.
28. INCB (International Narcotics Control Board). 2019. Single convention on narcotic drugs of 1961. https://www.incb.org/incb/en/narcotic-drugs/1961_Convention.html.

29. Wikipedia. 2019. Convention on psychotropic substances. https://en.wikipedia.org/wiki/Convention_on_Psychotropic_Substances.
30. UN (United Nations). 2019. *United Nations convention against illicit traffic in narcotic drugs and psychotropic substances*. New York: Office of Media Services.
31. Navconon. 2019. Alcohol history. https://www.narconon.org/drug-information/alcohol-history.html.
32. Staff. 2019. Alcohol: A short history. Foundation for a drug-free world. https://www.drugfreeworld.org/drugfacts/alcohol/a-short-history.html.
33. Yerby, N. 2019. Types of alcohol. Recovery worldwide, LLC. AlcoholRehabGuide.org.
34. Morris, S. 2019. US alcohol sales increased by 5.1% in 2018. *The Drink Business*, January 17.
35. Pollard, M. S., J. S. Tucker, and H. D. Green Jr. 2020. Changes in adult alcohol use and consequences during the COVID-19 Pandemic in the US. *JAMA Netw. Open* 3(9):e2022942.
36. WHO (World Health Organization). 2018. *Alcohol. Key facts*. Geneva: News Room.
37. WHO (World Health Organization). 2018. Harmful use of alcohol kills more than 3 million people each year, most of them men, September 21. Geneva: News Room.
38. CDC (Centers for Disease Control and Prevention). 2018. Fact sheets—alcohol use and your health. https://www.cdc.gov/alcohol/fact-sheets/alcohol-use.htm.
39. White, A. M., L-J. Castle, R. W. Hingson, and P. A. Powell. 2020. Using death certificates to explore changes in alcohol-related mortality in the United States, 1999 to 2017. *Alcoholism: Clinical and Experimental Research*, January 7.
40. Phillips, A. 2018. The environmental impact of alcohol. *The Environmental Magazine*, August 29.
41. Gale Encyclopedia of U.S. Economic History. 2000. Eighteenth amendment. Encyclopedia.com.
42. Schultz, H. W. 1981. Federal Alcohol Administration Act and internal revenue code—Alcohol excise taxes. In: *Food Law Handbook*. Springer, Dordrecht.
43. Hanson, D. J. 2019. Alcohol taxes: Federal & state. Alcohol problems and solutions. https://www.alcoholproblemsandsolutions.org/alcohol-taxes-on-beer-wine-spirits/.
44. ATF (Bureau of Alcohol, Tobacco, Firearms and Explosives). 2018. *Rules and regulations*. Washington, D.C.: U.S. Department of Justice.
45. WHO (World Health Organization). 2010. Global strategy to reduce harmful use of alcohol. https://www.who.int/substance_abuse/activities/gsrhua/en/.
46. Editors. 2019. Russia's alcohol policy: A continuing success story. *The Lancet*. October 5.
47. Sherk, A., G. Thomas, S. Churchill, and T. Stockwell. 2020. Does drinking within low-risk guidelines prevent harm? Implications for high-income countries using the International Model of Alcohol Harms and Policies. *J. Stud. Alcohol Drugs* 81(3):352–61.
48. Staff. 2020. Rethinking alcohol. Alcohol and your health. https://www.rethinkingdrinking.niaaa.nih.gov/how-much-is-too-much/is-your-drinking-pattern-risky/Drinking-Levels.aspx.

4 Firearms Violence

4.1 INTRODUCTION

The preceding chapters have described specific hazards in community physical environments and attendant policies to prevent their effects on human and ecological health. Examples of the hazards included climate change, air pollution, tobacco smoke, water contamination, food pathogens, and noise and light pollution, each of which has prompted the development of environmental policies intended for the prevention of effects on public health and other consequences.

In distinction, conditions in what might be called social environments are not specifically addressed in this book. Some examples of social conditions that can affect human health in particular could include racial discrimination, poverty, inadequate educational opportunities, unavailable health care, and social dysfunctions such as family breakups and instances of unlawful behavior. An additional consideration that makes this chapter about firearms different from the other chapters is the subject of acquisition of a known hazard. Individuals choose to possess firearms knowing that they can present a hazard if discharged. This personal assumption of risk is different from the other physical hazards in the environment discussed in this book. For instance, one does not choose to deliberately inhale polluted air or to consume known unsafe food products.

> From 1968 through 2015 there were 1,516,863 gun-related deaths on U.S. territory, with 96 fatalities on average each day in the U.S.

However, having said the foregoing about physical versus social environments, one must allow that the boundaries between the two environments are not always precise. For instance, food insecurity, as discussed in Volume 1, Chapter 4, is a subject of health concern, i.e., famine or malnutrition, within the physical environment of food availability. But there are also social environmental considerations such as food insecurity as a consequence of poverty, social status, and form of governance.

This chapter, firearms violence as a matter of public health, contains the greatest social dimension when compared to all the preceding chapters. Social issues include matters of firearms for an individual's self-defense, personal right to possess weapons, and governments' role in regulating the acquisition and use of firearms. These kinds of social considerations will be addressed, but not elaborated, in this chapter. Rather, the focus of this chapter will be on public health issues associated with firearms. And it is appropriate to state here that the chapter is confined to firearms in community settings and excludes the subject of firearms and other weapons assigned to military services and their utilization therein.

As a prelude to what follows in this chapter, and to perhaps to establish a public health soul for its content, from 1968 through 2015 there were 1,516,863 gun-related deaths on U.S. territory according to a compilation of data from several sources [1].

DOI: 10.1201/9781003212621-4

For sake of comparison, in 2018, the populations of Philadelphia, Pennsylvania, and Phoenix, Arizona were each approximately 1.5 million residents. And on average, 96 fatalities occur each day in the U.S. as a consequence of gun violence [2]. This same source states that on average 342 people in the U.S. are shot each day by some type of firearm, with 96 succumbing to their injuries. The latter number is approximately the same as the 102 fatalities that resulted in 2016 from vehicle crashes in the U.S.

In a second study, an epidemiological study of firearm injuries in the U.S. from 2009 to 2017 was conducted by University of Pennsylvania researchers [3]. Data on fatal injuries from the Centers for Disease Control and Prevention (CDC) were combined with national data on emergency department visits for nonfatal firearm injury from the Nationwide Emergency Department (ED) sample. Incidence, case fatality rate, and trends over time of firearm injury according to intent, age group, and urban-rural location were determined. From 2009 to 2017, there was a mean of 85.694 ED visits for nonfatal firearm injury and 34,538 deaths each year. An annual mean of 26, 445 deaths (76.6%) occurred outside of the hospital. Assault was the most common overall mechanism (38.9%), followed by unintentional injuries (36.9%) and intentional self-harm (19.6%).

> A firearm is "a weapon from which a shot is discharged by gunpowder – usually used of small arms."

Both forms of premature death due to firearms and vehicle crashes, respectively, represent preventable losses and therefore a matter of public health because the gist of public health is **prevention** (emphasis added) of disease and disability in at-risk populations.

4.2 DEFINITION

A standard dictionary definition of firearm is "a weapon from which a shot is discharged by gunpowder—usually used of small arms." This brief dictionary definition is quite distinct from the legal definition in U.S. law: Under 26 USCA § 861 (a), firearms is defined as "a shot gun or rifle having a barrel of less than eighteen inches in length, or any other weapon, except a pistol or revolver, from which a shot is discharged by an explosive if such weapon is capable of being concealed on the person, or a machine gun, and includes a muffler or silencer for any firearm whether or not such firearm is included within the foregoing definition" [4].

> Firearms came from the discovery of gunpowder. Experimenting with life-lengthening elixirs around A.D. 850, Chinese alchemists instead discovered gunpowder.

For this chapter, firearms will refer to handguns, rifles, and shotguns, the weapons most frequently used in the U.S. for personal protection and in instances of harm to human and ecological health.

4.3 PRÉCIS HISTORY OF FIREARMS

Firearms came from the discovery of gunpowder. Experimenting with life-lengthening elixirs around A.D. 850, Chinese alchemists instead discovered gunpowder. "Chinese scientists had been playing with saltpeter – a common name for the

Firearms Violence

powerful oxidizing agent potassium nitrate – in medical compounds for centuries when one industrious individual thought to mix it with sulfur and charcoal."

The result was a mysterious powder from which, observers remarked in a text dated from the mid-9th century, "smoke and flames result, so that (the scientists)] hands and faces have been burnt, and even the whole house where they were working burned down" [5]. Chinese alchemists soon learned to bind their powdery material onto projectiles that could be set afire and thrust into the sky, forming the first experiment in fireworks, a spectacular entertainment that is still enjoyed by millions of people globally.

In the 13th century, the science of gunpowder was passed along the ancient silk trade route to Europe and the Islamic world and where it became weaponized in many Middle Age skirmishes. "By 1350, rudimentary gunpowder cannons were commonplace in the English and French militaries, which used the technology against each other during the Hundred Years' War. The Ottoman Turks also employed gunpowder cannons with abandon during their successful siege of Constantinople in 1453. The powerful new weapon essentially rendered the traditional walled fortifications of Europe, impregnable for centuries, weak and defenseless" [5]. During this age, cannons became standard weapons used in warfare, with their progeny gradually giving rise to modern artillery.

In the mid-15th century, gunpowder was first inserted into the barrel of a handgun, in effect, downsizing from the use of gunpowder in cannons, resulting in guns becoming weaponry in the hands of individuals. Guns that could be carried and used by individuals gave rise to a new form of military group, armed infantry. While these early uses of gun-powdered weapons were primarily used by military organizations, armies and navies in particular, firearms gradually seeped into the possession of and for use by nonmilitary individuals. Homeowners, farmers, merchants, law enforcers, hunters, and mariners, concerned about the acquisition of food via hunting and protection of property and lives gradually acquired firearms, primarily rifles and handguns. And over time, firearms also became tools in the hands of persons with criminal intentions. As the quantity and quality of firearms increased with time, so did lives lost to their use. The loss of human lives has accompanied, for both good and ill, the global proliferation of firearms. And that loss translates and transacts into firearms as a matter of public health concern.

> The premature loss of human lives has accompanied, for both good and ill, the global proliferation of firearms. And that loss translates into firearms as a matter of public health concern.

4.4 FORMS OF FIREARMS

There are many forms and variations of current-day firearms available for nonmilitary purposes. The forms most available to the general public in the U.S. are handguns, rifles, and shotguns. Each of these types has different features and constructions. Further, weapons can be classified according to the size and number of projectiles (bullets, cartridges, shells) that can be fired from the weapon. The size of projectiles is stated in calibers, e.g., 22–38- or 45-caliber handguns. These cited

FIGURE 4.1 Example cache of firearms seized by the ATF. (ATF, 5/14/2014, https://www.dea.gov/press-releases/2014/05/12/dea-spd-atf-seize-cache-guns-magnolia-home.)

forms of modern firearms can also be manufactured so as to discharge a single projectile once, then require reloading, or operate in a semi- or fully automatic firing pattern. Illustrated in Figure 4.1 is a collection of firearms confiscated by a California police department. Noteworthy is the variety and size of the weapons.

Handguns are loaded by placing projectiles into the cylinders of a pistol, e.g., a six-shooter, or placed in a clip that fits into the hand grip of a pistol. Modifications or attachments to these forms of firearms include telescopic sights attached to rifles for the purpose of improving aiming at a target, silencers that can be attached to handguns for the purpose of muffling the sound of a discharge from a gun, and modified stocks to rifles that in effect convert a semi-automatic weapon to a fully automatic weapon. As will be discussed, these various forms of firearms are the subject of various firearm control policies implemented by federal, state, and local governments in the U.S. The following sections will provide data on the numbers of firearms possessed in the U.S. and globally.

4.4.1 Possession and Storage of Firearms in the U.S.

The number of firearms owned by U.S. civilians has been estimated by two sources of investigation, yielding considerable difference in findings, but with one common finding: the U.S. civilian population is the most heavily armed global population. Turning to the first of two surveys, according to the Small Arms Survey, a project of the Graduate Institute of International and Development Studies in Geneva, Switzerland, there are more than 393 million civilian-owned firearms in the U.S., or enough for every man, woman, and child to own one and still have 67 million guns left over [6]. The report, which draws on official data, survey data, and other measures for 230 countries, found that global firearm ownership is heavily concentrated in the U.S. In 2017, for instance, Americans made up 4% of the world's population but owned about 46% of the entire global stock of 857 million civilian firearms. The survey observed from their data that the number of civilian-owned firearms in the U.S. was 120.5 guns per 100 residents, meaning there were more firearms than people.

A different study of firearm possession in the U.S. was reported in 2016 by Harvard and Northeastern University researchers [7]. Researchers utilized the GfK Knowledge Panel, a nationally representative online panel used to poll Americans on a variety of subjects. Participants were asked to provide detailed information about their guns and their ownership habits, the types of guns and firearms in their homes, how they are stored, and why they keep them. The study yielded several results of relevance to public health issues. First, the responses revealed a fundamental shift in gun owners' attitudes. While previous surveys had found that U.S. gun owners possessed firearms primarily for hunting or sports activities, this survey revealed that firearms were being kept principally for self and family protection. Respondents also indicated that handguns had increased dramatically in popularity in comparison to results from prior studies.

> Survey results provide an estimate of about 265 million firearms were in private possession in the U.S. in 2016.

Further, the researchers found that about half of all firearms are concentrated in the possession of just 3% of American adults. Survey results provide an estimate of about 265 million firearms were in private possession in the U.S.

Complementing the acquisition of a firearm should be a policy of safe storage of the firearm. Safe storage laws promote responsible gun-owning practices by requiring gun owners to keep their firearms out of the reach of others, such as children or prohibited persons, who could use the weapon to deadly effect. The U.S. General Accounting Office has estimated that 31% of accidental deaths caused by firearms could be prevented by the addition of two devices: a child-proof safety lock and a "loading indicator," a safety device that indicates whether a firearm is loaded and still has a round of ammunition in its chamber. Massachusetts is the only U.S. state that requires firearms be stored with a locking device in place in all cases when they are not in use. Massachusetts' law also requires individuals to securely store firearms in a locked container or equipped with a tamper-resistant mechanical lock or other gun safety device [8].

The Massachusetts law on gun storage has been effective in preventing youth suicides using guns. Notably, guns are used in just 9% of youth suicides in Massachusetts, compared to 39% of youth suicides in the U.S. California, Connecticut, and New York also require that guns must be securely stored when around people who cannot legally possess them, which may include children. California also requires that all firearms sold in the state, including handguns and long guns sold by any party, must include a gun lock or other approved gun safety device; Connecticut, Massachusetts, and New Jersey require that gun locks must be included in the sale of a handgun by any party. Notably, along with New York, these three states have the lowest rates of youth suicide in the nation [8].

Another reason for safe storage of firearms is to deter their theft. Researchers estimate that more than 500,000 firearms are lost or stolen annually from private residences in the U.S. These unsecured firearms when taken by unauthorized persons help fuel episodes of gun violence that includes criminal activity, homicide, and nonlethal injuries [9].

4.4.2 POSSESSION OF FIREARMS GLOBALLY

At the end of 2017, according to data collected by the Small Arms Survey, there were approximately 1,013 million firearms in the 230 countries and autonomous territories of the world. Of this number, 84.6% were held by civilians, 13.1% by state militaries, and 2.2% by law enforcement agencies [9]. This total is greater than the Survey's previously published total of 875 million firearms in 2006, an increase of 15.7% for all identified firearms. According to the survey, much of this change was due to an increase of 32% in the estimated total of civilian-held firearms. Reported global totals for the law enforcement and military categories showed net decreases, mostly due to changes in estimating procedures used in the survey.

> At the end of 2017, there were approximately 1,013 million firearms in the 230 countries and autonomous territories of the world.

The survey estimated the U.S. population held the greatest number of firearms in 2017 [9]. The estimated total number of civilian-held legal and illicit firearms was approximately 395.3 million, a figure that exceeded the second-ranked country, India, with approximately 71 million firearms. When these numbers were converted to firearm rates, the U.S. stood atop the list, with 120.5 firearms per 100 residents, as shown in Table 4.1. The world's second-ranked country was Yemen, a state torn by civil war, where there were 52.8 firearms per 100 residents. Other countries' holdings of firearms are also itemized in Table 4.1. From this table it can be observed that other countries' gun holdings are one-fourth to one-sixth of those of the U.S. These differences in national rates of gun possession are attributable to variances in firearms policies (e.g., gun laws) and historical cultural factors (historical possession) distinct from those in the U.S.

TABLE 4.1
Civilian Firearms per 100 Residents in Top-Ranked Countries and Territories [6]

United States	120.5	Bosnia and Herzegovina	31.2
Yemen	52.8	Austria	30.0
Montenegro	39.1	Macedonia	29.8
Serbia	39.1	Malta	28.3
Canada	34.7	Switzerland	27.6
Uruguay	34.7	New Zealand	26.3
Cyprus	34.0	Kosovo	23.8
Finland	32.4	Sweden	23.1
Lebanon	31.9	Pakistan	22.3
Iceland	31.7	France, Germany, Iraq	19.6

4.5 HUMAN HEALTH IMPACTS OF FIREARMS

Firearms when discharged into humans intentionally or accidentally can result in injuries or death. In instances of warfare or other forms of deliberate violence, firearms that include pistols, rifles, cannons, mobile guns, and explosive devices are weapons used to injure or kill combatants. In instances of civilian use of firearms, injuries and fatalities can occur as consequences of accidental discharge of weapons, homicides, and suicides. The demographics of humans adversely affected by firearms range the full spectrum of human populations, from fetuses to elderly victims, and across all strata of human endeavors. In describing human fatalities caused by firearms, the following sections are framed as "premature deaths," because without a causal firearm, a person's life could have lasted longer. The same nomenclature would be appropriate for the death of a person who succumbed to an illness, disease, or other infirmity that in its absence would have led to a person's greater longevity. Or expressing this subject in a different vernacular, "He/she died before their time."

4.5.1 Premature Firearms-Related Deaths in the U.S.

For public health purposes, it is useful to relate the historical toll of premature deaths due to firearms. One source estimated the accrual of firearm-caused deaths from 1968 through 2015 of the U.S., noting, "Since 1968, when these figures were first collected, there have been 1,516,863 gun-related deaths on U.S. territory. Since the founding of the U.S., there have been 1,396,733 war deaths. That figure includes American fatalities in the Revolutionary War, extending through the Iraq war, as well as other conflicts, including those in Lebanon, Grenada, Panama, Somalia, and Haiti" [1]. In other words, the number of U.S. firearms-related deaths caused by civilian gun incidents exceeded all the deaths lost in all U.S. wars.

The same source comments that of the approximately 33,000 gun violence deaths that take place annually in the U.S., two-thirds are from suicide; further, more than 70,000 people are injured by guns each year in the U.S [1]. The source's data seem to pertain to the year 2015. However, two other sources cite different figures in regard to the annual number of fatalities and injuries in the U.S. due to gun violence. One source, University of Pennsylvania researchers, in a cross-sectional study of data from nationwide U.S. databases, from 2009 to 2017, found an annual average of 85,694 emergency department visits for nonfatal firearm injury and 34,538 deaths from firearm injury [3]. Although suicide accounted for 61% of deaths, self-harm accounted for only 3% of nonfatal injuries; assaults accounted for 35% of deaths and 41% of nonfatal injuries, and unintentional injuries accounted for 2% of deaths and 51% of nonfatal injuries.

> University of Pennsylvania researchers: Data from U.S. databases, from 2009 to 2017, showed an annual average of 85,694 emergency department visits for nonfatal firearm injury and 34,538 deaths from firearm injury.

Another source, the American Public Health Association (APHA), stated in 2019, "Gun violence is a leading cause of premature death in the U.S. Guns kill more than 38,000 people and cause nearly 85,000 injuries each year. As a longtime advocate for violence prevention policies, APHA recognizes a comprehensive public health approach to addressing this growing crisis is necessary" [10]. The basis for the APHA data is not given. For the purposes of this chapter, the most recently available data on mortality and morbidity attributable to gun violence are cited as summary statistics.

On an annual basis, data collected by the CDC in 2016 found there were 19,362 homicides in the U.S. Of this number, 14,415 were classified as firearm homicides, which equates to 74% of homicides committed in the U.S. [11]. The CDC also found a 31% increase in homicides involving firearms from 2014 to 2016. In 2014, 11,008 homicides involved a gun. The number rose to 14,415 by 2016 [12]. According to a different source, the Brady Campaign to Prevent Gun Violence, every day on average in the U.S. 47 children and teens are shot in murders, assaults, suicides, and suicide attempts, unintentional shootings, and police intervention. Of this number, eight die from gun violence [13]. This translates into 17,207 U.S. children and teens shot in gun violence each year, with 2,737 fatalities. The same source states that on average 124,760 people in the U.S. are victims of gun violence, with 35,141 fatalities annually. Both the number of annual deaths and nonfatal injuries attributable to gun violence, together with the proportion due to suicides, represent a challenge in terms of public health impacts.

Yet another source, the Gun Violence Archive, provides the data given in Table 4.2. This source states that it is "a not for profit corporation formed in 2013 to provide free online public access to accurate information about gun-related violence in the U.S. GVA will collect and check for accuracy, comprehensive information about gun-related violence in the U.S. and then post and disseminate it online" [14]. According to the data in Table 4.2 for the years 2016–2018, there was a mean of approximately 59,000 total incidents of gun violence in the U.S. These incidents caused approximately 15,000 fatalities, with approximately 30,000 non-fatalities. According to the Archive, there was a mean of 356 mass shootings during the span of 2016–2018. It should be noted that suicides occurring from the use of firearms are not included in the Archive's database. Inclusion of suicides would have brought their data closer that cited by the CDC in the preceding paragraph.

TABLE 4.2
Number of Gun Violence Events in the U.S. [14]

Year	2016	2017	2018	Mean
Total incidents	58,963	61,929	57,067	59,320
Number deaths	15,099	15,656	14,708	15,154
Number injuries	30,642	31,247	28,169	30,019
Children (0–11 years) deaths or injuries	671	732	667	690
Teens (12–17 years) deaths or injuries	3,136	3,247	2,833	3,072
Mass shootings	382	346	340	356

Source: Gun Violence Archive [14]; deaths/injuries exclude suicides; data based on four or more persons shot and/or killed in a single incident, at the same general time and location; shooter's death not counted in count of fatalities (FBI definition).

Firearms Violence

4.5.2 Premature Firearms-Related Deaths Globally

According to the Global Burden of Disease Study, there were an estimated 251,500 deaths in 2016 from firearm injuries worldwide [15]. Six countries (Brazil, Colombia, Guatemala, Mexico, United States, Venezuela) accounted for 50.5% of the deaths. Globally, the majority (64%) of firearm injury deaths in 2016 was homicides, numbering some 161,000 deaths; additionally, 27% were firearm suicide deaths, numbering 67,500, and 9% were unintentional firearm deaths, numbering 23,000. From 1990 to 2016, there was no significant decrease in the estimated global age-standardized firearm homicide rate, whereas firearm suicide rates decreased globally at an annualized rate of 1.6%, but in 124 of 195 countries and territories included in this study, these levels were either constant or significant increases were estimated. There was an annualized decrease of 0.9% in the global rate of age-standardized firearm deaths from 1990 to 2016. Aggregate firearm injury deaths in 2016 were highest among persons aged 20–24 years, where for men, an estimated 34,700 deaths, and for women, an estimated 3,580 deaths had occurred. The study noted that the number of firearms by country was associated with higher rates of firearm suicide.

4.5.3 Children and Adolescent Premature Deaths

A study by the University of Michigan of the deaths of 20,360 U.S. children and teenagers in 2016 found that firearms were the second leading cause of premature death. (The leading cause was motor vehicle crashes, causing 4,074 deaths.) The number of premature deaths caused by firearms was 3,143, which translates to about eight deaths each day. Homicides were the cause of 59% of deaths, with 35% attributable to suicides [16].

The relation between youth suicide rates and firearm ownership was investigated in 2017 by researchers at the Boston University School of Public Health [17]. Researchers studied 10 years of teenage suicide rates in the U.S. and found that 4.6 million young people in the U.S. reside in homes with at least one unloaded, unlocked firearm. This study takes advantage of the availability of state-level data from the Youth Risk Behavior Surveillance System (YRBSS) on risk behaviors that might influence suicide, including rates of severe negative affect, suicidal planning, and suicide attempts. This study models the relationship at the state level between household gun ownership and suicide rates among youth aged 10–19 years during the period 2005–2015. The researchers noted that household gun ownership was statistically positively associated with the overall youth suicide rate. For each 10%-point increase in household gun ownership, the youth suicide rate increased by 26.9%.

During the 2005–2015 study period, the investigators found "the overall youth suicide rate ranged from a high of 15.2 per 100,000 youth in Alaska to a low of 2.6 per 100,000 youth in New Jersey … The estimated household gun ownership in 2004 ranged from a high of 65.5% in Wyoming to a low of 10.2% in Hawaii. In the 10 states with the highest youth suicide rates, the average household gun

> Researchers have noted that household gun ownership is statistically associated with the overall youth suicide rate in the U.S.

ownership was 52.5%, compared with a household gun ownership rate of 20.0% in the 10 states with the lowest youth suicide rates. The household gun ownership level in 2004 was highly correlated with the overall youth suicide rate, explaining 55% of the variance across the 50 states" [18,19].

Because firearms are a leading cause of premature death of children, the American Academy of Pediatrics in 2012 (reaffirmed in 2016) articulated the following policy statement: "The absence of guns from children's homes and communities is the most reliable and effective measure to prevent firearm-related injuries in children and adolescents. Adolescent suicide risk is strongly associated with firearm availability. Safe gun storage (guns unloaded and locked, ammunition locked separately) reduces children's risk of injury. Physician counseling of parents about firearm safety appears to be effective, but firearm safety education programs directed at children are ineffective. The American Academy of Pediatrics continues to support a number of specific measures to reduce the destructive effects of guns in the lives of children and adolescents, including the regulation of the manufacture, sale, purchase, ownership, and use of firearms; a ban on semi-automatic assault weapons; and the strongest possible regulations of handguns for civilian use" [20]. This statement, because its intent is to prevent children's death caused by firearms, is a statement of the core public health principle: prevention of disease and disabilities in at-risk populations.

Perspective: The number of persons in the U.S. who were gun violence victims is a subject of some debate. This is due to differences in the data collection systems used and sometimes the bias of the persons or organizations that collect and interpret the data on gun violence. As with other data of public health significance, data collected and evaluated and subsequently published in peer-reviewed journals is generally the most reliable for purposes of public health policy-making. However, gun violence data collected and evaluated by organizations pro and con firearms possession should be reviewed and decisions formed on the quality and objectivity of data made available to the public.

4.5.4 Mass Casualty Shootings

In the U.S., mass casualty shootings are the most common and most closely tracked type of mass casualty events [21]. The Congressional Research Service defines mass shootings as events where more than four people are killed with a firearm "within one event, and in one or more locations in close proximity." Congress uses the term "mass killings" and describes these events as "three or more killings in a single incident." The FBI uses the term "Active Shooter," which it defines as "an individual actively engaged in killing or attempting to kill people in a populated area." Nongovernmental organizations, including *Mother Jones*, *USA Today*, and the Stanford Mass Shootings in America data project, use various combinations of these definitions [21]. It is therefore important to understand the definition of what constitutes a definition of gun violence used in any database being considered for analysis.

A comprehensive analysis of mass shooting in the U.S. was conducted by investigative reporters for *The Washington Post*, published in 2018 [22]. The investigators prefaced their report by observing "Public mass shootings account for a tiny fraction of the country's gun deaths, but they are uniquely terrifying because they occur without warning in the most mundane places. Most of the victims are chosen not for what they have done but simply for where they happen to be." The investigators identified 161 shootings in which four or more people were killed by a lone shooter (two shooters in a few cases). The shootings occurred from August 1966 through the end of 2018. Their analysis did not include shootings tied to gang disputes or robberies that went awry, and it did not include domestic shootings that took place exclusively in private homes. The shootings caused 1,148 fatalities, where 189 were children and adolescents. Regarding the weapons used by shooters, 308 guns were involved, with 172 having been obtained legally and 58 illegally obtained by the shooters.

The Washington Post's investigators found there were 165 shooters. Some of these were persons known to have violent tendencies or criminal pasts. Others seemed largely fine until they attacked. All but three were male. The vast majority of shooters were between the ages of 20 and 49. More than half of them – 91– died at or near the scene of the shooting, often by suicide. Shootings in schools and houses of worship, while often standing out in the public's attention, were a relatively small portion of mass shootings. More common were those in offices and retail establishments such as restaurants and stores. Mass public shootings occurred in 42 states and the District of Columbia. California had 25 public mass shootings, which was more than any other state [22].

> The U.S. had more mass shootings – and more people cumulatively killed or injured – than ten other affluent nations combined.

A second source of data on mass shootings in the U.S. is maintained by the Associated Press, in collaboration with *USA Today* and Northeastern University. According to their data, there were 41 mass killings in the U.S. in 2019, which were defined as instances when four or more people were killed, excluding the perpetrator. More than 210 people were left dead in the mass killings, 33 of which were mass shootings [23]. Research dating back to the 1970s shows there was no other year with as many mass killings. When mass shootings rather than mass killings were taken into account, there were 410 mass shootings – defined as any incident in which four people were shot except the shooter – in the U.S.

A question sometimes arises as to where the U.S. ranks globally in regard to the number of mass shootings in comparison to other countries. A comprehensive review of mass shootings was conducted by a fact checking organization [24]. They reviewed and cited the research of two academic researchers at the State University of New York in Oswego and the Texas State University. These two researchers had analyzed mass shootings in 11 countries, covering the period from 2000 to 2014. In addition to the U.S., they examined mass shootings in Australia, Canada, China, England, Finland, France, Germany, Mexico, Norway, and Switzerland. The investigators found that the U.S. had more mass shootings – and more people cumulatively killed or injured – than the other ten nations combined. While part of this is because the U.S. has a much bigger population than all but China, when adjusted for population, the U.S. ranked in the upper half of their list of 11 countries, ranking higher than Australia, Canada, China, England, France, Germany, and Mexico. The U.S. did rank lower than three countries (Norway, Finland, and Switzerland) but those countries have populations so small that only one or two mass-casualty events can produce a relatively high per capita rate.

Another question arises as to how the U.S. states rank in the number of mass shootings in regard to the nature of their gun possession policies. To explore this question, researchers from Columbia University, New York University, Boston University, and the University of Pennsylvania analyzed states' mass shooting rates, the permissiveness of their firearm laws, and levels of gun ownership from 1998 to 2015 [25]. Data for mass shootings came from the FBI's Uniform Crime Reporting System, which defines a mass shooting as an event in which four or more people, not including the shooter, were shot and killed. To measure the permissiveness of gun laws, the researchers utilized a 100-point scale from the *Traveler's Guide to the Firearms Laws of the Fifty States*, a database compiled by legal professionals as a reference guide for gun owners traveling between states. The investigators noted, "A 10 percent increase in gun ownership was associated with an approximately 35 percent higher rate of mass shootings after adjusting for key factors." In other words, the researchers observed that where there are more guns, there are more mass shootings. And where gun laws are weaker, there are more mass shootings.

There are some caveats to the study. The biggest one: It found correlation, not causation. It's possible other factors not accounted for in the study, besides levels of gun ownership or the permissiveness of state gun laws, are driving higher mass shooting rates. And the researchers also raised concerns that the FBI's mass shooting data could be incomplete since some U.S. states sometimes fail to consistently report to the FBI system.

In closing this section, it is important to cite in memory of the victims the deadliest mass shooting in U.S. history at the time of the attack [26]. On the night of October 1, 2017, a gunman used an assortment of firearms to fire on a crowd attending the final night of a country music festival in Las Vegas, killing 58 people and injuring more than 800. The gunman, a 64-year-old retired man who lived in Mesquite, Nevada, targeted the crowd of concert-goers on the Las Vegas strip from the 32nd floor of the Mandalay Bay hotel. He had checked into the hotel several days before the massacre. The gunman began firing at the crowd at 10:05 p.m. using an arsenal of 23 guns, 12 of which were upgraded with bump stocks – a tool used to fire semi-automatic guns in rapid succession. Within the 10-minute assault period, he was able to fire more than 1,100 rounds of ammunition.

4.5.4.1 School Shootings

Especially noteworthy as gun violence incidents are school shootings, given the loss of life and nonfatal injuries sustained by children who were victims. An investigative report by *CNN News* staff reviewed media reports and a variety of databases to ascertain how school shootings in the U.S. compare in frequency with those of other countries [27]. The study covered data for the period from January 1, 2009, to May 21, 2018. A school shooting was included in the CNN analysis if: (1) the shooting involved at least one person being shot (excluding the shooter) and (2) the shooting occurred on school grounds. Using these two criteria, for the span of time considered, 288 school shootings occurred in the U.S. In comparison, other countries selected for comparison and their numbers of school shootings were as follows: Mexico (8), Brazil (2), Canada (2), France (2), Germany (0), Japan (0), Italy (0), and the UK (0). These data convey the troubling message that school shootings in the U.S. were 57 times the total of the other six G7 countries (Canada, France, Germany, Japan, Italy, and the UK) and 29 times the number of the U.S.'s nearest neighbors (Mexico, Canada).

Many U.S. school shootings can be classified as mass casualty shootings. The investigative reporting magazine *Mother Jones* has been collecting and reporting on mass casualty shootings for several years [28]. In particular, the magazine's staff has collected an important database on such episodes of firearm-based violence. The database covers the period from the year 1982 through 2018, listing the details for 107 instances of mass shootings wherein four or more fatalities occurred. Their published data permit one to glean some important details pertinent to the public health issue of firearms. For instance and of relevance to public health, the sums of deaths and injuries from firearms in the 107 episodes of mass shootings over the span of 36 years are 875 and 1,327, respectively.

Particularly troubling to the American public are the number of mass shootings that occur in educational institutions. A review of the *Mother Jones* database reveals 15 instances of such mass shootings [28]. Of this number, ten occurred in elementary or high school settings. Shown in Table 4.3 are the ten mass shootings in these kinds of schools. It is interesting to note in the table that home of the perpetrator was the source of four weapons used in school shootings. And although not noted in the table, four of the 10 school shootings involved perpetrators with alleged mental health issues. The remaining six school shootings were committed by current or former students with anger or disaffection issues. While these numbers of fatalities and injuries in mass school shootings in the U.S. are relatively small in comparison to the numbers of deaths and injuries suffered through other forms of annual civilian firearm-related casualties, they are nevertheless troubling in their own right in that school shootings create a psychosocial impact unequaled by other incidents of mass shootings. The victims are children, a community's most valuable resource, accosted in a place of presumed safe environment.

4.5.4.2 Psychosocial Impacts of School Shootings

Mass school shootings have consequential public health impacts on families of shooting victims, survivors of the shooting, community medical providers, and emergency response personnel. As observed by a George Washington University psychiatrist, "Mass shootings are a first-line traumatic event that can potentially trigger posttraumatic stress disorder (PTSD) in people who are directly exposed, as life and limb are under direct and violent threat. Children, in particular, are even more vulnerable; multiple studies have shown that childhood trauma has more lifelong and pervasive effects on young developing psyches, both in terms of their psychologic worldview, and their physiological systems that handle stress and anxiety" [29].

The National Center for PTSD estimates that 28% of people who have witnessed a mass shooting event will develop post-traumatic stress disorder and about a third develop acute stress disorder [30]. The same source states that mass shooting survivors might be at greater risk for mental health difficulties compared with people who experience other forms of trauma, e.g., natural disasters such as hurricanes. Further, "most survivors show resilience. But others – particularly those who

> 28% of people who have witnessed a mass shooting event will develop PTSD and about a third develop acute stress disorder.

TABLE 4.3
Mass Shootings in the U.S. Involving Elementary and High School Students [28]

Event	Location	Date	Fatalities	Injured	Weapon	Source
Santa Fe H.S.	Santa Fe, TX	5/18/2018	10	13	Shotgun, handgun	Home
Marjory Stoneman Douglas H.S.	Parkland, FL	2/14/2018	17	14	Semi-automatic rifle	Pawn shop
Marysville-Pilchuck H.S.	Marysville, WA	10/24/2014	5	1	Pistol	Home
Sandy Hook Elementary	Newtown, CT	12/14/2012	27	2	Pistol, rifle	Home
Amish School	Lancaster County, PA	10/2/2006	6	5	Pistol, rifle, shotgun	Local stores
Columbine H.S.	Littleton, CO	4/20/1999	13	24	Shotguns, carbine, handgun	Local stores; gun show
Thurston H.S.	Springfield, OR	5/21/1998	4	25	Pistol (2), semi-automatic rifle	Home
Westside Middle School	Jonesboro, AR	3/24/1998	5	10	Semi-automatic pistol (2)	Home
Lindhurst H.S.	Olivehurst, CA	5/1/1992	4	10	Rifle, shotgun	Gun shop
Stockton schoolyard	Stockton, CA	1/17/1989	6	29	Semi-automatic pistol	Trading post

Data from *Mother Jones*, with permission; mass school shootings with three or more fatalities

Firearms Violence

believed their lives or those of their loved ones were in danger or who lack social support – experience ongoing mental health problems, including post-traumatic stress, depression, anxiety and substance abuse."

Community psychologists advise making counseling available for survivors, parents of survivors, and parents of victims. In addition, they stress the importance of mental health promotion efforts that help the community heal together. Ensuring victims are aware that support is available and accessible to them – even if they never take advantage of it – can help the healing process. Of note, memorial events, particularly those that are student and community initiated and led, are most helpful to survivors in terms of recovering after a mass violence event [29].

A health care provider has observed that survivors of a mass shooting incident are not the only persons to be adversely affected by mental health issues. Many studies have found elevated rates of PTSD among first responders to such events. Emergency medical responders, law enforcement personnel, hospital physicians and nurses, and even some news personnel can be adversely affected by their experience in responding to the fatalities and injuries among a shooting's victims. "Many of the men and women who respond to these tragedies have become heroes and victims at once. Some firefighters, emergency medical providers, law enforcement officers and others say the scale, sadness and sometimes sheer gruesomeness of their experiences haunt them, leading to tearfulness and depression, job burnout, substance abuse, relationship problems, even suicide" [31]. This source also notes that emergency responders and associated personnel are themselves victims of an accumulation of stress in responding to gun shootings and other traumatic incidents. Employers of emergency response personnel have been slow to provide counseling and other services for the purpose of preventing mental health disorders among their responding staff.

Perspective: Mass casualty shootings, in a public health sense, are like a form of infectious disease that appears unexpectedly, leading to commercial and social media broadcasting awareness of outcomes that in turn engender fear in public forums. And as with public health, mass shootings will bring attention by affected individuals, e.g., school administrators or public health specialists, with how to prevent further occurrences. Although much more research is needed on the circumstances of mass shootings, there is some indication that mental health issues are a factor in some shooters' actions. And for some school mass shootings, disaffected current or former students were the shooters. In such situations, recognition of signs of abnormal mental or psychological effects in individuals should forge appropriate kinds of interventions, e.g., medical diagnosis or psychological counseling of students and parents. And there is the issue of access to firearms, a subject addressed in the hazard interventions section of this chapter. Put simply, no access to firearms obviates shooters' use of them.

4.6 ECOLOGICAL HEALTH IMPACTS OF FIREARMS

While the impacts of firearms on human health are serious, humans' use of firearms has also had substantive consequences on ecological health and quality, both for ill and perhaps some good outcomes. Concerning good outcomes, as human populations formed into social groups, firearms became purposeful for hunting food supplies, in particular, hunting feral animals such as deer, birds, bison, and squirrels. This need for firearms in support of gathering food supplies was particularly acute as the American society moved westward from the eastern states. While this hunting-gathering of feral food supplies gradually decreased as agrarian-based settlements became established, the use of firearms in managing wildlife populations remains. This is exemplified by hunting seasons established by many U.S. states in their management of feral animals that lack nature's predators to keep their numbers in check. Perhaps the premier example is deer hunting seasons in many U.S. states.

> Two examples of species severely impacted in the U.S. by firearms are the passenger pigeon and the American bison.

In distinction to some beneficial impacts of firearms derived from managing some ecological systems, there are a number of deleterious impacts. One sentinel impact has been the impact of firearms on the reduction of the numbers of some species of animals. Two examples of species severely impacted in the U.S. are the passenger pigeon and the American bison. The last passenger pigeon, Martha, died in captivity at the Cincinnati, Ohio, Zoo in 1914. A bird that had once existed in the billions was eliminated by uncontrolled commercial hunting of the bird for food and "sport." The American bison almost suffered the same fate as millions of bison were killed by migrating European settlers moving westward, traversing the Great Plains of North America. While the Plains Indians had relied on bison as a source of food and clothing, taking only what was needed, European settlers slaughtered bison herds for both food and "sport." Bison were considered as being in the way of westward expansion and were killed in great numbers. Only in the 20th century were bison herds reestablished in Yellowstone National Park and on Plains Indians' lands [32].

Another ecological impact of firearms is the consequence of lead-containing ammunition on wildlife and the environment. Lead is a toxic, poisonous element, as discussed in Volume 1, Chapter 6. Hunters who use lead-containing ammunition for shooting wildlife such as ducks and other birds create an adverse ecological impact. This stems from the presence of lead in the tissues of game, which is in turn consumed by foraging animals, for example, eagles and coyotes, or lead in game that died directly from lead poisoning. In 1991, conservationists and biologists estimated that about 2 million ducks in the U.S. succumbed annually from ingesting spent lead pellets [33]. On January 19, 2017, President Obama's Fish and Wildlife Service issued a ban on lead in ammunition and fishing tackle used on federal lands and waters in order to protect birds and fish from lead poisoning. However, this ban was overturned by the Trump administration's incoming U.S. Interior Secretary on March 2, 2017, his first day on the job, in a nod to hunters and fishermen [34].

In addition to the federal ban, 34 U.S. states have expanded restrictions beyond the 1991 federal ban on the use of lead ammunition to hunt waterfowl. Further, in 2013, California enacted the first statewide phase-out of lead-containing ammunition for all forms of hunting [35]. Beginning July 1, 2019, lead ammunition was banned for hunting wildlife anywhere in California. This was the final phasing in

> 34 U.S. states have expanded restrictions beyond the 1991 federal ban on the use of lead ammunition to hunt waterfowl.

of the law California passed in 2013. The law was enacted in large part to protect the threatened California condor.

Internationally, in November 2020, the European parliament voted to ban lead shot from all wetlands in the EU in a decision that is expected to pave the way for phasing out all toxic ammunition [34]. The ban will ensure that any wildfowl or waterbirds are shot with non-toxic steel ammunition after scientific studies found that 1 million waterbirds in the EU are killed by lead poisoning each year. The parliament's vote will allow the European Commission to introduce the new regulations by the end of 2020.

Many manufacturers are developing nontoxic ammunition. As of April 2015, the U.S. Fish and Wildlife Service had approved 13 nontoxic varieties for hunting. Ammunition made of steel, copper, and bismuth, among the most common and effective nontoxic materials, is available at major outfitters throughout the U.S. and is widely available online [33]. According to the U.S. Humane Society, despite the overwhelming evidence that lead ammunition is toxic to wildlife and presents an environmental legacy, it still remains widely used by hunters, resulting in an estimated 10–20 million nontarget animals in the U.S. dying from lead poisoning each year [33].

Firing ranges are a further source of environmental sources of lead of consequence to ecological systems. Outdoor firing ranges can pose significant environmental risks due to contamination from the materials associated with gun use. These contaminants include lead, copper, zinc, antimony, and even mercury, all of which can sink into the soil and sometimes leach into groundwater and surface water if not appropriately waste managed [35]. There are 16,000–18,000 firing ranges in the U.S. [35]. Discharge of lead-laden dust and gases is a consequence of shooting guns in firing ranges. Persons working in indoor ranges and frequent customers can be exposed to lead through breathing lead-contaminated air. Outdoor ranges can pollute the air, soil, and water is appropriate waste remediation measures are not taken.

At the federal level, the EPA has issued Best Management Practices that recommend site practices for firing ranges in order to minimize the impact of lead on the environment. The EPA does not enforce its recommendations through criminal prosecutions, but implementation relies on civil law suits brought either by the government or by private parties who claim injury from toxic contamination [36]. However, U.S. state governments can establish environmental protection regulations for firing ranges.

> Outdoor firing ranges can pose significant environmental risks due to contamination from the materials associated with gun use.

Perspective: While the damage of firearms to ecological systems may not weigh as heavily in socioeconomic costs as the damage in lives prematurely lost or in nonlethal injuries, the costs are nonetheless significant. History has shown that damage to the environment and to the ecological systems that humans share with other living entities has adverse consequences. Soil and water supplies and feral animals contaminated with metals, especially lead, as a consequence of spent ammunition left after firearms' discharges have the potential to become sources of human exposure, thus becoming a matter of public health. Substitution of ammunition constructed of materials with fewer or no toxic properties would lessen the impact of firearms on ecological systems.

4.7 SOCIOECONOMIC IMPACTS OF FIREARMS

In addition to the mortality and morbidity caused by improper use of firearms, there are socioeconomic consequences as well. The social consequences are difficult to quantify, but persons who are lost to premature death or suffer from disabling injuries are unable to contribute to social relationships. Perhaps the greatest social impact is the imprimatur of violence that accompanies frequent reports of firearm-associated social disruption. A public that is too often informed of episodes of gun violence runs the risk of becoming inured to the violence.

As to the economic consequence of improper use of firearms, researchers from Johns Hopkins University quantified the financial burden of gun violence in the U.S. by examining data from 150,930 gun-violence patients who visited U.S. emergency rooms between the years 2006 and 2014 [37]. Their findings indicated that the annual cost for victims of gun violence in the U.S. is an average of $2.8 billion in emergency room and inpatient charges. But if victims' lost wages were factored in, the financial burden rose to $45 billion each year. For sake of comparison, the 2018 budget for the National Institutes of Health (NIH), the premier biomedical research agency in the U.S., was approximately $39 billion. Put in a public health context, the annual financial cost to the U.S. of gun violence exceeds the annual cost of NIH's research on finding causes and cures for diseases such as cancer and neurological disorders.

> Johns Hopkins University: The annual cost for victims of gun violence in the U.S. is an average of $2.8 billion in emergency room and inpatient charges. But if victims' lost wages were factored in, the financial burden rose to $45 billion each year.

4.8 COVID-19 IMPACTS ON FIREARM ISSUES

The global pandemic of COVID-19 coronavirus, as discussed in Chapter 1 and elsewhere, affected social, health, and economic systems in unprecedented ways. Because the virus was primarily transmitted from person to person, social distancing became the principal public health method for containing the spread of the virus. This was achieved by promoting 'shelter-in-home' policies, keeping 6 or more feet away from other persons, and the wearing of face masks in public places. A companion social event to the pandemic was public demonstration in several U.S. cities, protesting police shootings of persons of color. An outcome of the demonstrations was pressure on public officials to re-examine policy policies. The presence of the virus, together with issues of policing, produced a period of some uncertainty in many U.S. residences. A feeling of the need for self-protection evolved in the minds of many U.S. residents.

This period of angst manifested in increased sales of weapons. Gun sales increased 95% while ammunition sales increased 139% compared to the same period in 2019, according to a survey by the National Shooting Sports

> During the COVID-19 pandemic gun sales increased 95% in the U.S. while ammunition sales increased 139% compared to the same period in 2019.

Foundation (NSSF) [38]. U.S. retailers nationwide reported a record 10.3 million firearm transactions in the first half of 2020, with approximately 92% reporting an increase in ammunition sales and 87% reporting an increase in firearm sales. The survey noted that the demographics of customers who purchased guns and ammo "consisted of 55.8% White males, 16.6% White females, 9.3% Black males, 5.4% Black females, 6.9% Hispanic males, 2.2% Hispanic females, 3.1% Asian males and 0.7% Asian females" [38].

The effects of actions to mitigate COVID-19, in particular, social distancing, appear to have affected criminal activity in the U.S. People are staying home more and going out less, which helps explain why crime rates have mostly declined across the U.S. during the coronavirus pandemic, according to research from the University of Pennsylvania [39]. Across almost all of the 25 cities examined, crime fell more than 23% below the average of the same time period in the previous 5 years according to the researcher. The cities that saw the biggest drop in crime rates were Baltimore, Boston, Chicago, Philadelphia, San Francisco, and Washington D.C.

> COVID-19: Across almost all of the 25 cities examined, crime fell more than 23% below the average of the same time period in the previous 5 years, but gun violence increased. More than 41,500 people died by gun violence in 2020 nationwide.

In contrast, more than 41,500 people died by gun violence in 2020 nationwide, which is a record, according to the independent data collection and research group Gun Violence Archive. That included more than 23,000 people who died by suicide [40]. A rise in shootings accounts for the surge in many of the cities. Shootings are up 95% from the same time last year in New York, up 67% in Philadelphia, and up 34% in Atlanta, according to local police departments. Those factors include surging gun sales, increasing tension between police departments and the communities they police, an extended summer violence spike, and the disruption of school, social services and outreach and intervention programs as a consequence of COVID-19.

4.9 GOVERNMENT POLICIES TO CONTROL FIREARMS

Government policies to control the possession, use, and societal consequences of firearms have evolved in response to the impacts of firearms on social structures. The social structures include healthcare, community sustainability, family cohesion, and penal policies. Firearms control policies have developed across all levels of government in response to the public's concern about adverse effects of firearms on health and well-being. This section presents government firearms control policies existent in the U.S. at the federal and state levels. A discussion of personal policies for firearms control concludes the section.

4.9.1 MAJOR U.S. FEDERAL POLICIES TO CONTROL FIREARMS

It is not wrong to assert that the U.S. is anchored in the use of firearms. The European colonists who immigrated to what was a new land for them were accompanied by their firearms – rifles, pistols, and cannons. These firearms were deemed essential for purposes of hunting wild game and for personal and community protections against unwelcome intruders. In time, the colonists used their muzzle-loading rifles and other firearms as weapons of war against England during what became known as the American War of Revolution. The country's subsequent growth in population, geography, economy, and global influence was always accompanied by increases in quantity and quality of firearms. The complexity of the national growth eventually resulted in the U.S. federal government's enacting policies on firearms, gradually assuming a significant role in regard to the manufacture, sale, and use of firearms.

As noted by the Congressional Research Service, Congress has broad authority pursuant to the Commerce Clause to enact laws in areas that may overlap with traditional state jurisdiction [41]. As such, Congress has passed complex statutory provisions that regulate the possession, receipt, transfer, and manufacture of firearms and ammunition. Generally, courts have upheld the validity of firearms laws pursuant to Congress's commerce power. However, courts have been confronted with the question of whether federal laws can be applied to intrastate possession and intrastate transfers of firearms, or whether such application exceeds the authority of Congress. The federal government's authorizing role in regard to firearms is addressed in Amendment II of the U.S. Constitution, as discussed in the following section.

4.9.1.1 Second Amendment to the U.S. Constitution

The U.S. Constitution is the document composed by the nation's founding leaders that came into force in 1789. It is the supreme statement of law for the U.S. It includes stipulating the three-element structure and authorities of the federal government, relationships with the U.S. states, and limits of authority of the federal Congress, executive branch, and the judiciary. The Constitution has been amended 27 times in order to meet the needs of a nation that has profoundly changed since the 18th century when the Constitution was first written. In general, the first 10 amendments, known collectively as the Bill of Rights, offer specific protections of individual liberty and justice and place restrictions on the powers of government. One can assert that the amendments were developed and made law in both order of importance and year of adoption. If that be true, Amendment I states, "Congress shall make no law respecting an establishment of religion, or prohibiting the free exercise thereof; or abridging the freedom of speech, or of the press; or the right of the people peaceably to assemble, and to petition the government for a redress of grievances." It is not hyperbole to observe that this amendment forms the bedrock of individuals' freedoms in the U.S.

> Amendment II states, "A well-regulated Militia, being necessary to the security of a free State, the right of the people to keep and bear Arms, shall not be infringed."

If the amendments to the Constitution are more or less in descending order of societal importance, then the second amendment reflects the political reality that the framers of the Constitution were committed to the right of some individuals to bear firearms. For the purposes of this chapter, the Second Amendment is relevant to the subject of firearms as a matter of public health. Amendment II states, "A well-regulated Militia, being necessary to the security of a free State, the right of the people to keep and bear Arms, shall not be infringed." The Second Amendment originally applied only to the federal government, leaving the states to regulate weapons as they saw fit. However, these 27 words, perhaps not artfully written at the time, have engendered legal and societal disputes and arguments for many years, and especially in the 20th and early 21st centuries.

Until early in the 21st century, the judiciary treated the Second Amendment almost as a dead letter, i.e., essentially ignored it [42]. However, that dormancy changed with the U.S. Supreme Court case in 2008 of District of Columbia v. Heller, where the Court invalidated a federal law that forbade nearly all civilians from possessing handguns in Washington, D.C. A 5–4 majority of the court's justices ruled that the language and history of the Second Amendment showed that it protects a private right of individuals to have arms for their own defense, not a right of the states to maintain a militia. But the court ruled only that the right applied inside the home, for self-defense. A passage in the Heller decision said some restrictions were presumptively constitutional. The four dissenters argued that even if the Second Amendment did protect an individual right to have arms for self-defense; it should be interpreted to allow the government to ban handguns in high-crime urban areas [42]. This decision by the U.S. Supreme Court was hailed by U.S. gun advocates as an endorsement of individuals' right to bear arms within the strictures of applicable gun control laws, as will be subsequently described.

4.9.1.2 The National Firearms Act, 1934

The National Firearms Act (NFA) is the first piece of U.S federal gun control legislation. The act was passed by Congress on June 26, 1934, and soon thereafter signed into law by President Franklin Delano Roosevelt (D-NY). This act imposed a tax on the making and transfer of those firearms defined by the Act, as well as a special (occupational) tax on persons and entities engaged in the business of importing, manufacturing, and dealing in NFA firearms. The law also required the registration of all NFA firearms with the Secretary of the Treasury. Firearms subject to the 1934 Act included shotguns and rifles having barrels less than 18 inches in length, certain firearms described as "any other weapons," machineguns, and firearm mufflers and silencers [43].

While the NFA was enacted by Congress as an exercise of its authority to tax, the NFA had an underlying purpose unrelated to revenue collection. As the legislative history of the law discloses, its underlying purpose was to curtail, if not prohibit, transactions in NFA firearms. Congress found these firearms to pose a significant crime problem because of their frequent use in crime, particularly the gangland crimes of that era such as what became known as the St. Valentine's Day Massacre. This was the 1929 Valentine's Day murder by a Chicago criminal gang, using machineguns, of seven members of a rival Chicago gang. The brutality of the massacre caught the public's attention and led to demands for more effective gun control and associated removal of gang activities. Congress responded with the enactment of the NFA.

> As structured in 1934, the NFA imposed a duty on persons transferring NFA firearms, as well as mere possessors of unregistered firearms, to register them with the Secretary of the U.S. Treasury.

As structured in 1934, the NFA imposed a $200 making and transfer tax on most NFA firearms. This was considered quite severe at the time and thought to be an adequate policy for carrying out Congress' purpose of discouraging or eliminating transactions in these firearms. The $200 tax has not changed since 1934. Further, the NFA imposed a duty on persons transferring NFA firearms, as well as mere possessors of unregistered firearms, to register them with the Secretary of the U.S. Treasury.

If the possessor of an unregistered firearm applied to register the firearm as required by the NFA, the Treasury Department could supply information to U.S. state authorities about the registrant's possession of the firearm. State authorities could then use the information to prosecute the person whose possession violated State laws. For these reasons, the U.S. Supreme Court in 1968 held in the Haynes case that a person prosecuted for possessing an unregistered NFA firearm had a valid defense to the prosecution – the registration requirement imposed on the possessor of an unregistered firearm violated the possessor's privilege from self-incrimination under the Fifth Amendment of the U.S. Constitution. The Supreme Court's Haynes decision in effect made the 1934 Act virtually unenforceable.

4.9.1.3 Gun Control Act, 1968

This Legislation regulated interstate and foreign commerce in firearms, including importation, "prohibited persons," and licensing provisions. The Gun Control Act imposed stricter licensing and regulation on the firearms industry, established new categories of firearms offenses, and prohibited the sale of firearms and ammunition to felons and certain other prohibited persons. It also imposed the first federal jurisdiction over "destructive devices," including bombs, mines, grenades, and other similar devices [44]. The GCA was signed into law by President Lyndon B. Johnson (D-TX) on October 22, 1968.

This Act amended the NFA to cure the constitutional flaw pointed out in the Supreme Court's Haynes case [45]. First, the requirement for possessors of unregistered firearms to register was removed. Indeed, under the amended law, there is no mechanism for a possessor to register an unregistered NFA firearm already possessed by the person. Second, a provision was added to the law prohibiting the use of any information from an NFA application or registration as evidence against the person in a criminal proceeding with respect to a violation of law occurring prior to or concurrently with the filing of the application or registration. In 1971, the Supreme Court reexamined the NFA in the Freed case and found that the 1968 amendments cured the constitutional defect in the original NFA. Title II of the GCA also amended the NFA's definitions of "firearm" by adding "destructive devices" and expanding the definition of "machinegun."

4.9.1.4 Omnibus Crime Control and Safe Streets Act, 1968

The Omnibus Crime Control and Safe Streets Act of 1968 was enacted June 19, 1968, by Congress and signed into law by President Lyndon B. Johnson (D-TX) [46]. This act is the federal wiretapping and electronic eavesdropping statute. The act permits federal and state law enforcement officers to use wiretapping and electronic eavesdropping under strict limitations. The act (1) prohibits the unauthorized, nonconsensual interception of 'wire, oral, or electronic communications' by government agencies as well as private parties; (2) establishes procedures for obtaining warrants to authorize wiretapping by government officials; and (3) regulates the disclosure and use of authorized intercepted communications by investigative and law enforcement officers. The act prohibits the intentional actual or attempted interception, use, disclosure, or procurement of any other person to intercept or endeavor to intercept any wire, oral, or electronic communication.

4.9.1.5 Firearm Owners' Protection Act, 1986

This Act amended the NFA's definition of "silencer" by adding combinations of parts for silencers and any part intended for use in the assembly or fabrication of a silencer [47]. The Act also amended the Gun Control Act of 1968 to prohibit the transfer or possession of machine guns. Exceptions were made for transfers of machineguns to, or possession of machine guns by, government agencies, and those lawfully possessed before the effective date of the prohibition, May 19, 1986. The bill was signed into law on May 19, 1986, by President Ronald Reagan (R-CA).

4.9.1.6 Undetectable Firearms Act, 1988

In the 1980s, guns were being manufactured with plastic components and were therefore a problem for gun detecting security devices at airports and other places of surveillance. In particular, some pistols had frames and grips made from lightweight polymer, leading to public criticism that their relative lack of metal content meant they might be able to slip past airport metal detection and be suitable for use by terrorists. Reacting to this concern, Congress enacted the U.S. Undetectable Firearms Act of 1988. This act makes it illegal to manufacture, import, sell, ship, deliver, possess, transfer, or receive any firearm that is not as detectable by walk-through metal detection as a security exemplar containing 3.7 oz. (105 g) of steel, or any firearm with major components that do not generate an accurate image before standard airport imaging technology. The act was signed into law by President Ronald Reagan (R-CA) on November 10, 1988 [48].

4.9.1.7 Gun-Free School Zones Act, 1990

At the federal level, there have been two laws enacted that address guns in or near schools [49]. In 1990, Congress enacted the Gun-Free School Zones Act (GFSZA), signed into law by President George W. Bush (R-TX), which made it a crime under federal law to knowingly possess a firearm within 1,000 feet of a school zone (public, private, or parochial). Concerns related to school violence and rising crime rates led to the passage of this law. The act was challenged, however, and the U.S. Supreme Court found it unconstitutional. The court ruled that the law exceeded Congress' authority to regulate this type of activity (under the Commerce Clause), and in so doing, it overturned the conviction of a high school senior who brought a handgun to school. The law was later amended to apply only to guns that have either moved in or otherwise affected interstate commerce.

There are some exceptions to the GFSZA. The act allows guns to be permitted on school grounds for individuals who are licensed to possess guns. Additionally, this act allows firearms in school zones if the unloaded weapon is locked in a container or rack in a motor vehicle, if the weapon is used for a school program, or if the weapon used is part of a contract with the school and an individual (or an employer of the individual) [49].

Congress made another attempt at protecting students in schools when it enacted the Gun-Free Schools Act in 1994 (later amended in 2002), which requires states to enact statutes mandating expulsion for students who possess a firearm on school grounds. Significantly, the act also allows states to grant authority to school officials to modify the expulsion of a particular student on a case-by-case basis. Since these acts were passed, as of 2020, there has been no major federal legislation enacted that directly addresses guns in schools. There are, unfortunately, few studies about which gun and safety policies work in schools, as the federal government refuses to fund new research related to gun violence.

4.9.1.8 Brady Handgun Violence Prevention Act, 1993

In 1981, James Brady, who served as press secretary for President Ronald Reagan (R-CA), was shot in the head by an assassin during an attempt on the president's life outside a hotel in Washington, D.C. Reagan himself was shot in his left lung but recovered and returned to the White House within 2 weeks. Brady, the most seriously injured in the attack, was momentarily pronounced dead at a local hospital. He survived but suffered a debilitating brain injury from which he only partially recovered in later years.

During the 1980s, Brady and his wife, Sarah Kent Brady, became leading crusaders for gun-control legislation and in 1987 their advocacy efforts succeeded in getting a bill introduced into Congress. The Brady Bill, as it became known, was vigorously opposed by guns advocacy organizations and many members of Congress, who, in reference to the Second Amendment to the U.S. Constitution, questioned the constitutionality of regulating the ownership of arms. But in 1993, with the support of President Bill Clinton (D-AR), an advocate of gun control, the Brady Bill became law.

> The Brady Handgun Violence Prevention Act of 1993 requires federally licensed firearms dealers, but not private sellers, to perform background checks on prospective handgun purchasers.

Prior to the adoption of the Brady Act, gun sales were governed by the "honor system," i.e., a dealer was permitted to rely upon the representations of a purchaser regarding his or her eligibility to own firearms. However, the Brady Handgun Violence Prevention Act of 1993 amended the Gun Control Act to require federally licensed firearms dealers – but not private sellers – to perform background checks on prospective handgun purchasers to ensure that the firearm transfer would not violate federal or state law [50].

In order to comply with the Brady Act, the FBI created the National Instant Criminal Background Check System (NICS), a centralized catalog of records comprising three separate national databases. Among other things, the NICS contains information about individuals' criminal and mental health histories and any civil orders entered against them that might affect their eligibility to purchase or possess a gun, such as domestic violence restraining orders. Since February 28, 1994, when the federal background check requirement became effective, and

2015, the FBI asserts that more than 3 million people have been denied a firearm transfer or permit through the FBI's background check system. Of those denials, 61% were based on an individual's status as a convicted felon (43%) or fugitive from justice (19%) [51].

4.9.1.9 Law Enforcement Officers Safety Act, 2004

The Law Enforcement Officers Safety Act of 2004 was signed into law by President George W. Bush (R-TX) on July 24, 2004. As summarized by the Congressional Research Service, "The law exempts qualified current and former law enforcement officers employed by local, state, or federal law enforcement agencies from state laws prohibiting the carrying of concealed firearms. The act does not supersede federal law that governs the carrying of firearms onto aircraft, federal buildings, federal property, and national parks, nor does it limit the laws of any states that permit private persons or entities to prohibit or restrict the possession of concealed firearms on their property, nor does it supersede state laws that prohibit or restrict the possession of firearms on any state or local government property, installation, building, base, or park" [52].

4.9.1.10 Protection of Lawful Commerce in Arms Act, 2005

The Protection of Lawful Commerce in Arms Act (PLCAA) was enacted by Congress in 2005. As commented by the Congressional Research Service, the PLCAA generally shields licensed manufacturers, dealers, and sellers of firearms or ammunition, as well as trade associations, from any civil action "resulting from the criminal or unlawful misuse" of a firearm or ammunition, but lists six exceptions where civil suits may be maintained. (The six exceptions are specific matters of civil law and generally provide limited access in ability to sue a commercial weapons enterprise.)

This act was introduced in response to litigation brought by municipalities, e.g., Cincinnati, Ohio, and victims of shooting incidents against federally licensed firearms manufacturers and dealers, some of whom were located outside the state where the injuries occurred. Consequently, most lawsuits brought after the enactment of this law have been dismissed notwithstanding the exceptions that would permit a civil suit to proceed against a federal firearms licensee [53]. It is interesting to compare this industry-wide protection against litigation noted in Volume 1, Chapter 1 where some municipalities and concerned individuals have sued the fossil fuel industry (coal, oil, gas) for failing to protect the climate. The course of this climate change litigation has in 2019 yet to run its course through U.S. federal courts.

An interesting challenge to the Protection of Lawful Commerce Act occurred in November 2019. The U.S. Supreme Court cleared the way for relatives of Sandy Hook Elementary School shooting victims (Table 4.3) to sue the Remington Arms Company, the maker of the rifle used in the massacre [54]. The court said that it would not hear an appeal by Remington of a ruling by Connecticut's Supreme Court that allowed a lawsuit brought by the families of the victims to go forward. The Sandy Hook families' suit used one of six narrow exemptions to the 2005 law to argue that Remington violated Connecticut's Unfair Trade Practices Act. The suit said the gun maker recklessly marketed the Bushmaster AR-15-style rifle.

4.9.1.11 Bureau of Alcohol, Tobacco, and Firearms

The only federal government organization bearing the word 'firearms' in its title is the Bureau of Alcohol, Tobacco, Firearms and Explosives (ATF), which is a federal law enforcement organization within the U.S. Department of Justice. Its responsibilities include the investigation and prevention of federal offenses involving the unlawful use, manufacture, and possession of firearms and explosives; acts of arson and bombings; and illegal trafficking and tax evasion of alcohol and tobacco products. The ATF also regulates via licensing the sale, possession, and transportation of firearms, ammunition, and explosives in interstate commerce.

As background, in the early 1950s, the Bureau of Internal Revenue was renamed "Internal Revenue Service" (IRS), and its Alcohol Tax Unit (ATU) was given the additional responsibility of enforcing federal tobacco tax laws. At this time, the name of the ATU was changed to the Alcohol and Tobacco Tax Division (ATTD). In 1968, with the passage of the Gun Control Act, the agency changed its name again, this time to the Alcohol, Tobacco, and Firearms Division of the IRS and first began to be referred to by the initials "ATF." In Title XI of the Organized Crime Control Act of 1970, Congress enacted the Explosives Control Act, which provided for close regulation of the explosives industry and designated certain arsons and bombings as federal crimes. The Secretary of the Treasury was made responsible for administering the regulatory aspects of the new law and was given jurisdiction over criminal violations relating to the regulatory controls. These responsibilities were delegated to the ATF division of the IRS. The Secretary of the Treasury and the U.S. Attorney General were given concurrent jurisdiction over arson and bombing offenses [55].

On July 1, 1972, the ATF was officially established as an independent bureau within the U.S. Treasury Department. This transferred the responsibilities of the ATF division of the IRS to the new Bureau of Alcohol, Tobacco, and Firearms, which brought the organization into a new era where federal firearms and explosives laws addressing violent crime became the primary mission of the agency [55]. However, taxation and other alcohol issues remain principal duties because the ATF is responsible for collecting federal taxes on alcohol and tobacco sales.

Perspective: Through actions of Congress, the federal government has authority under the previously described laws and their attendant regulations to control some aspects of firearms possession in the U.S. Sales of firearms to felons are banned. Firearms dealers are required to register with the Treasury Department. Possession of certain firearms, e.g., machine guns, is restricted and destructive devices such as grenades are banned. Background checks are required of persons who desire to purchase firearms. Gun-free school zones are mandated. These federal statutes overarch firearms control policies of the U.S. states and territories, as illustrated in the following section.

4.9.1.12 Trump and Biden Firearms Executive Actions

Firearms violence in the U.S. has been the subject of voiced concern by both U.S. President Donald Trump (R-NY) and Joseph Biden (D-DE). Statements of concern were precipitated by events of firearms violence, usually a mass shooting. But in point of fact, U.S. presidents have rather limited authority to change gun control laws

Firearms Violence

and regulations. At the federal level, the principal authority for firearms control is vested with Congress, implied by material in this chapter that discusses federal laws pertaining to firearms.

But the preceding comments are not meant to suggest that neither president did anything regarding responses to firearms violence. In February 2018, President Trump signed a memorandum instructing U.S. Attorney General Jeff Sessions to regulate the use of bump stocks, effectively banning the use of the devices that can allow rifles to mimic automatic weapons. In April 2021, President Biden instructed the U.S. Department of Justice to tighten regulations on sales of so-called "ghost guns," which are untraceable firearms assembled from kits that can be purchased via internet sales.

4.9.2 U.S. State and Local Firearm Control Policies

In addition to complying with federal laws and regulations bearing on control of firearms, U.S. states govern how their residents may purchase, own, and use firearms, including which types of guns and weapons are prohibited. An illustration of one state's gun laws follows for the state of Georgia. As summarized by a legal source, Georgia's gun laws stipulate the following control of firearms within the state [56].

"Legal Arms: Some types of weapons are illegal in Georgia, and most are military-style firearms. Fully automatic machine guns are illegal, as well as sawed-off shotguns or rifles. Those are firearms that once had a long barrel, but have had the barrel shortened for ease of use. The difference between a fully automatic machine gun and a semi-automatic machine gun is that a fully automatic machine gun can fire multiple shots with one pull of the trigger. With a semi-automatic gun, the trigger must be pulled for every shot."

Dangerous weapons, including bazookas, mortars, and hand grenades, are also illegal, given the extreme damage they cause, and the fact that they are often too unwieldy for personal defense or hunting. Silencers are illegal as well.

Waiting Periods: Georgia does not have a waiting period to purchase a firearm but does require a 60-day waiting period before receiving a license to carry a handgun.

People Prohibited From Owning Firearms: Certain people are not allowed to own firearms. Those include Convicted felons, People convicted of drug crimes, Fugitives, People going through felony proceedings, People convicted of carrying a deadly weapon in a school zone or public gathering, People less than 18 years old, Anyone hospitalized for mental issues, or drug or alcohol abuse" [56].

Some states, like California, require a waiting period before purchasing a firearm. One of the justifications is to allow an angry person to cool down before making a bad decision. While each U.S. state and territory has legislated laws and other policies for the purpose of firearm possession and uses, both criminal and otherwise.

> California has implemented controls on the sale and distribution of ammunition used in firearms.

Perspective: Georgia's state laws on control of firearms are considered mild in comparison to other states, e.g., New York or California, where waiting periods and registration of firearms are more stringent. However, Georgia's gun laws are typical of most states with large numbers of hunters and gun sports enthusiasts.

4.9.2.1 "Red Flag" Gun Seizure Statutes

Five U.S. states (California, Connecticut, Indiana, Oregon, and Washington) have enacted what are called "red flag" gun seizure statutes [57]. The operative policy behind this kind of statute is concordant with the public health principle of prevention of disease and disability. Red flag policies allow police to **temporarily** (emphasis added) remove guns from people deemed by a judge to be dangerous to themselves or others. Police are alerted by a family member or a concerned citizen that a person in possession of a firearm might have the intent to harm themselves or others. Police must deem the possession to be a potential threat, and then petition a judge to issue a gun seizure action. Police removal of a weapon is temporary, typically one to 2 weeks, as decreed by the court. In theory, temporary removal of a firearm permits the firearm's owner to resolve any personal issues that might have precipitated a suicide or a criminal act.

Connecticut was the first state to pass a red flag statute in 1999, in response to a mass shooting at a Connecticut state lottery office. The authorities in the statute were seldom utilized until 2007 when a mass shooting occurred on the campus of Virginia Tech University, Blacksburg, Virginia, when 49 persons were shot, 32 fatally. Overall, Connecticut has seen 1,519 cases of red flag actions, with about two-thirds of those occurring post-2012 [57]. The effectiveness of red flag policies in the prevention of gun violence, including suicides, is unknown, but some law enforcement agencies opine the policy has been effective.

4.9.2.2 California's Ammunition Control Statute

In addition to gun control policies, one state, California, has implemented an approach to reducing gun violence that goes beyond gun control. The state has implemented controls on the sale and distribution of ammunition used in firearms. As background, the purchase of ammunition remains subject to far fewer federal restrictions than the weapons that fire them. Purchasing ammunition typically requires no form of identification, is handed over with no questions asked and, in most of the U.S, can be ordered online and delivered to domiciles [58]. In contrast, gun dealers have to keep detailed sales records of firearms and generally have to be licensed.

But in California, local laws require ammunition dealers to keep detailed records of sales. In January 2019, a state restriction on internet sales went into effect, preventing online buyers from having ammunition delivered directly to their homes. Instead, ammunition must be shipped to a licensed dealer so that the police can run a background check on the purchaser [58]. In July 2019 California began requiring stores to conduct point-of-purchase background checks on ammunition buyers. However, background checks will not be required for people purchasing ammunition at gun ranges for on-site use.

4.9.2.3 Retail Sales of Assault Weapons

In lieu of any firearms control policies by the U.S. federal government, and in recognition of mass shootings in the U.S., some retail stores have decided to forego sales of assault rifles and other firearms materials. One large retailer, Dick's Sporting Goods, made the decision to pull all AR-15 assault rifles off the stores' shelves and had all $5-million dollars of AR-15 rifles destroyed [59]. The company's decision was met with plaudits by organizations opposed to gun violence and by derision by gun rights groups. In a similar decision, Walmart's decided that after "selling through our current inventory commitments," it would stop selling certain short-barrel rifle ammunition and all handgun ammunition [60].

4.9.3 INTERNATIONAL GOVERNMENT POLICIES TO CONTROL FIREARMS

Government policies to control firearms and their public health consequences vary between countries. This variance ranges from relatively mild gun control policies in the U.S. to quite restrictive policies in some other nations, e.g., China and India. In addition to national policies on firearms, the UN has issued a protocol against firearms trafficking.

The Protocol against the Illicit Manufacturing of and Trafficking in Firearms, their Parts and Components and Ammunition (Firearms Protocol) is the only legally binding international instrument to counter the illicit manufacturing of and trafficking in firearms, their parts and components and ammunition. It was adopted by resolution 55/255 of May 31, 2001, at the fifty-fifth session of the General Assembly of the United Nations and it entered into force on July 3, 2005 [61].

The Firearms Protocol provides a framework for States (e.g., France) to control and regulate licit arms and arms flows, prevent their diversion into the illegal circuit, facilitate the investigation and prosecution of related offences without hampering legitimate transfers. "The Firearms Protocol aims at promoting and strengthening international cooperation and developing cohesive mechanisms to prevent, combat and eradicate the illicit manufacturing of and trafficking in firearms, their parts and components and ammunition (firearms). By ratifying or acceding to the Firearms Protocol, States make a commitment to adopt and implement a series of crime-control measures that aim at: (a) establishing as criminal offence the illicit manufacturing of and trafficking in firearms in line with the Protocol's requirements and definitions; (b) adopting effective control and security measures, including the disposal of firearms, in order to prevent their theft and diversion into the illicit circuit (c) establishing a system of government authorizations or licensing intending to ensure legitimate manufacturing of, and trafficking in, firearms; (d) ensuring adequate marking, recording and tracing of firearms and effective international cooperation for this purpose" [61].

The Protocol was signed by 52 parties and the treaty entered into force on July 3, 2005. As of October 2018 it has 116 parties, including 115 states and the European Union. Notably, three of the top four arms manufacturers in the world—the U.S., Russia, and France—have not signed the protocol [62].

> The UN Firearms Protocol is the only legally binding international instrument to counter the illicit manufacturing of and trafficking in firearms. The U.S., Russia, and France are not signatories to the Protocol.

Perspective: The stated intent of the UN Protocol "aims at promoting and strengthening **international cooperation** and developing cohesive mechanisms to prevent, combat and eradicate the illicit manufacturing of and trafficking in firearms (emphasis added), their parts and components and ammunition (firearms)." Given this intent, it is disappointing in a public health sense that commercial interests, i.e., firearms manufacturers, have apparently persuaded some government policymakers to forego the Protocol.

4.9.4 U.S. Federal Policies on Firearms Research

Among the leading causes of death in the U.S., firearms claim about as many lives as car crashes. As previously stated, more than 38,000 people in the U.S. die by guns every year, the majority of them suicides – an average of more than 100 gun deaths every day. Yet the federal government spends only about 1.6% as much on gun violence research as it does on research into traffic crashes and other leading causes of death. The lack of federal government research pertaining to gun violence and its public health consequences is the result of Congressional actions and directives. The absence of federal research was not always the case.

In the mid-1990s, Congress zeroed out the budget for gun violence prevention at the CDC after some of its studies linked home gun ownership with higher rates of firearm deaths. In response, gun possession advocates persuaded their allies in Congress to take action. Led by Congressman Jay Dickey (R-AR), a provision was added to a 1996 spending bill declaring that "[n]one of the funds made available in this title may be used, in whole or in part, to advocate or promote gun control." The CDC has interpreted that rule, known as the Dickey Amendment, as an outright ban on most gun research [63].

> In the mid-1990s, Congress zeroed out the budget for gun violence prevention research at the CDC.

However, as noted in a review of the Dickey Amendment's origin and impact, "Congress reached a compromise in passing an omnibus spending bill in March 2018. The Dickey Amendment has not gone away, but a report accompanying the spending bill clarifies that the amendment does not prohibit federal funding of research on the causes of gun violence. This compromise may help reduce the Dickey Amendment's chilling effect on gun violence research, but it remains to be seen whether more funding will actually be devoted to such research" [63].

The CDC was not the only federal agency affected. In 2011, Congress added a similar clause to appropriations legislation for the National Institutes of Health (NIH). However, due to a directive from the Obama administration, the NIH continued to provide funding for gun research. That NIH research funding faded as the Obama administration left office. In 2017, the NIH discontinued its funding program that specifically focused on firearm violence. While firearms researchers can still apply for funding through more general NIH funding opportunities, funding for gun research appears uncertain [64].

With the absence of federal government research on gun violence, research on this subject has been left to academic investigators and organizations with interests in firearms issues. An example of such research is a comprehensive review of gun violence in the U.S. performed by the Rand Corporation. Following a comprehensive review of policies pertaining to gun possession and gun violence data, the researchers concluded, "The best evidence showed that child-access prevention laws could reduce accidental and self-inflicted gun deaths and injuries among youth. The researchers also found moderate evidence that gun restrictions for people with mental illness could reduce violent crime; and moderate evidence that stand-your-ground laws might increase homicide rates. They found moderate evidence as well that dealer background checks reduce suicides and gun homicides, but inconclusive evidence for expanded background checks by private sellers" [65].

The Rand Corporation investigators strongly endorsed further research on gun violence, causes, and policies. This Rand recommendation for research on gun violence was preceded by a similar recommendation in 2016 from a coalition of 141 medical groups that urged Congress to restore funding for gun violence research [66]. The coalition's spokes group was Doctors for America, an advocacy group for health care in the U.S. The medical group voiced their support for research on accidental shootings, firearms suicides, and the impact of state policies on the rate of deaths related to firearms. This recommendation from a coalition of medical groups and some public health organizations were helpful for influencing a Congressional budget decision in December 2019 that provides $25 million to study gun violence [67]. The CDC and the NIH will each receive $12.5 million to study gun violence and ways to prevent it. This bipartisan budget decision marks the first time in more than 20 years that Congress has appropriated money for gun violence research.

4.10 FIREARMS HAZARD INTERVENTIONS

Firearms are hazardous to human and ecological health by their very design and intended purpose. The intended purpose is to discharge a metal cartridge or other form of ammunition for the purpose of propelling a projectile at a desired target. As discussed in this chapter, this sequence of firearm events can be deliberate or unintentional. And as with other chapters in this book, because firearms can represent a physical hazard in the environment, efforts to intervene in their occurrence comport with public health practices and values. Given the content of this chapter, the following interventions are suggested.

- Governments at all levels are obliged to establish and enforce firearms control policies that are appropriate for the societal conditions at hand. This assertion is in concert with a primary duty of government: protection and well-being of citizenry.
- Individuals who possess firearms are obligated to educate themselves on the use, distribution, and storage of them. Possession of a firearm should be restricted for persons less than the age of maturity. For example, the state of Washington bans anyone less than age 21 years old from purchasing assault rifles [68].
- Individuals who possess firearms have a societal obligation to safely store and secure them. This is particularly necessary whenever children might

have access to the possessed firearms. One source notes that nearly 2 million U.S. children reside with unlocked, unsecured, loaded guns [69].
- The most secure method to avoid unintended discharges of firearms in a household is to have no firearms therein.
- Possession of firearms should be denied to persons with known mental health disorders. Similarly, school students and other persons with disaffections with real or perceived authority should be referred to counseling and access to firearms negated.
- Gun clubs and firearm support organizations should adopt policies of providing public education about the prevention of gun violence.
- Commercial enterprises that profit from sales of firearms should have the ethical responsibility for practicing business behaviors that do not contribute to anti-social uses of firearms. As an example, in September 2019 Walmart announced that it would formally end handgun sales, discontinue sales on certain types of ammunition and ask customers not to openly carrying firearms in the company's stores [70].
- Research on the causes and circumstances of gun violence should be adequately supported at the federal level, with commensurate funding of academic research on the causes and prevention of gun violence.

4.11 SUMMARY

Humankind resides in and is sustained by its physical environment. We are sustained by air, water, food, and other products of nature. But within our physical environment are hazards to our health and well-being. Chemical and physical pollutants in air, water, and food can produce adverse human and ecological health effects. These kinds of hazards arise from anthropogenic activities. Certainly, the environmental hazard discussed in this chapter, firearms, is a product of anthropogenic effort. As presented herein as a public health concern, from 1968 through 2015, there were 1,516,863 gun-related deaths on U.S. territory. Annually, approximately 38,000 gun violence deaths occur in the U.S., with two-thirds occurring from suicide. Further, more than 85,000 people in the U.S. are nonfatally injured by guns annually. Globally, in 2016 there were an estimated 251,500 deaths from firearm injuries, with the majority (64%) of firearm injury deaths homicides. For the practice of public health, elimination or marked reduction of a hazard to health is the norm, for example, regulations to limit the amount of pollution permitted in public water supplies. Therefore, to prevent or markedly reduce firearm violence would necessitate elimination or substantive reduction to access to firearms [71]. In those countries where access to firearms is tightly controlled, e.g., China and India, premature loss of life due to firearms is infrequent. But for those countries, e.g., U.S., where civilian access to firearms is relatively easy, prevention of gun violence will remain a challenge to the public's health and well-being. The public health calculus is quite evident – fewer firearms possessed means less gun violence expressed.

REFERENCES

1. Chalabi, M. 2017. How bad is US gun violence? These charts show the scale of the problem. *The Guardian*, October 5.
2. Brady Center. 2019. *Key gun violence statistics.* Washington, DC: Brady Campaign to Prevent Gun Violence.
3. Kaufman, E. J., D. J. Wiebe, R. A. Xiong, et al. 2021. Epidemiologic trends in fatal and nonfatal firearm injuries in the US, 2009–2017. *JAMA Intern. Med.* 181:237–44.
4. Anonymous. 2018. 2019 Firearms law and legal definition. *US Legal.* https://definitions.uslegal.com/.
5. Whipps, H. 2008. How gunpowder changed the world. *Live Science*, April 6.
6. Ingraham, C. 2018. There are more guns than people in the United States, according to a new study of global firearm ownership. *The Washington Post*, June 19.
7. Masters K. 2016. Why a new survey from Harvard and Northeastern is the most authoritative assessment of American gun ownership in 20 years. *The Trace*, January 21.
8. Anonymous. 2019. *Safe storage.* Giffords Law Center. https://lawcenter.giffords.org/gun-laws/policy-areas/child-consumer-safety/safe-storage/.
9. Karp, A. 2018. *Briefing Paper. Estimating Global Civilian-Held Firearms Numbers. Small Arms Survey.* Canberra: Australian Government, Department of Foreign Affairs and Trade.
10. APHA (American Public Health Association). 2019. *Gun violence.* Washington, DC: APHA.
11. NCHS (National Center for Health Statistics). 2017. *Assault or homicide.* Atlanta, GA: Centers for Disease Control and Prevention.
12. Fox, M. 2018. Homicides using guns up 31 percent, CDC finds. *NBC News*, July 27.
13. Brady Center. 2019. *Gun violence statistics.* Washington, DC: The Brady Center to Prevent Gun Violence.
14. Gun Violence Archive. 2019. About. Washington, DC. https://www.gunviolencearchive.org
15. Naghavi, M., L. B. Marczak, M Kutz, et al. 2018. Global mortality from firearms, 1990–2016. *JAMA* 320(8):792–814.
16. Cunningham, R. M., M. A. Walton, and P. M. Carter. 2018. The major causes of death in children and adolescents in the United States. *N. Engl. J. Med.* 379:2468–75.
17. Azrael, D., J. Cohen, C. Salhi, and M. Miller. 2018. Firearm storage in gun-owning households with children: Results of a 2015 national survey. *J. Urban Health* (Online, May 20). https://www.thetrace.org/wp-content/uploads/2018/05/Firearm-Storage-in-Households-with-Children_JUH.pdf.
18. Knopov, A., R. J. Sherman, J. R. Raifman, et al. 2019. Household gun ownership and youth suicide rates at the state level, 2005–2015. *Am. J. Prev. Med.* 56:335–342.
19. Bandlamudi, A. 2019. Youth suicide rates are higher in states with high gun ownership, according to a new study. *KERA News*, January 17. https://www.keranews.org/post/youth-suicide-rates-are-higher-states-high-gun-ownership-according-new-study.
20. Council on Injury, Violence, and Poison Prevention Executive Committee. 2012. Firearm-related injuries affecting the pediatric population. *Pediatrics* 130(5):e1416–23.
21. Anonymous. 2018. *Mass casualty shootings. 2018 National crime victims' rights week reference guide: Crime and victimization fact sheets.* The National Center for Victims of Crime. https://ovc.ncjrs.gov/ncvrw2018/info_flyers/fact_sheets/2018NCVRW_MassCasualty_508_QC.pdf.
22. Berkowitz, B., D. Lu, & C. Alcantara. 2018. The terrible numbers that grow with each mass shooting. *The Washington Post*, April 22.
23. Daugherty, O. 2019. U.S. mass killings reach record in 2019, largely due to shootings. *The Hill*, December 28.

24. PolitiFact Staff. 2018. The facts on mass shootings, guns in the United States. *PolitiFact*, May 18. https://www.politifact.com/truth-o-meter/article/2018/may/18/facts-mass-shootings-guns-united-states/.
25. Lopez, G. 2019. Study: Where gun laws are weaker, there are more mass shootings. *Vox*, March 8.
26. Editors. 2019. Gunman opens fire on Las Vegas concert crowd, wounding hundreds and killing 58. *History.com*, September 30.
27. Grabow, C. and L. Rose. 2018. The US had 57 times as many school shootings as the other major industrialized nations combined. *CNN News*, May 21.
28. Follman, M., G. Aronsen, and D. Pan. 2018. US mass shootings, 1982–2018: Data from Mother Jones' investigation. *Mother Jones*, November 19.
29. Fagan, C. 2018. The impact of mass school shootings on the mental health of survivors: What parents need to know. *PSYCOM*, November 25. https://www.psycom.net/mental-health-wellbeing/school-shooting-survivor-mental-health/.
30. Novotney, A. 2018. What happens to the survivors? *J. Am. Psycholog. Assoc.* 49(8):36.
31. de Marco, H. 2018. The other victims: First responders to horrific disasters often suffer in solitude. *Kaiser Health News*, July 6.
32. Anonymous. 2019. America's Plains Indians bring back bison to their lands. *Deutsche Welle*, January 8.
33. Anonymous. 2019. *Lead ammunition: Toxic to wildlife, people and the environment.* The Humane Society of the United States. https://www.humanesociety.org/resources/lead-ammunition-toxic-wildlife-people-and-environment.
34. Barkham, P. 2020. EU to ban use of lead shot by wetland bird hunters. *The Guardian*, November 26.
35. Laidlaw, M. A. S., G. Filippelli, H. Mielke, et al. 2017. Lead exposure at firing ranges – A review. *Environ. Health.* 16:34. DOI: 10.1186/s12940-017-0246-0.
36. EPA (Environmental Protection Agency). 2005. *Best management practices for lead at outdoor shooting ranges.* New York: Division of Enforcement and Compliance Assistance, RCRA Compliance Branch.
37. Rubin, M. 2017. The cost of US gun violence has finally been calculated – At $2.8 billion a year. *Quartz*, October 3.
38. Manfred, L. 2020. Gun, ammo sales surging amid coronavirus pandemic, civil unrest: Survey. *Fox News*, July 22.
39. Wolford, B. 2020. Coronavirus has 'massive' impact on US crime rates: Here are some of the big changes. *Miami Herald*, July 21.
40. Hauck, G. 2020. 'They're not forgotten': America's other epidemic killed 41,000 people this year. *USA Today*, December 18.
41. Chu, V. S. 2013. *Congressional authority to regulate firearms: A legal overview.* Washington, DC: Congressional Research Service.
42. Lund, N. and A. Winkler. 2018. The second amendment. Matters of Debate. https://constitutioncenter.org/interactive-constitution/amendments/amendment-ii.
43. ATF (Bureau of Alcohol, Tobacco, Firearms). 2018. *Rules and regulations. The National Firearms Act of 1934.* Washington, DC: U.S. Department of Justice.
44. ATF (Bureau of Alcohol, Tobacco, Firearms). 2016. *Gun Control Act (GCA) of 1968.* Washington, DC: U.S. Department of Justice.
45. ATF (Bureau of Alcohol, Tobacco, Firearms). 2018. *National Firearms Act. Title II of the Gun Control Act (GCA) of 1968.* Washington, DC: U.S. Department of Justice.
46. Anonymous. 2019. Omnibus Crime Control and Safe Streets Act. *US Legal.* https://uslegal.com/.
47. ATF (Bureau of Alcohol, Tobacco, Firearms). 2018. *Firearm Owners' protection Act.* Washington, DC: U.S. Department of Justice.

48. Wikipedia. 2019. Undetectable Firearms Act. https://en.wikipedia.org/wiki/Undetectable_Firearms_Act.
49. NASSP (National Association of Secondary School Principals). 2019. Gun policies and schools. https://www.nassp.org/2019/03/01/legal-matters-march-2019/.
50. Editors. 2018. Brady Bill signed into law. This day in history. *History.com*, November 30. https://www.history.com/this-day-in-history/brady-bill-signed-into-law.
51. Anonymous. 2018. *Background check procedures*. Giffords Law Center. https://lawcenter.giffords.org/gun-laws/policy-areas/background-checks/background-check-procedures/.
52. CRS (Congressional Research Service). 2006. *Law Enforcement Officers Safety Act of 2004*. Washington, DC: American Law Division.
53. Chu, V. S. 2012. *The protection of lawful commerce in Arms Act: An overview of limiting tort liability of gun manufacturers*. Washington, DC: Congressional Research Service.
54. Chappell, B. 2019. Supreme Court allows Sandy Hook families' case against Remington arms to proceed. *NPR*, November 12.
55. ATF (Bureau of Alcohol, Tobacco, Firearms and Explosives). 2018. *Rules and regulations*. Washington, DC: U.S. Department of Justice.
56. Anonymous. 2019. *Georgia gun control laws*. Atlanta, GA: McLendon Law Firm, LLC.
57. Johnson, K. 2018. States mull 'red flag' gun seizures from people deemed dangerous. *The New York Times*, February 23.
58. Urbina, I. 2018. California tries new tack on gun violence: Ammunition control. *The New York Times*, September 9.
59. Shumway, J. 2019. Customers at local Dick's sporting goods react to CEO's decision to destroy $5m worth of assault rifles. *CBS Pittsburgh*, October 8.
60. Corkery, M. 2019. Walmart to limit ammunition sales and discourage 'open carry' of guns in stores. *The New York Times*, September 3.
61. UNODC (United Nations Office of Drugs and Crime). 2019. The firearms protocol. http://www.unodc.org/unodc/en/firearms-protocol/the-firearms-protocol.html.
62. Anonymous. 2019. Protocol against the illicit manufacturing of and trafficking in firearms. *Wikipedia*. https://en.wikipedia.org/wiki/Protocol_against_the_Illicit_Manufacturing_of_and_Trafficking_in_Firearms#cite_note-ratifications-2.
63. Rostron, A. 2018. The Dickey amendment on federal funding for research on gun violence: A legal dissection. *Am. J. Public Health* 108(7):865–7.
64. Wallace, L. 2017. Here's the reason why there is so little scientific research on guns in the US. *Science Alert*. https://www.sciencealert.com/why-so-little-gun-control-research-usa-questions-answers?perpetual=yes&limitstart=1.
65. Anonymous. 2018. *More research could help prevent gun violence in America*. Santa Monica: The Rand Corporation.
66. Masters, K. 2016. 141 Medical groups urge Congress to restore funding for gun violence research. *The Trace*, April 6.
67. Hellmann, J. 2019. Congress reaches deal to fund gun violence research for first time in decades. *The Hill*, December 16.
68. Associated Press. 2019. Washington bans anyone under 21 from buying assault rifles. Associated Press, January 1.
69. Schaechter, J. 2018. Guns in the home. *Healthy Children.org*. https://www.healthychildren.org/English/safety-prevention/at-home/Pages/Handguns-in-the-Home.aspx.
70. Gstalter, M. 2019. Walmart to discontinue sales of some ammunition, asks customers not to open carry. *The Hill*, September 3.
71. Fisher, M. and J. Keller. 2017. What explains U.S. mass shootings? International comparisons. *The New York Times*, November 7.

5 Noise and Light Pollution

5.1 INTRODUCTION

This chapter addresses two forms of environmental pollution that in some regard are nontraditional in the suite of hazards in humans' physical environment. As discussed in some of the previous chapters, air, water, and food pollution are long-standing recognized hazards in our physical environment, likely dating back to primitive times and early human development. Moreover, as time passed, new forms of environmental hazards emerged, for example, during the Industrial Age when workers' and communities' exposure to toxic substances commenced on a large scale. In the 20th century, as discussed in this book's chapters on air pollution, water contamination, toxic substances, and transportation, new hazards emerged as threats to human and ecological health.

Two of these newer hazards are noise and light pollution. One might question why these two forms of energy are being called "new." After all, noise has accompanied human development over eons of time whenever someone complained that some particular acoustic sound was somehow offensive. Perhaps some cave-dwelling ancestor thought the sound of constructing stone weapons was too noisy. And what is "new" about light? Have not humans always relied on each day's arrival of the Sun's rays as the key environmental element to provide the light for determining social, commercial, and personal lifestyles? All the foregoing questions are quite reasonable to ask. In response, what is "new" is the proliferation of sources of noise and light and the effects on human and ecological health. Put differently, it is relatively "new" to consider excess, novel sources of noise and light as matters of pollution. A description of these two forms of environmental pollution, their sources, human and ecological effects, and policies to control them therefore form the content of this chapter. The two sources of pollution will be considered separately in this chapter, commencing with a discussion of noise pollution.

5.2 NOISE POLLUTION

Noise can be considered acoustic energy that is deemed personally offensive, i.e., bothersome or unwelcome. It is a matter of personal value. What might be considered noise in one setting could be considered pleasing in a different setting. For example, a trumpet solo at a music concert could be received as quite pleasant by concert goers, but the same performance given in an apartment building at an early hour of the morning could be heard as noise by some apartment dwellers. Governments up until the 1970s viewed noise as a "nuisance" rather than an environmental problem. But commencing in 1972, the U.S. enacted federal legislation for purpose of controlling noise sources that presented hazards to public health. Later, other governments undertook similar policy actions, as will be subsequently described herein.

Before proceeding, some terms should be defined, specifically, sound pressure level (SPL) and dB SPL. One parameter of the acoustic (sound) wave that is generally used to assess the exposure of humans to sound is the SPL. This metric is expressed in μPa or Pa, which is pressure expressed in pascal units, where 1

> Sound levels of 65 dBA are annoying to most individuals, constant or repeated exposure to levels of 90 dBA or greater can lead to hearing loss.

pascal (Pa) equals one newton per square meter [1]. The human ear's audible SPLs range from 20 μPa (hearing threshold) up to 20 Pa (pain threshold), a range with a scale of 1:10,000,000. Since using such a large scale is not practical, a logarithmic scale in decibels (dB) was introduced. A dB of SPL (dB SPL) is defined as $20 \log_{10} p_1/p_0$ where p_1 is the measured SPL of a given sound and p_0 is a reference value of 20 μPa, which corresponds to the lowest hearing threshold of a young, healthy ear. In the logarithmic scale, the range of a human ear's audible sound is from 0 dB SPL (hearing threshold) to 120–140 dB SPL (pain threshold). A-weighted decibels, abbreviated dBA, or dBa, or dB(a), are an expression of the relative loudness of sounds in air as perceived by the human ear. In the A-weighted system, the decibel values of sounds at low frequencies are reduced, compared with unweighted decibels, in which no correction is made for audio frequency. This correction is made because the human ear is less sensitive at low audio frequencies, especially less than 1,000 Hz, than at high audio frequencies [2].

Turning to sound levels of health concern, as noted by the Congressional Research Service, while sound levels of 65 dBA are annoying to most individuals, constant or repeated exposure to levels of 90 dBA or greater can lead to hearing loss [3]. Another source, the World Health Organization's (WHO's) guidelines for community noise recommend less than 30 dBA in bedrooms during the night for a sleep of good quality and less than 35 dBA in classrooms to allow good teaching and learning conditions. The WHO guidelines for night noise recommend less than 40 dBA of annual average outside of bedrooms to prevent adverse health effects from night noise [4]. To limit the public's exposure to potentially harmful sound levels, the U.S. federal government sets and enforces uniform noise standards for aircraft and airports, interstate motor carriers and railroads, workplace activities, medium and heavy-duty trucks, motorcycles and mopeds, portable air compressors, federal highway projects, and federal housing projects. U.S. state, territorial, and local governments determine the extent to which other sources are regulated, including commercial, industrial, and residential activities. Shown in Table 5.1 are sound levels associated with various noise sources [5].

> The WHO guidelines for night noise recommend less than 40 dBA of annual average outside of bedrooms to prevent adverse health effects.

Following a presentation of noise pollution definition and sources of noise pollution, the adverse effects of noise pollution on human and ecological health will be discussed in the following sections, together with federal and illustrative state and local policies to control sources of noise.

TABLE 5.1
Sound Levels Generated by Various Sources of Noise [5]

Sound Level	dBA
Quiet library, soft whispers	30
Quiet office, library, quiet residential area	40
Large office, rainfall, refrigerator	50
Normal conversation, electric toothbrush, sewing machine	60
Freeway traffic, power lawn mower, TV audio	70
Noisy restaurants, handsaws, blender	80
Constant exposure to the following sound levels can lead to hearing loss	
Tractor, truck, shouted conversation	90
Motorcycle, snowmobiles, subway	100
Chain saw, ambulance sirens, band concerts	120
Airplane taking off, stock car races, bicycle horn	140
Rocket launching from a pad, shotgun	180

Source: Center for Hearing and Communication.

5.2.1 DEFINITION

It is important at this juncture to define noise pollution. One standard dictionary's definition is "annoying or harmful noise (as of automobiles or jet airplanes) in an environment" [6]. A different source declares noise pollution to be "noise, such as that from traffic, that upsets people where they live or work and is considered to be unhealthy for them" [7]. While these two definitions differ somewhat in words and breadth in specifics, both contain the same and pertinent element of health, where the former definition mentions 'harmful' and the latter mentions "unhealthy." For the purposes of this chapter, the latter definition is preferred and therefore chosen for use.

> One source declares noise pollution to be "noise, such as that from traffic, that upsets people where they live or work and is considered to be unhealthy for them."

5.2.2 SOURCES OF NOISE POLLUTION

Broadly speaking, noise pollution has two sources: industrial and non-industrial [8]. The industrial source includes the noise from various industries and big machines working at a very high speed and high noise intensity. Non-industrial sources of noise include the noise created by transport/vehicular traffic and the neighborhood noise generated by various noise pollution, which can be divided into two categories, natural and anthropogenic. Most leading noise sources will fall into the following categories: roads traffic, aircraft, railroads, construction, industry, noise inside buildings, and consumer products, e.g., kitchen appliances [8]. Shown in Figure 5.1 is an airplane landing at an airport, thereby constituting a noise source.

FIGURE 5.1 Noise pollution nearing a city airport. (2021. U.S. Department of Transportation Federal Aviation Administration, Aircraft Noise, FAA-2021–0037-001.)

5.2.3 Effects of Noise Pollution on Human Health

It is relatively well known that excessive exposure to noise can damage the organs of hearing. This damage can be temporary or permanent, depending on such factors as intensity and duration of the noise. Impaired hearing is a public health problem. The CDC has reported that trouble with hearing is the third most commonly reported chronic health condition in the U.S. [9]. They estimate that about 40 million Americans ages 20–69 years old have hearing loss in one or both ears that might be due to noise exposure. The CDC further observed that studies have shown that untreated hearing loss is associated with anxiety, depression, loneliness, and stress. In addition to hearing loss, chronic noise exposure has been associated with worsening of heart disease, increased blood pressure, and other adverse health effects. In a separate analysis, University of Michigan public health researchers used data from several noise sources to estimate that 104 million Americans had annual noise levels greater than 70 dBA (equivalent to a continuous average exposure level over 24 hours) in 2013 and were at risk of noise-induced hearing loss [10]. The researchers opined that tens of million more Americans might be at risk of heart disease and other noise-induced adverse health effects.

> University of Michigan researchers estimated that 104 million Americans had annual noise levels greater than 70 dBA in 2013 and were at risk of noise-induced hearing loss.

The WHO has analyzed the health and economic impacts of noise pollution, finding, "Exposure to prolonged or excessive noise has been shown to cause a range of health problems ranging from stress, poor concentration, productivity losses in the workplace, and communication difficulties and fatigue from lack of sleep, to more serious issues such as cardiovascular disease, cognitive impairment, tinnitus and hearing loss… The studies analyzed environmental noise from planes, trains and vehicles, as well as other city sources, and then looked at links to health conditions such as cardiovascular disease, sleep disturbance, tinnitus, cognitive impairment in children, and annoyance" [11].

The WHO team used the health information to calculate the disability-adjusted life-years or DALYs – basically the healthy years of life – lost to "unwanted" human-induced noise dissonance. The agency found that at least 1 million healthy years of life are lost each year in Europe alone due to noise pollution, a figure that did not include noise from industrial workplaces. Based on their health and economic analyses, the WHO ranked traffic noise second among environmental threats to public health following air pollution. The agency also noted that while other forms of pollution are decreasing, noise pollution is increasing [11].

The effects of noise pollution on the human cardiovascular system are a matter of great significance, given the knowledge that heart disease is the most prevalent cause of death globally. And as will be subsequently described, the consequences of noise exposure to the prevalence of hypertension have consequential economic impacts. The effects of environmental noise on the cardiovascular system were the subject of an extensive review in 2018 by researchers at the Johannes Gutenberg University, Germany [12]. Excerpts from this comprehensive review follow.

"The association between traffic noise and coronary heart disease has been studied extensively. The most recent meta-analysis from 2015 included studies on road traffic and aircraft noise, and found a 6% significant increase in risk for every 10 dBA increase in traffic noise (day-evening-night equivalent noise level A-weighted), starting as low as 50 dBA. In 2011, a large cohort study found a 14% significant higher risk of incident stroke for every 10 dBA increase in road traffic noise (day-evening-night equivalent noise level A-weighted). This result was later confirmed by two large studies from London on aircraft and road traffic noise. Both daytime and nighttime aircraft noise greater than 55 dBA significantly increased risk for stroke hospitalization with, respectively, 8% and 29%, when compared with levels less than 50 dBA in a population of 3.6 million people living around London's Heathrow airport, which suggests that nighttime noise may be especially hazardous. Similarly, based on 8.6 million residents of London, road traffic noise was found to significantly increase risk for stroke hospitalization" [12].

> Studies on road traffic and aircraft noise found a 6% significant increase in risk in coronary heart disease for every 10 dBA increase in traffic noise.

"Traffic noise may also result in heart failure and atrial fibrillation. Two large population studies of 0.75 and 4.41 million persons, respectively, found both road traffic and aircraft noise to significantly increase the risk for heart failure, ranging from 2% to 7% increase in risk per 10-dBA rise, depending on study and type of exposure.

The most comprehensively studied risk factor for CVD in a noise context is hypertension. In 2012, a meta-analysis of 24 studies showed that a 5-dBA rise in road traffic noise was associated with a significant odds ratio for prevalent hypertension of 1.034. The meta-analysis was based on cross-sectional studies, which limits the interpretation regarding causality. However, studies on incident hypertension are emerging, largely showing that traffic noise is associated with hypertension, and thereby supporting the cross-sectional findings. Also, stronger associations with hypertension have been observed for indoor noise as compared with outdoor noise.

In further support, a large population study found aircraft, railway, and road traffic noise to be associated with hypertensive heart disease.

In conclusion, more and more large studies of high quality find that traffic noise is associated with coronary heart disease and stroke, as well as with major risk factors for CVD, most importantly hypertension and metabolic disease" [12].

An interesting study considered the potential cardiovascular effects of joint exposure to air pollution and noise pollution. Researchers at the Swiss Tropical and Public Health Institute, Basel, Switzerland, looked at the combined effects of air pollution and transportation noise for heart attack mortality by considering all deaths that occurred in Switzerland between 2000 and 2008. Analyses that only included fine particulates ($PM_{2.5}$) suggested that the risk for a heart attack rises by 5.2% per 10 µg/m^3 increase in the long-term concentration at home. Studies that also accounted for road, railway, and aircraft noise revealed that the risk for a heart attack attributable to fine particulates in fact increases considerably less; 1.9% per 10 µg/m^3 increase. These findings indicate that the negative effects of air pollution might have been overestimated in studies that failed to concurrently consider noise exposure. The investigators reported "Our study showed that transportation noise increases the risk for a heart attack by 2.0 to 3.4% per 10 decibels increase in the average SPL at home. Strikingly, the effects of noise were independent from air pollution exposure" [13]. These important findings will require further elaboration in similar studies of joint exposure to air and noise pollution.

Additionally, noise in workplaces has long been recognized as a hazard to workers' hearing and other adverse health effects. The Occupational Safety and Health Administration (OSHA) estimates that 22 million U.S. workers are exposed to potentially damaging noise at work each year, causing an estimated $242 million spent annually on workers' compensation for hearing loss disability [14]. The OSHA sets legal limits on noise exposure in the workplace. These limits are based on a worker's time-weighted average over an 8-hour day. The OSHA's permissible exposure limit (PEL) is 90 dBA for all workers for an 8-hour day. The OSHA standard uses a 5 dBA exchange rate. This means that when the noise level is increased by 5 dBA, the amount of time a person can be exposed to a certain noise level to receive the same dose is cut in half [15].

> The OSHA estimates that 22 million U.S. workers are exposed to potentially damaging noise at work each year.

The OSHA's noise standard (29 CFR 1910.95) requires employers to have a hearing conservation program in place if workers are exposed to a time-weighted average (TWA) noise level of 85 dB (dBA) or higher over an 8-hour work shift. The OSHA's authorities to set noise level standards derive from the Occupational Safety and Health Act of 1970.

5.2.4 Effects of Noise Pollution on Ecological Health

The effects of noise pollution on ecological systems have been investigated most often in studies of bird populations. In one investigation reported in 2009 by University of Colorado ecologists of nesting behavior of bird populations, investigators reported the following: "The change in the avian community is in line with earlier studies implicating the negative influence of noise on birds; however, we provide the first evidence of this trend while simultaneously controlling for confounding stimuli and potential noise-caused detection biases. This is the strongest evidence to date that noise negatively influences bird populations and communities, and acoustic masking may be a dominant mechanism precluding many birds from breeding in noisy habitats. Because noise also indirectly facilitates reproductive success, species intolerant of noise may suffer from not only exclusion from noisy habitats that might be otherwise suitable but also higher rates of nest predation relative to species inhabiting noisy areas" [16]. It is interesting to note that the investigators observed that noise had indirectly facilitated birds' reproductive success, likely owing to noise having provided a stimulus for keeping prey birds and other animals away.

> Noise negatively influences bird populations and communities, and acoustic masking may be a dominant mechanism precluding many birds from breeding in noisy habitats.

The same ecologists further commented: "Using observations, vegetation surveys and pollen transfer and seed removal experiments, we found that effects of noise pollution can reverberate through communities by disrupting or enhancing these ecological services. Specifically, noise pollution indirectly increased artificial flower pollination by hummingbirds, but altered the community of animals that prey upon and disperse Pinus edulis seeds, potentially explaining reduced P. edulis seedling recruitment in noisy areas. Despite evidence that some ecological services, such as pollination, might benefit indirectly owing to noise, declines in seedling recruitment for key-dominant species such as P. edulis may have dramatic long-term effects on ecosystem structure and diversity" [17].

In addition to studies on the effects of noise pollution on avian species, other research has reported adverse ecological effects of noise on other species, as summarized by the Australian Academy of Science. They noted that a study published in 2010 found that traffic noise decreased the foraging efficiency of an acoustic predator, the greater mouse-eared bat. Successful foraging bouts decreased and search times increased dramatically with proximity to the highway. As the animals being hunted by the bats are themselves predators, the study noted that 'the noise impact on the bats' foraging performance will have complex effects on the food web and ultimately on the ecosystem stability'. Noise pollution could potentially interfere with other acoustic predators, such as owls, in a similar fashion [18].

The Australian academy further observed that in another species impacted by noise pollution, researchers had noted a different outcome for the black-tailed prairie dog in free-ranging colonies in Colorado. The dogs were exposed to simulated traffic noise from a series of speakers, similar to that which would be heard if a real highway were 100 meters from the colony. The dogs did not leave their homes, but the researchers did note a distinct change in their behavior during times of traffic noise broadcast. Three of the most significant changes in the dogs' social behavior were the following: The number of prairie dogs above ground declined by 21%. The proportion of individuals foraging declined by 18%. Vigilance (looking out for predators) increased by 48%. The researchers concluded that road noise can alter key survival behaviors and that "these findings highlight that the presence of animals in a location is no guarantee of population and ecological integrity. The investigations also commented that while noise pollution might not necessarily drive animals away from a site, it may alter their established behaviors and be having a less-obvious negative effect on their physical well-being" [18].

Interestingly, noise made by humans is increasingly disrupting life below the surface, with many marine animals being affected. Rising levels of intense underwater sound are produced by a range of sources – shipping traffic, industrial noise from oil and gas exploration, seismic surveys, military sonar, and others. This cacophony can present a range of problems for marine species, many of which rely on hearing as their primary sense for mating, hunting, and communicating. Between 1950 and 1975, ambient noise over low frequencies at one location in the Pacific Ocean increased by about 10 dB, representing a 10-fold increase [18].

> Noise made by humans is increasingly disrupting life below the surface, with many marine animals being affected.

In 2000, whales of four different species stranded themselves on beaches in the Bahamas after U.S. sonar-system testing in the area. A government report after the incident established that the mid-frequency sonar was the cause of the strandings. Of 40 recorded instances of mass strandings of Cuvier's beaked whales since the 1960s, 28 have occurred at the same time and place as naval maneuvers or the use of active sonar. For example, noise has been shown to reduce humpback whale communication, with less "song" during periods of noise, even when the origin of the noise is 200 kilometers away. Both right and blue whales have been found to increase the level of vocalizations when exposed to sound sources in their vocal range. In effect, they need to "shout" to allow themselves to be heard [18].

> In 2000, whales of four different species stranded themselves on beaches in the Bahamas after U.S. sonar-system testing in the area.

Perspective: The material cited in this section indicates that noise pollution is significantly impacting ecological systems, with birds, terrestrial animals, and marine life all adversely affected in ways disruptive to their natural life style and life cycle. Control of noise sources such as marine noise-making sources is the route to reducing the ecological effects of environmental noise pollution.

5.2.5 ECONOMIC IMPACTS OF NOISE POLLUTION

The human health consequences of exposure to noise pollution are both individually personal as well as societal impactful. Adverse health effects in general can bring disability, pain, and social disruption, resulting in economic impacts in the form of medications, healthcare, and emergency services. As to environmental noise pollution, the adverse health impacts of exposure to excessive levels of noise include hearing loss, stress, sleep disruption, annoyance, and cardiovascular disease. And as previously cited, estimates in 2013 suggested that more than 100 million Americans are exposed to unhealthy levels of noise. Given the pervasive nature and significant health effects of environmental noise pollution, the corresponding economic impacts were the subject of two major economic studies [19].

A study in 2014 by University of Michigan public health researchers developed a new approach to estimate the impact of environmental noise on the prevalence and cost of key components of hypertension and cardiovascular disease in the U.S. [19]. The researchers assert that by placing environmental noise in context with comparable environmental pollutants, this approach can inform public health law, planning, and policy. The effects of hypothetical national-scale changes in environmental noise levels on the prevalence and corresponding costs of hypertension and coronary heart disease were estimated, with the caveat that the national-level U.S. noise data exposure estimates were more than 30 years old in 2014. The analyses suggested that a 5-dB noise reduction scenario would reduce the prevalence of hypertension by 1.4% and coronary heart disease by 1.8%. The annual economic benefit was estimated at $3.9 billion.

> University of Michigan researchers estimated a 5-dB noise reduction scenario would reduce the prevalence of hypertension by 1.4% and coronary heart disease by 1.8%.

In a separate study conducted in the UK, researchers focused on the economic impacts of hypertension (HT) associated with environmental noise exposure [20]. Their study aims were to identify key HT-related health outcomes and to quantify and monetize the impact on health outcomes attributable to environmental noise-related HT. A reiterative literature review identified key HT-related health outcomes and their quantitative links with HT. The health impact of increases in environmental noise greater than recommended daytime noise levels (55 dBA) was quantified in terms of quality-adjusted life years and then monetized. A case study evaluated the cost of environmental noise, using published data on health risks and the number of people exposed to various bands of environmental noise levels in the United Kingdom (UK).

The researchers commented, "Three health outcomes were selected based on the strength of evidence linking them with HT and their current impact on society: Acute myocardial infarction (AMI), stroke, and dementia. In the UK population, an additional 542 cases of HT-related AMI, 788 cases of stroke, and 1,169 cases of dementia were expected per year due to daytime noise levels ≥55 dBA. The cost of these additional cases was valued at around £1.09 billion, with dementia accounting for 44%. The methodology is dependent on the availability and quality of published data and the resulting valuations reflect these limitations. The estimated intangible cost provides an insight into the scale of the health impacts and conversely the benefits that the implementation of policies to manage environmental noise may confer" [20].

5.3 POLICIES TO CONTROL NOISE POLLUTION

Policies to control the sources of noise and means to regulate noise pollution began to appear as legislative actions in the U.S. and elsewhere, commencing in the 1970s. In the U.S., federal government policies appeared as enacted statutes for a relatively brief period, succeeded by Congress's transferring noise control authority to the U.S. states and territories. As will be described, some regional and national governments elsewhere have also adopted policies to control noise pollution, often delegating the responsibility to lower levels of government such as provinces and municipalities.

5.3.1 U.S. Federal Policies to Control Noise Pollution

The U.S. federal government was active in the 1970s in enacting legislation purposed for controlling environmental sources of noise. The two principal acts were the Noise Control Act of 1972 and the Quiet Communities Act of 1978, although in 1982 Congress shifted federal noise control policy by transferring the primary responsibility of regulating noise to U.S. states and local governments. However, both acts were never rescinded by Congress but currently remain as enacted legislation [21]. But both acts are essentially unfunded by Congress, which means executive branch agencies are unable to implement any programs related to the two noise statutes. Commencing in 1972 the EPA coordinated all federal noise control activities through its Office of Noise Abatement and Control. Later, the EPA ended the office's funding in 1982. However, federal regulations on environmental noise control continue, as described in a subsequent section, following a description of federal noise control statutes.

5.3.1.1 Aircraft Noise Abatement Act, 1968

The Aircraft Noise Abatement Act of 1968 requires the Federal Aviation Administration (FAA) to develop and enforce safe standards for aircraft noise [22]. In developing these standards, the FAA generally follows the noise restrictions established by the International Civil Aviation Organization (ICAO). Federal noise regulations define aircraft according to four classes: Stage 1, Stage 2, Stage 3, and Stage 4. Stage 1 aircraft are the loudest, and Stage 4 are the quietest. All Stage 1 aircraft have been phased out of commercial operation, and all unmodified Stage 2 aircraft over 75,000 pounds were phased out by December 31, 1999, as required by the Airport Noise and Capacity Act of 1990.

Stage 3 aircraft must meet separate standards for runway takeoffs, landings, and sidelines, ranging from 89 to 106 dBA depending on the aircraft's weight and its number of engines. Stage 4 standards are stricter and require a further reduction of 10 dBA overall relative to Stage 3 standards. The Stage 4 standards are relatively new and are based on standards that the ICAO adopted in June 2001 (referred to as "Chapter 4" in ICAO parlance). The FAA proposed the Stage 4 standards in December 6, 2003, and finalized them in July 7, 2005, adopting the ICAO standards by reference.

The Stage 4 standards apply to newly manufactured airplanes for which a new design is submitted for airworthiness certification on or after January 1, 2006. As the majority of aircraft designed in recent years are already quite enough to attain the Stage 4 standards, some have commented that the impact of the stricter standards on most aircraft manufacturers may be less significant than otherwise. At this time, existing Stage 3 aircraft also will be allowed to continue operation.

5.3.1.2 Noise Control Act, 1972

This legislation was signed into law by U.S. President Richard Nixon (R-CA) on October 28, 1972. The act gave the EPA the primary role for controlling environmental noise. It had been submitted to Congress in 1971 as part of the Nixon Administration's environmental "package." The Act established "a national policy to promote an environment for all Americans free from noise that jeopardizes their health and welfare" [23]. The Act also served to: (1) establish a means for effective coordination of Federal research and activities in noise control; (2) authorized the establishment of Federal noise emission standards for products distributed in commerce;

> The two principal federal acts for noise control are the Noise Control Act of 1972 and the Quiet Communities Act of 1978. Both are unfunded by Congress.

and (3) provided information to the public respecting the noise emission and noise reduction characteristics of such products [24]. The Act also directed the EPA to coordinate the programs of all U.S. federal agencies relating to noise research and noise control. In response to these responsibilities under the Act, the EPA established its Office of Noise Abatement and Control. Although the EPA's noise control authorities withered post-1982, it is worthwhile for historical purposes to cite here the agency's early accomplishments under the Noise Control Act of 1972 (NC Act) [25].

Terms of the NC Act gave the EPA the responsibility for coordinating all Federal programs in noise research and control. The EPA was to be consulted by other Federal agencies prior to publishing new regulations on noise. The EPA also had the authority to set standards for any product or class of products that have been identified as a major source of noise. They would be based on criteria that the EPA was required to develop before proposing any standards. Categories of equipment covered by the legislation included construction, transportation (including recreational vehicles), motors or engines, and electrical and electronic.

An early action by the EPA under its NC Act authorities was to establish noise levels useful to provide a basis for state and local governments' judgments in setting noise standards [26,27]. The EPA's review of published literature led the agency in April 1974 to state the following: "The document identifies a 24-hour

> EPA declared a 24-hour exposure level of 70 dB as the level of environmental noise which will prevent any measurable hearing loss over a lifetime.

exposure level of 70 decibels as the level of environmental noise which will prevent any measurable hearing loss over a lifetime. Likewise, levels of 55 decibels outdoors and 45 decibels indoors are identified as preventing activity interference and annoyance. These levels of noise are considered those which will permit spoken conversation and other activities such as sleeping, working and recreation, which are part of the daily human condition" [25].

The levels are not single event or "peak" levels. Instead, they represent averages of acoustic energy over periods of time such as 8 or 24 hours, and over long periods of time such as years. For example, occasional higher noise levels would be consistent with a 24-hour energy average of 70 dB, so long as a sufficient amount of relative quiet is experienced for the remaining period of time.

Noise levels for various areas are identified according to the use of the area. Levels of 45 decibels are associated with indoor residential areas, hospitals and schools, whereas 55 decibels is identified for certain outdoor areas where human activity takes place. The level of 70 decibels is identified for all areas in order to prevent hearing loss" [26,27].

Under the NC Act, the Federal Aviation Agency (FAA) retained authority to set aircraft noise regulations, but the EPA was required to recommend to FAA any regulations it deemed necessary. To fulfill this responsibility, the EPA, working with representatives of the Departments of Commerce (including the National Bureau of Standards), Defense, Transportation, and Housing and Urban Development, was to make a major study of existing airport flight and operational control regulations. The results of this study were to be contained in a special report to the Congress and in proposed regulations to the FAA.

In the consumer area, the EPA had authority to label products as to their noise-generating characteristics or their effectiveness in reducing noise. The standard-setting and labeling authorities applied to both domestic and imported products. Manufacturers or importers of non-conforming or mislabeled products were subject to fines of up to $25,000 per day for each violation and to imprisonment of up to 1 year. Manufacturers were required to issue warrants that their regulated products are in compliance at the time of sale. They were also required to maintain records and provide information, including supplying products coming off the assembly line for testing, to the EPA if requested. The EPA was also authorized under the NC Act to conduct research, technical assistance, and public information and certify low-noise emission products for purchase by the Federal Government. The authorities given to the EPA under the NC Act ended in 1982 when Congress shifted noise control responsibilities to the U.S. states and territories.

> The authorities given to the EPA under the NC Act ended in 1982 when Congress shifted noise control responsibilities to the U.S. states and territories.

5.3.1.3 Quiet Communities Act, 1978

The Quiet Communities Act of 1978 (QC Act) amended the Noise Control Act of 1972 to require the Administrator of the EPA to: (1) disseminate educational materials on the public health effect of noise and the most effective means of noise control; (2) conduct or finance specified research projects on noise control; (3) administer a nationwide Quiet Communities Program designed to assist local governments in controlling levels; and (4) provide technical assistance to U.S state and local governments in implementing noise control programs [28]. The QC Act also amended the Federal Aviation Act to establish a time limit of 90 days following a public hearing on aircraft noise control regulations within which the FAA must publish such regulations as prescribed by the EPA or give notice in the *Federal Register* that it does not intend to publish the regulations as prescribed.

The QC Act further: (1) Prescribed a civil penalty for violators of the Noise Control Act; (2) Permitted a state or local official to petition the EPA Administrator to adopt more stringent regulations for noise control than originally prescribed, if needed to protect public health or welfare; (3) Authorized appropriations to carry out the purposes of this Act; (4) Made technical amendments to the Solid Waste Disposal Act of 1976; (5) Directed the Secretary of Transportation and the Administrator of the EPA to jointly study the aircraft noise effects from an airport on communities located in a state other than the state in which such airport is located; (6) Set forth criteria for the selection of the airport to be studied; and (7) Required that a report on such study be submitted to Congress [28].

5.3.1.4 Quiet Communities Act, 2015, 2016, 2017

One member of Congress has promoted legislation to revive the Quiet Communities Act [29]. Her proposed legislation would reestablish EPA's Office of Noise Abatement and Control and would specify other purposes for federal control of noise pollution. The Congresswoman's attempts have been unsuccessful to date, owing to lack of support from other members of Congress. Moreover, no companion bill in the U.S. Senate has been developed in support of re-establishment of a federal noise control program. As a consequence, control of noise pollution resides as a matter of policy with those U.S. states, territories, and municipalities that have developed such policies, as discussed in the subsequent subsection.

Perspective: The Noise Control Act of 1972 and the Quiet Communities Act of 1978 were the two federal policies intended for the purpose of controlling noise pollution sources in the general environment. Both statutes authorized the EPA and other federal agencies to develop and implement various actions to control noise from exceeding levels known to adversely affect human health. And although both acts remain on the roles of authorized federal statutes, neither act has been funded by Congress, nor actively sought for reactivation by any presidential administration since 1982, the year when Congress and the administration shifted issues of noise control from federal policies to U.S. state, territorial, and municipal governments. And although noise pollution is now recognized as a hazard to human and ecological health at noise levels exceeding designated dB levels, it is unlikely that a federal re-entry into control of noise pollution will occur in the near term.

5.3.1.5 What Sources of Noise are Subject to Federal Regulation?

The Noise Control Act of 1972 and several other federal laws required the federal government to set and enforce uniform noise standards for aircraft and airports, interstate motor carriers and railroads, workplace activities, medium and heavy-duty trucks, motorcycles and mopeds, portable air compressors, federal highway projects, and federal housing projects. The Noise Control Act also required federal agencies to comply with all federal, state, and local noise requirements. Most federal noise standards focus on preventing hearing loss by limiting exposure to sounds of 90 dBA and greater. However, some are stricter and focus on limiting exposure to quieter levels that are annoying to most individuals and can diminish one's quality of life. Federal noise standards and the agencies that set and enforce them are discussed in the following narratives [30].

> Most federal noise standards focus on preventing hearing loss by limiting exposure to sounds of 90 dBA and greater.

Aircraft and airports: As previously noted, the Aircraft Noise Abatement Act of 1968 requires the Federal Aviation Administration (FAA) to develop and enforce safe standards for aircraft noise. In developing these standards, the FAA generally follows the noise restrictions established by the International Civil Aviation Organization (ICAO). Federal noise regulations define aircraft according to three classes: Stage 1, Stage 2, and Stage 3. Stage 1 aircraft are the loudest, and Stage 3 are the quietest. All Stage 1 aircraft have been phased out of commercial operation in the U.S., and all unmodified Stage 2 aircraft of more than 75,000 pounds were phased out by December 31, 1999, as required by the Airport Noise and Capacity Act of 1990. Stage 3 aircraft must meet separate standards for runway takeoffs, landings, and sidelines, ranging from 89 to 106 dBA depending on an aircraft's weight and its number of engines. The ICAO has adopted stricter Stage 4 aircraft noise standards, which are quieter by 10 dBA than the current Stage 3 standards. However, the Stage 4 standards must go through the federal rule-making process before they could be applied in the U.S., and the FAA has not proposed such standards to date.

Interstate motor carriers: The Noise Control Act of 1972 required the EPA to develop noise standards for motor carriers engaged in interstate commerce, and it authorized the Federal Highway Administration to enforce the standards. All commercial vehicles more than 10,000 pounds are subject to standards for highway travel and stationary operation, but the standards do not apply to sounds from horns or sirens when operated as warning devices for safety purposes, e.g., emergency responding equipment. For highway travel, the standards range from 81 to 93 dBA, depending on the speed of the vehicle and the distance from which the sound is measured. The standards for stationary operation are similar and range from 83 to 91 dBA, depending on the distance from the vehicle. The standards apply at any time or condition of highway grade, vehicle load, acceleration, or deceleration.

Interstate railroads: The Noise Control Act of 1972 also required the EPA to establish noise standards for trains and railway stations engaged in interstate commerce, and it authorized the Federal Railroad Administration to enforce them. The standards do not apply to sounds from horns, whistles, or bells, when operated as warning devices for safety purposes. There are separate standards for locomotives, railway cars, and railway station activities such as car coupling. For locomotives built before 1980, noise is limited to 73 dBA in stationary operation and at idle speeds and is limited to 96 dBA at cruising speeds. The standards for locomotives built after 1979 are stricter, and limit noise in stationary operation and at idle speeds to 70 dBA and at cruising speeds to 90 dBA. Noise from railway cars must not exceed 88 dBA at speeds of 45 miles per hour (mph) or less, and must not surpass 93 dBA at speeds greater than 45 mph. Noise from car coupling activities at railway stations is limited to 92 dBA.

> Noise from railway cars must not exceed 88 dBA at speeds of 45 miles per hour (mph) or less, and must not surpass 93 dBA at speeds greater than 45 mph.

Workplace activities: The Occupational Safety and Health Act of 1970 required the Occupational Safety and Health Administration (OSHA) to develop and enforce safety and health standards for workplace activities. To protect workers, OSHA established standards that specify the duration of time that employees can safely be exposed to specific sound levels. At a minimum, constant noise exposure must not exceed 90 dBA averaged over 8 hours. The highest sound level to which workers can constantly be exposed is 115 dBA, and exposure to this level must not exceed 15 minutes within an 8-hour period. The standards limit instantaneous exposure, such as impact noise, to 140 dBA. If noise levels

> OSHA: The highest sound level to which workers can constantly be exposed is 115 dBA, and exposure to this level must not exceed 15 minutes within an 8-hour period.

exceed these standards, employers are required by OSHA to provide hearing protection equipment that will reduce sound levels to acceptable limits.

Other regulated sources of noise: The Noise Control Act of 1972 directed the EPA to set and enforce noise standards for transportation, construction, and electrical equipment, and motors or engines. Under this authority, the EPA established standards for motorcycles and mopeds, medium and heavy-duty trucks more than 10,000 pounds, and portable air compressors. The standards for motorcycles only apply to those manufactured after 1982 and range from 80 to 86 dBA, depending on the model year and whether the motorcycle is designed for street or off-road use. Noise from mopeds is limited to 70 dBA. The standards for trucks more than 10,000 pounds only apply to those manufactured after 1978 and range from 80 to 83 dBA depending on the model year. These standards are separate from those for interstate motor carriers. Noise from portable air compressors is limited to 76 dBA. The Noise Control Act also authorized the EPA to require labels for products that reduce noise. Under this authority, the EPA established Noise Reduction Ratings for hearing protection devices that require manufacturers to identify the level of sound from which the device protects the user.

There also are noise standards for federal highway projects and federal housing projects. The Federal-Aid Highway Act of 1970 required the Federal Highway Administration to develop standards for highway noise levels that are compatible with different land uses. The law prohibits the approval of federal funding for highway projects that do not incorporate measures to meet these standards, which range from 52 to 75 dBA depending on land use. Under general authorities provided by the Housing and Urban Development Act of 1968, there also are standards for federal housing projects located in noise-exposed areas. The standards limit interior noise to a daily average of 65 dBA. More specific details about the key federal noise control acts follow.

Perspective: Although the two principal federal statutes on control of noise pollution, the Noise Control Act and the Quiet Communities Act, are legislatively dormant, other federal statutes or residual actions from the Noise Control Act service the act of noise control of specific sources of noise. Of particular note are federal standards for control of noise from aircraft and airports, interstate motor carriers, and interstate railroads, each of which can be a significant source of noise in a community.

5.3.2 International Government Policies to Control Noise Pollution

Noise pollution increases with such sources as industrialization, urbanization, population growth, technology development and application, and increased entertainment sources. These kinds of sources of noise pollution have stimulated many countries to adopt policies to control the sources and the effects on human and ecological health.

5.4 NOISE POLLUTION HAZARD INTERVENTIONS

Noise pollution has become a modern-day environmental physical hazard to human and ecological health, dependent on the characteristics of the sound energy and methods in place to control the noise sources. A number of hazard interventions have been implemented for purpose of controlling noise pollution. The following actions are interventions that an individual could consider for controlling noise pollution.

- Purchase and operate equipment in workplaces, private residences, and community settings that are designed for sound control, e.g., office equipment and home appliances.
- Maintain all sound-producing equipment, e.g., chain saws, in good condition so as to avoid excess noise production.
- Support local policies that provide individuals with access to obtaining relief from noise polluting sources, e.g., music concerts that produce sound levels in excess of those permitted by local ordinances.
- Advocate for state and federal policies that are purposed for noise pollution control.
- Work with local public health departments for purpose of educating the public about the adverse health effects of noise pollution that exceeds permitted levels.
- Advocate that local education sources include classroom material pertaining to the importance of controlling noise pollution sources.
- Take personal responsibility for keeping sound levels to an acceptable level as determined by local health authorities.
- Choose to reside in areas with non-hazardous noise pollution levels.

5.5 LIGHT POLLUTION

Humans and their antecedents have always experienced a light-dark cycle each day. Light was provided by the forces of nature. The Sun gave us periods of light between the hours of sunrise and sunset. Human activities during periods of light focused on activities such as hunting food, seeking shelter, and practicing tribal social affairs. The dark periods provided time for rest and social activities such as the development of communication skills. Over time, humans sought to provide light during periods of darkness by creating artificial sources of light such as building fires and creating burning torches. Perhaps these new sources of light were bothersome to some of our primordial ancestors, creating sleep disturbances, for example. Maybe a bone or stone was thrown for purpose of eliminating a source of perceived light pollution.

Setting aside the experience of our primordial ancestors, one might speculate that the evolution of light pollution was a consequence of Thomas A. Edison's invention and patenting in 1879 of the first commercially successful electric light bulb. Edison's invention was an artificial source of light. It soon changed humans' light-dark experience. In 1879 Edison's incandescent light bulbs first illuminated a New York street, an event that began the modern era of electric lighting [31].

> In 1879 Edison's incandescent light bulbs first illuminated a New York street, an event that began the modern era of electric lighting.

As a brief digression, it is interesting to reflect on the growth of municipal lighting from one street in New York City to lighting whole cities such as Los Angeles, an illumination that can be observed 200 miles above Earth. Edison's accomplishment

Noise and Light Pollution

was transformative. Lighted streets reduced crime rates, gave business owners extended hours of commerce, provided sports complexes with evening and nighttime events, and residential electric lights that replaced candles, oil lamps, and lanterns as sources of home light. The passage of time has seen technological advances in electric lights from Edison's incandescent bulbs to light-emitting diodes (LEDs), which are now considered the future of lighting due to a lower energy requirement to run, a lower monthly price tag, and a longer life than traditional incandescent light bulbs. One source estimates human illumination of the planet is growing in range and intensity by about 2% a year [32].

5.5.1 Definition

According to the Rensselaer Polytechnic Institute, "Light pollution is an unwanted consequence of outdoor lighting and includes such effects as sky glow, light trespass, and glare.

- Sky glow is a brightening of the sky caused by both natural and human-made factors. The key factor of sky glow that contributes to light pollution is outdoor lighting. For much of Earth's history, our remarkable universe of stars has been visible in the darkness of the night sky. But increasing urbanization, combined with the excessive and inefficient use of light, has created a kind of pollution that obscures the stars from view and leads to numerous other disturbances.
- Light trespass is light being cast where it is not wanted or needed, such as light from a streetlight or a floodlight that illuminates a neighbor's bedroom at night making it difficult to sleep.
- Glare can be thought of as objectionable brightness. It can be disabling or discomforting. There are several kinds of glare, the worst of which is disability glare because it causes a loss of visibility from stray light being scattered within the eye. Discomfort glare is the sensation of annoyance or even pain induced by overly bright sources... Discomfort and even disability glare can also be caused by streetlights, parking lot lights, floodlights, signs, sports field lighting, and decorative and landscape lights" [33].

5.5.2 Sources of Light Pollution

"**Urban sky glow**: the brightening of the night sky over inhabited areas. It is the "glow" effect that can be seen over distant populated areas. This light that escapes up into the sky is created by the combination of all light reflected off of what is being illuminated, from all of the badly directed light in that area, and from that light that is scattered (redirected) by the atmosphere itself from reaching the ground." Illustrated in Figure 5.2 is the light pollution typical of a modern city.

> Light trespass is light falling where it is not intended, wanted, or needed.

Light trespass: light falling where it is not intended, wanted, or needed. Street lighting, for example, should light streets and sidewalks, not shine into peoples'

FIGURE 5.2 A city's light pollution at night. (2021. NASA, 10/9/13. Night lights can disrupt wildlife, https://earthobservatory.nasa.gov/images/145767/night-lights-can-disrupt-wildlife.)

bedroom windows or illuminate rooftops or tree branches. Also known as spill light, light trespass occurs whenever light shines beyond the intended target and onto adjacent properties.

Glare: the sensation produced by luminance within the visual field that is sufficiently greater than the luminance to which the eyes are adapted. It creates an unnerving, oppressive, annoying, discomforting feeling that can cause a loss in our visual performance, visibility and can be dangerous. High levels of glare can decrease visibility for the elderly, drivers of motor vehicles, and astronomers. It is easily recognized by when a viewer's pupils will close down in its presence. This makes dimmer objects harder to see, and that increases the danger.

Uplight: wasted light, pure and simple. Light that goes directly up into the night sky is "lost in space" and serves no useful purpose (though the most often used for self-centered vain purposes). Uplight is the bane of astronomers and the occasional stargazer because atmospheric scattering artificially brightens the night sky, making distant celestial light sources difficult or impossible to see. Uplight often results from light fixtures which also produce glare and light trespass" [33].

> High levels of glare can decrease visibility for the elderly, drivers of motor vehicles, and astronomers.

5.5.3 Effects of Light Pollution on Human Health

In 2016 the Council on Science and Public Health (CSPH) of the American Medical Association (AMA) comprehensively reviewed the published literature on the effects of light pollution on human and ecological health [34]. Concerning human health, the CSPH reported: "Much has been learned over the past decade about the potential adverse health effects of electric light exposure, particularly at night. The core concern is the

> The potential adverse health effect of electric light exposure, particularly at night, is disruption of circadian rhythmicity.

disruption of circadian rhythmicity. With waning ambient light, and in the absence of electric lighting, humans begin the transition to nighttime physiology at about dusk; melatonin blood concentrations rise, body temperature drops, sleepiness grows, and hunger abates, along with several other responses."

"A number of controlled laboratory studies have shown delays in the normal transition to nighttime physiology from evening exposure to tablet computer screens, backlit e-readers, and room light typical of residential settings. These effects are wavelength and intensity dependent, implicating bright, short wavelength (blue) electric light sources as disrupting transition. These effects are not seen with dimmer, longer wavelength light (as from wood fires or low wattage incandescent bulbs). In human studies, a short-term detriment in sleep quality has been observed after exposure to short wavelength light before bedtime. Although data are still emerging, some evidence supports a long-term increase in the risk for cancer, diabetes, cardiovascular disease and obesity from chronic sleep disruption or shiftwork and associated with exposure to brighter light sources in the evening or night" [34].

"Electric lights differ in terms of their circadian impact. Understanding the neuroscience of circadian light perception can help optimize the design of electric lighting to minimize circadian disruption and improve visual effectiveness. White LED streetlights are currently being marketed to cities and towns throughout the country in the name of energy efficiency and long-term cost savings, but such lights have a spectrum containing a strong spike at the wavelength that most effectively suppresses melatonin during the night. It is estimated that a "white" LED lamp is at least 5 times more powerful in influencing circadian physiology than a high-pressure sodium light based on melatonin suppression. Recent large surveys found that brighter residential nighttime lighting is associated with reduced sleep time, dissatisfaction with sleep quality, nighttime awakenings, excessive sleepiness, impaired daytime functioning, and obesity. Thus, white LED street lighting patterns also could contribute to the risk of chronic disease in the populations of cities in which they have been installed. Measurements at street level from white LED street lamps are needed to more accurately assess the potential circadian impact of evening/nighttime exposure to these lights" [34].

In a meta-analysis of 126 published papers on light pollution, University of Exeter scientists found that light pollution impacted hormone levels, breeding cycles, activity patterns, and vulnerability to predators across a broad range of species [32]. In all the animal species examined, they found reduced levels of melatonin – a hormone that regulates sleep cycles – as a result of artificial light at night. Behavioral patterns were also disturbed in both nocturnal and diurnal creatures. For instance, rodents, which mostly forage at night, were active for a shorter duration, while birds started singing and searching for worms earlier in the day.

> In a 2019 cohort study of 43,722 women, artificial light at night while sleeping was significantly associated with increased risk of weight gain and obesity.

The association of light pollution with obesity was further explored in an investigation by NIEHS epidemiologists [35]. In a 2019 cohort study of 43,722 women, artificial light at night (ALAN) while sleeping was significantly associated with increased risk of weight gain and obesity, especially in women who had a light or a television on in the room while sleeping. Associations did not appear to be explained by sleep duration and quality or other factors influenced by poor sleep. The women included in the analysis had no history of cancer or cardiovascular disease, were not shift workers, daytime sleepers, or were not pregnant at baseline. Data were analyzed from September 1, 2017, through December 31, 2018. Compared with no ALAN, sleeping with a television or a light on in the room was associated with gaining 5 kg or more, which was a BMI increase of 10% or more.

> NIEHS: Artificial light at night while sleeping was significantly associated with women's increased risk of weight gain and obesity.

In a different kind of study, Harvard Medical School researchers found an association between living in areas with high amounts of ambient nighttime light and slightly increased odds for breast cancer in younger women who smoke [36]. The researchers used nighttime satellite images and records of night shift work to help figure out the amount of nighttime light to which each woman might have been exposed. The investigators noted that earlier research had suggested that high levels of exposure to light at night disrupt the body's internal clock. In turn, that might lower levels of a hormone called melanin which, in turn, might boost the risk of breast cancer. To test their theory, the investigators tracked almost 110,000 U.S. women, followed as part of a long-term study of nurses from 1989 to 2013. The investigators found that breast cancer levels in premenopausal women who currently smoked or had smoked in the past grew by 14% if they were in the 20% deemed to have had the most exposure to outdoor light at night. Furthermore, as levels of outdoor nighttime light went up, so did the likelihood of breast cancer for this subgroup of women.

5.5.4 Effects of Light Pollution on Ecological Health

Concerning the effects of light pollution on ecological health, the Council on Science and Public Health (CSPH) reported, "The detrimental effects of inefficient lighting are not limited to humans; 60% of animals are nocturnal and are potentially adversely affected by exposure to nighttime electrical lighting.

> 60% of animals are nocturnal and are potentially adversely affected by exposure to nighttime electrical lighting.

Many birds navigate by the moon and star reflections at night; excessive nighttime lighting can lead to reflections on glass high-rise towers and other objects, leading to confusion, collisions, and death. Many insects need a dark environment to procreate, the most obvious example being lightning bugs that cannot 'see' each other when light pollution is pronounced. Other environmentally beneficial insects are attracted to blue-rich lighting, circling under them until they are exhausted and die.

Unshielded lighting on beach areas has led to a massive drop in turtle populations as hatchlings are disoriented by electrical light and sky glow, preventing them from reaching the water safely. Excessive outdoor lighting diverts the hatchlings inland to their demise. Even bridge lighting that is 'too blue' has been shown to inhibit upstream migration of certain fish species such as salmon returning to spawn. One such overly lit bridge in Washington State now is shut off during salmon spawning season.

Recognizing the detrimental effects of light pollution on nocturnal species, U.S. national parks have adopted best lighting practices and now require minimal and shielded lighting. Light pollution along the borders of national parks leads to detrimental effects on the local bio-environment. For example, the glow of Miami, FL extends throughout the Everglades National Park. Proper shielding and proper color temperature of the lighting installations can greatly minimize these types of harmful effects on our environment" [34].

An international agency, UNESCO (United Nations Educational, Scientific and Cultural Organization), characterized the adverse effects of light pollution on ecological systems as the following: "Light pollution, in particular, has been shown to have a widespread, negative impact on many different species. Scientific evidence for this impact in migratory birds, hatchling sea turtles, and insects is striking, because of the large-scale mortality that has occurred as a result of artificial night lighting. Light pollution can confound animal navigation (many species use the horizon and stars for orientation), alter competitive interactions, mutualisms, and reproduction behavior, change the natural predator-prey relationship and even affect animal physiology. Amphibians are well-studied in this sense, as well as a number of nocturnal or crepuscular mammals such as bats, some primates, many rodents and marsupials, which all suffer from what is now called "biological photopollution." Disturbing data on light pollution effects on flora and phytoplankton are also being obtained. This is because many plants time their development, growth and flowering behavior by measuring the seasonally changing length of the night, which is impossible when there is light pollution" [37].

The adverse effects of light pollution on insects was the subject of a comprehensive review of published literature by ecologists at Washington University in St. Louis who reported, "Artificial light at night can affect every aspect of insects' lives, the researchers said, from luring moths to their deaths around bulbs, to spotlighting insect prey for rats and toads, to obscuring the mating signals of fireflies" [38]. The researchers cited recent research in the UK that found greater losses of moths at light-polluted sites than dark ones. Further, vehicle headlights pose a deadly moving hazard, and this fatal attraction has been estimated to result in 100 billion insect deaths per summer in Germany according to the reviewers.

5.5.5 Effects of Light Pollution on the Night Sky

One of the most noticeable effects of light pollution is the increase in night sky brightness. The light wasted upward by the nighttime outdoor lighting lights up atmospheric particles and molecules producing a luminous background that obstructs the vision of the night sky [39]. Satellite surveys of night skies in Europe and North America have produced data that characterize the impact of light pollution, particularly lights in metropolitan areas, on the night sky. The *First World Atlas of Night Sky Brightness* shows the effects of light pollution on the night sky globally.

> One of the most noticeable effects of light pollution is the increase in night sky brightness.

The Italian investigators who developed the atlas found the following: "Based on radiance-calibrated high-resolution satellite data and on accurate modeling of light propagation in the atmosphere, it provides a nearly global picture of how mankind is proceeding to envelop itself in a luminous fog. Comparing the Atlas with the U.S. Department of Energy (DOE) population density data base, we determined the fraction of population who are living under a sky of given brightness.

> About one-tenth of the World population, more than 40% of the U.S. population, and one-sixth of the EU population no longer view the heavens because of the sky brightness.

About two-thirds of the World population and 99% of the population in the U.S. (excluding Alaska and Hawaii) and EU live in areas where the night sky is above the threshold set for polluted status. Assuming average eye functionality, about one-fifth of the World population, more than two-thirds of the U.S. population and more than one half of the European Union population have already lost naked eye visibility of the Milky Way. Finally, about one-tenth of the World population, more than 40% of the U.S. population, and one-sixth of the EU population no longer view the heavens with the eye adapted to night vision because of the sky brightness" [40].

An increase in night sky brightness causes problems for amateur and professional astronomy, cultural disruptions in societies that function with night sky observations, and aesthetic consequences, e.g., loss of vision of the Milky Way. For these and other reasons, as will be subsequently discussed, policies have been proposed to promote dark skies uncontaminated by light pollution.

Perspective: The findings from the CSPH led the American Medical Association (AMA) to articulate the following recommendation: "Recognizing the detrimental effects of poorly-designed, high-intensity LED lighting, the AMA encourages communities to minimize and control blue-rich environmental lighting by using the lowest emission of blue light possible to reduce glare. The AMA recommends an intensity threshold for optimal LED lighting that minimizes blue-rich light. The AMA also recommends all LED lighting should be properly shielded to minimize glare and detrimental human health and environmental effects, and consideration should be given to utilize the ability of LED lighting to be dimmed for off-peak time periods" [41].

5.6 POLICIES TO CONTROL LIGHT POLLUTION

The American Medical Association accepted a proposal for an anti-light pollution policy and passed it unanimously. Resolution 516 asked the American Medical Association to: (1) advocate that all future outdoor lighting be of energy-efficient designs to reduce waste of energy and production of greenhouse gasses that result from this wasted energy use; (2) support light pollution reduction efforts and glare reduction efforts at both the national and state levels; and (3) AMA support efforts to ensure all future streetlights be of a fully shielded design or similar non-glare design to improve the safety of our roadways for all, but especially vision impaired and older drivers [41].

5.6.1 U.S. FEDERAL POLICIES

There are no U.S. federal government policy statements or statutes that are specific to the control of light pollution, as was the case with noise pollution. The U.S. unstated policy is to concede considerations of light pollution to the U.S. states and territories.

5.6.2 U.S. State and Local Policies

The National Conference of State Legislatures (NCSL) notes that "at least 18 states, the District of Columbia, and Puerto Rico have laws in place to reduce light pollution. The majority of states that have enacted so-called 'dark skies' legislation have done so to promote energy conservation, public safety, aesthetic interests, and astronomical research capabilities. Municipalities in a number of states have also been active on this issue, adopting light pollution regulations as part of their zoning codes" [42].

> At least 18 states, the District of Columbia, and Puerto Rico have laws in place to reduce light pollution.

Most state laws are limited to outdoor lighting fixtures installed on the grounds of a state building or facility or on a public roadway. The most common dark skies legislation requires the installation of shielded light fixtures that emit light only downward. Replacement of unshielded with fully shielded lighting units often allows for the use of a lower wattage bulb, resulting in energy savings. Other laws require the use of low-glare or low-wattage lighting, regulate the amount of time that certain lighting can be used, and the incorporation of Illuminating Engineering Society guidelines into state regulations.

Shown in Table 5.2 are illustrative light pollution control statutes from three U.S. states (Delaware, Florida, New Mexico), which were selected here as examples of different variations of light control policies [42]. It is interesting in a policy sense to compare the specifics of these three states' approaches to control of light pollution. The Delaware policy gives emphasis to energy conservation and minimization of the installation of light fixtures. The Florida code emphasizes ecological protection of marine turtles through the protection of nesting habitats from light pollution sources. The New Mexico policy makes reference to the state's Night Sky Protection Act and emphasizes the conservation of energy used for light sources. This state's statute also gives specific values for prohibiting light fixtures that exceed specific limits, e.g., 70 watts light sources.

5.7 LIGHT POLLUTION HAZARD INTERVENTIONS

Environmental responsibility requires energy efficiency and conservation. The International Dark-Sky Association has offered several methods to reduce light pollution in communities and private residences [43]. Their recommendations include the following:

- Installing quality outdoor lighting could cut energy use by 60–70%, save billions of dollars and cut carbon emissions.
- Outdoor lighting should be fully shielded and direct light down where it is needed, not into the sky.
- Fully shielded fixtures can provide the same level of illumination on the ground as unshielded ones, but with less energy and cost.
- Unnecessary indoor lighting – particularly in empty office buildings at night – should be turned off.

TABLE 5.2
Illustrative Examples of Three States' Statutory Policies on Control of Light Pollution [42]

State	Code	Statute
Delaware	Del. Code Ann. tit. 7, §§7101a et seq.	"Prohibits the use of state funds to install or replace a permanent outdoor lighting fixture unless (1) the fixture is designed to maximize energy conservation and minimize light pollution, (2) the fixture emits only as much light as necessary for the intended purpose, (3) a cutoff luminaire is used when the output is more than 1,800 lumens and (4) in the case of roadway lighting, the purpose of an additional light cannot be achieved by lowering the speed limit, installing reflective markers, etc. Exceptions may apply"
Florida	Fla. Stat. §161.163; Fla. Admin. Code §§62b-55.001 et seq.	"Contains a model lighting ordinance to guide local governments in developing policies that protect hatching marine turtles from the adverse effects of artificial lighting, provide overall improvement in nesting habitat degraded by light pollution and increase successful nesting activity and production of hatchlings"
New Mexico	N.M. Stat. Ann. §§74-12-1 et seq	"The Night Sky Protection Act regulates outdoor lighting fixtures to preserve the state's dark sky while promoting safety, conserving energy and protecting the environment for astronomy. Requires all outdoor lighting fixtures to be shielded, except incandescent fixtures of 150 watts or less or other sources of 70 watts or less. Prohibits outdoor recreational facilities from using lighting after 11:00 p.m. Provides for a fine of up to $25 for any person, firm or corporation in violation of the law. Exceptions may apply"

- New lighting technologies can help conserve energy.
- LEDs and compact fluorescents can help reduce energy use and protect the environment, but only warm-white bulbs should be used.
- Dimmers, motion sensors, and timers can help to reduce average illumination levels and save energy.

5.8 SUMMARY

As presented in this chapter, noise and light pollution are relatively recent arrivals in the physical environment and accordingly policy developments in place or in development have too been recent. Both sources of pollution have evolved as humans' numbers and their technologies and social arrangements, e.g., megacities, have evolved. And while nether loud noises and startling light displays are novel in human experience, perhaps commencing with claps of thunder and bolts of lightning, the global proliferation of noise and light sources is relatively novel. With the expansion of noise and light sources into daily life of persons in developed countries has come adverse human and ecological health effects.

As noted by the WHO, exposure to prolonged or excessive noise has been shown to cause a range of health problems ranging from stress, poor concentration, productivity losses in the workplace, and communication difficulties and fatigue from lack of sleep, to more serious issues such as cardiovascular disease, cognitive impairment, tinnitus, and hearing loss. And light pollution has been related by the AMA as a disruption of circadian rhythmicity, with some evidence supporting a long-term increase in the risk for cancer, diabetes, cardiovascular disease, and obesity from chronic sleep disruption or shiftwork and associated with exposure to brighter light sources in the evening or night. Both sources of pollution also have caused documented adverse effects on ecological systems, primarily disruptions of normal habitats and reproductive patterns in birds. Polices to control noise and light sources of pollution have been largely assigned to local levels of government, e.g., states and provinces and municipalities. Research is wanting as to the suite of potential adverse human, ecological, and economic effects of these two forms of pollution.

REFERENCES

1. Anonymous. 2018. Personal music players & hearing. *GreenFacts.* greenfacts@cogeneris.eu. (https://ec.europa.eu/health/scientific_committees/opinions_layman/en/hearing-loss-personal-music-player-mp3/l-3/11-health-effects-sound.htm)
2. Rouse, M. 2018. A-weighted decibels (dBA, or dBa, or dB(a)). *WhatIs.com.* https://whatis.techtarget.com/definition/A-weighted-decibels-dBA-or-dBa-or-dBa.
3. Bearden, D. M. 2000. *Noise abatement and control: An overview of federal standards and regulations.* Report to Congress RS20531. Washington, DC: Congressional Research Service.
4. WHO (World Health Organization). 2009. *Noise: Data and statistics.* Copenhagen: Regional Office for Europe.
5. Anonymous. 2018. *Common environmental noise levels.* New York: Center for Noise and Communication.
6. Anonymous. 2018. Noise pollution definition. *Merriam-Webster Dictionary.* https://www.merriam-webster.com/dictionary/noise%20pollution.
7. Anonymous. 2018. Noise pollution definition. *Cambridge English Dictionary.* https://dictionary.cambridge.org/us/dictionary/english/noise-pollution.
8. Miglani D. Noise pollution: Sources, effects and control. *Legal Service India.* http://www.legalserviceindia.com/articles/noip.htm.
9. CDC (Centers for Disease Control and Prevention). 2017. New vital signs study finds noise-related hearing loss not limited to work exposure. *Media Relations.* February 7. https://www.cdc.gov/media/releases/2017/p0207-hearing-loss.html.
10. Hammer, M. S., T. K. Swinburn, and R. L. Neitzel. 2014. Environmental noise pollution in the United States: Developing an effective public health response. *Environ. Health Perspect.* 122:115–9.
11. Kim, R. H. 2011. *New evidence from WHO on health effects of traffic-related noise in Europe.* Copenhagen: WHO Regional Office for Europe.
12. Münzel, T, F. P. Schmidt, S. Steven, et al. 2018. Environmental noise and the cardiovascular system. 2018. *J. Am. Coll. Cardiol.* 71(6):688–97.
13. Héritier, H., D. Vienneau, M. Foraster, et al. 2019. A systematic analysis of mutual effects of transportation noise and air pollution exposure on myocardial infarction mortality: A nationwide cohort study in Switzerland. *Eur. Heart J.* 40(7):598–603.
14. OSHA (Occupational Safety and Health Administration). 2018. *Occupational noise exposure: Overview.* Washington, DC: U.S. Department of Labor.

15. OSHA (Occupational Safety and Health Administration). 2018. *Laboratory safety noise. OSHA fact sheet.* Washington, DC: U.S. Department of Labor.
16. Francis, C. D., C. P. Ortega, and A. Cruz. 2009. Noise pollution changes avian communities and species interactions. *Curr. Biol.* 19(16):1415–9.
17. Francis, C. D., N. J. Kleist, C. P. Ortega, and A. Cruz. 2012. Noise pollution alters ecological services: Enhanced pollination and disrupted seed dispersal. *Proc. Royal Soc. B.* DOI: 10.1098/rspb.2012.0230.
18. Anonymous. 2018. *Noise pollution and the environment.* Canberra: Australian Academy of Science.
19. Swinburn, T. K., M. S. Hammer, and R. L. Neitzel. 2015. Valuing quiet: An economic assessment of US environmental noise as a cardiovascular health hazard. *Am. J. Prev. Med.* 49(3):345–53.
20. Harding, A. H., G. A. Frost, E. Tan, et al. 2013. The cost of hypertension-related ill-health attributable to environmental noise. *Noise Health* 15(67):437–45.
21. EPA (Environmental Protection Agency). 2018. *History: Noise and the Noise Control Act.* Washington, DC: Office of Media Relations.
22. Bearden, D. M. 2006. *Noise abatement and control: An overview of federal standards and regulations.* Washington, DC: Congressional Research Service.
23. Anonymous. 2018. Noise Control Act of 1972. *BallotPedia.* https://ballotpedia.org/Noise_Control_Act_of_1972.
24. EPA (Environmental Protection Agency). 2018. *Summary of the Noise Control Act.* Washington, DC: Office of Media Relations.
25. EPA (Environmental Protection Agency). 1972. *EPA to launch noise control program. EPA Press Release,* November 6. Washington, DC: Office of Media Relations.
26. EPA (Environmental Protection Agency). 1974. *EPA identifies noise levels affecting health and welfare. Press Release,* April 2. Washington, DC: Office of Media Relations.
27. EPA (Environmental Protection Agency). 1974. *Information on levels of environmental noise requisite to protect public health and welfare with an adequate margin of safety.* Washington, DC: Office of Noise Abatement and Control.
28. Govtrack. 1978. S. 3083 (95th): Quiet Communities Act. https://www.govtrack.us/congress/bills/95/s3083/summary.
29. House of Representatives. 2015. H.R. 3384. A bill to reestablish the Office of Noise Abatement and Control in the Environmental Protection Agency; and for other purposes, Washington, DC.
30. Bearden, D. M. 2003. *Noise abatement and control: An overview of federal standards and regulations.* Report to Congress RS20531. Washington, DC: Congressional Research Service.
31. Palermo, E. 2017. Who invented the light bulb? *LiveScience,* August 16.
32. Sanders, D., E. Frago, R. Kehoe, et al. 2021. A meta-analysis of biological impacts of artificial light at night. *Nat. Ecol. Evol.* 5(1):74–81.
33. Lighting Research Center. 2007. *What is light pollution? NLPIP lighting,* Vol. 7(2). Troy: Rensselaer Polytechnic Institute.
34. CSPH (Council on Science and Public Health). 2016. *Human and environmental effects of light emitting diode (LED) community lighting.* Report 2-A-16. Chicago, IL: American Medical Association.
35. Mark Park, Y.-M., A. J. White, C. L. Jackson, et al. 2019. Association of exposure to artificial light at night while sleeping with risk of obesity in women. *JAMA Intern. Med.* Published online June 10. DOI:10.1001/jamainternmed.2019.0571.
36. Staff. 2017. Could urban lighting raise breast cancer risk for some women? *HealthDay,* August 17.

37. UNESCO (United Nations Educational, Scientific and Cultural Organization). 2009. *Starlight reserve concept.* https://www3.astronomicalheritage.net/images/astronomicalheritage.net/documents/Starlight_Reserve_Concept-En.pdf.
38. Owens, A. C. S., P. Cochard, J. Durrant, et al. 2020. Light pollution is a driver of insect declines. *Biological Conservation* 241:January 108259.
39. Cinzano, P., F. Falchi, C. D. Elvidge. 2001. The first World Atlas of the artificial night sky brightness. *Mon. Not. R. Astron. Soc.* 328:689–707.
40. Cinzano, P. 2018. Light pollution in Italy. http://www.lightpollution.it/cinzano/en/index.html.
41. AMA (American Medical Association). 2016. *AMA adopts guidance to reduce harm from high intensity street lights.* Chicago, IL: Office of Media Relations.
42. NCSL (National Conference of State Legislatures). 2018. States shut out light pollution. http://www.ncsl.org/research/environment-and-natural-resources/states-shut-out-light-pollution.aspx.
43. IDA (International Dark-Sky Association). 2018. *Who we are.* Tucson, AZ: International Headquarters.

6 The Built Environment

6.1 INTRODUCTION

Communities are where people live together in close proximity. For urban communities, this means quite a large number of people, while for more rural communities, that number can be quite small. Regardless of size, there are many environmental health consequences of people residing together. Additionally, the many modifications we make to the environment – whether buildings or the structures that allow us to move between them – can add additional hazards and greater risk to human and ecosystem health. The broad area of concern over the modified environment in a community is often called *the built environment*. This encompasses the physical elements and structures where people spend their days [1]. This not only includes buildings and green spaces but also sidewalks, traffic, watersheds, powerlines, and internal environment between members of the community [2]. As illustrated in Figure 6.1, there are several determinants of health and well-being in human habitation, including the built environment, which is the subject of this chapter.

The study of the built environment transcends academic disciplines, with fields like architecture, engineering, and sociology all learning from each other. Disciplines like public health and urban planning can both trace their roots to similar concerns regarding the built environment in the 19th century: the conditions for many living in urban centers, and their rapidly decreasing quality of life [3]. As the 20th century

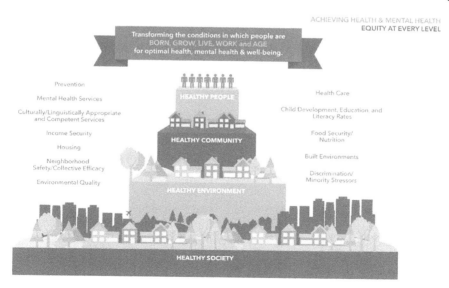

FIGURE 6.1 Model of the determinants of health and well-being in human habitation. (Let's Get Health California. https://letsgethealthy.ca.gov/sdoh/)

DOI: 10.1201/9781003212621-6

progressed and those initial concerns (communicable diseases, sanitation, overcrowding, and injury prevention) no longer commanded popular attention, the two fields began to drift apart. Public health shifted toward preventing childhood diseases and public health delivery systems, while urban planning focused more on housing and development. However, a central irony of public health and urban planning, as will be discussed later in this chapter, is that the decisions made at the time to manage poor sanitation and overcrowding had unintended consequences that contributed to many modern issues facing community health [3].

> The broad area of concern over the modified environment in a community is often called the built environment.

Today's urban planners and public health professionals are once again working together, along with many other disciplines, to address built environment issues of physical activity, health promotion, and sustainability. As described by Frank and Engelke [1], this requires a multidisciplinary perspective. For example, policies that make street networks denser could help increase physical activity and decrease overall vehicle emissions. However, the resulting, centrally located population may be exposed to traffic congestion and the toxicants that come with that exposure, which can create additional health problems.

The context of these issues will be explored in this chapter, first by detailing how the urban planning approach to the built environment has evolved, and how it is currently being managed. This will be done by first introducing key terms and concepts that are used in the built environment. Next, the history of built environment in the U.S. will be described as a survey of how these issues developed through internal and external forces.

6.2 TERMS AND CONCEPTS

Prior to discussing policies and actions that attend the built environment, it is important to define key terms and concepts that are used throughout this chapter.

6.2.1 LAND-USE POLICY TOOLS

Central to how the built environment is designed and how it is now being used to improve human and ecological health are several policy tools – almost all of which are available exclusively to local and regional government organizations. The policy tools include the following:

- **Comprehensive plans**: These are official documents that create protocols for making changes to land use and transportation. They prepare capital improvement programs and for determining the rate, timing, and location of future growth.
- **Zoning codes**: These codes are the fundamental tools in urban planning. Codes regulate how land can be used. Zoning can also control most of the

> Zoning codes are the fundamental tool in urban planning

physical attributes of any proposed structure – what it looks like, how tall it can be, where it can sit on the land, and what can be around it.
- **Building codes**: These have been used for centuries. They determine the material and methods used in constructing the building – its bulk, scale, and style. They can also dictate what type of material can be used (or not used), and how it is to be used.
- **Subdivision regulations**: This tool, along with zoning and building codes, carry out the comprehensive plan of a community. These manage how large parcels of land can be divided into units for development. In contrast to zoning, subdivision regulations dictate the quality of the development. They contribute to how streets and systems (e.g., sewage and water) are laid out in the context of the surrounding community. Subdivision regulations tend to be more permanent than zoning codes. Development patterns are set by this policy tool and can have long-lasting impacts on taxes, overcrowding of schools, and environmental issues such as overextended sewage and water resources.

6.2.2 LAND-USE POLICY ACTORS

The planning of how land is to be used is necessarily difficult and often complex. Experience over time has shown that the reliance on various policy actors can be vital for successful land use planning. The most prominent of these kinds of policy actors include the following:

City planning commissions: In many circumstances, this organization has the greatest influence over a community's-built environment. These commissions are traditionally seen as apolitical, citizen-led organizations that impact how their recommendations are heeded by the political city government. Following the model for planning set in 1929, these groups are responsible for creating ordinances through (1) commissioning studies, (2) creating and adopting master plans, (3) advising on zoning and subdivision regulations, (4) advising on capital improvements, and (5) responding to queries from elected officials.

Regional Planning Commissions: As communities developed, and as services provided by the government (e.g., transportation) and resources available to a region (e.g., water) crossed city lines, regional planning commissions emerged to manage an area's development. Metropolitan Planning Organizations (MPOs): MPOs are responsible for planning and programing transportation funds in a metro area. The Federal-Aid Highway Act of 1962 mandated – in exchange for federal highway funds – that cities create these groups, and the act charged them to take on "continuing, comprehensive, and cooperative" planning [4]. Specifically, they have five core functions: (1) frame regional decision-making in the metro area, (2) determine and evaluate alternative transportation options, (3) prepare and maintain the region's long-range transportation plan, or metropolitan transportation plan (MTP), (4) develop 4-year transportation improvement plans (TIPs), and (5) engage the public in the development of TIPs [5]. MPOs consist of local elected officials and state agency representatives.

> The Federal-Aid Highway Act of 1962 mandated – in exchange for federal highway funds – that cities create Metropolitan Planning Organizations.

MPOs historically were seen as a rubber stamp for state agencies, but with the passage of the Intermodal Surface Transportation Efficiency Act (ISTEA) of 1991 and the Transportation Equity Act for the 21st Century (TEA-21) of 1998, MPOs were given greater authority and a broader scope (e.g., considering other modes of transportation beyond automobiles, and environmental concerns). This new power gave members of the MPT greater authority to affect land use and environmental health issues in a metro region. Beyond this federal charge, some states have added additional responsibilities to MPOs or given them a more specific charge. For example, Virginia mandates land use and transportation planning, and California mandates additional considerations to climate change [5a]. One of those considerations is that the state became the first U.S. state to require solar panels on new homes. Effective January 1, 2020, solar panels will be required on all new homes up to three stories, although California regulators have evidenced willingness to cooperate with local housing developers as to the location of solar sources of energy [6]. This policy comports with California's plan to decrease greenhouse gas emissions, with a 2045 goal to transition to a fully renewable energy grid devoid of fossil fuels.

Region Council of Governments (COGs): In contrast to the federally mandated MPOs, COGs were traditionally formed through a grassroots-level mobilization. There are more than 500 COGs in the U.S., and their specific charges can cover air quality protection, waste management, water quality planning, and transportation modeling. COGs and MPOs serve together as a regional medium for built environment decision-making. States differ in how they organize COGs and MPOs. While most are separate entities, some regions join them into one organization. For example, central Ohio merges the COG and MPO under what is called Ohio's Mid-Ohio Regional Planning Commission. Additionally, some COGs are established by state statutes while some operate as nonprofits. Historically, they are seen as a body lacking clear statutory direction and have no direct accountability to the citizens of the region. Additionally, COGs tend to represent the interests of suburbs over urban centers, as their vote apportionment is typically based on member jurisdiction and not population [7].

> The Federal-Aid Highway Act of 1962 mandated that cities create Metropolitan Planning Organizations.

6.3 HISTORY OF U.S. PLANNING PRACTICES

Planning practices have existed for thousands of years. The customs and precedents established early still influence how cities are laid out today. Gridded streets were widely used in ancient Roman times and were carried over to the U.S. with the first colonial communities and cities (e.g., Boston, Philadelphia, Savannah). The grid principle spread to the U.S. frontier, as grid systems fit nicely with land surveys and were also widely adopted by the U.S. through The Land Ordinance Act of 1785, which mandated land be parceled via a rectilinear land survey [8].

Similarly, nuisance laws were widely established for several centuries before the Industrial Age. Early planning documents demonstrate the premise that certain noxious facilities be placed on the edges of towns. In the Middle Ages, it was common for shops like leather tanning to be forced away from city centers [8]. This concept was further codified into English Common law through public nuisance laws. It was then carried over to the British colonies. For example, a 1692 regulation in the Massachusetts colony designated that slaughterhouses be confined to specific areas. U.S. cities continued these traditions, offering the first uses of zoning for these particular situations [8].

> A 1692 regulation in the Massachusetts colony designated that slaughterhouses be confined to specific areas.

6.3.1 Industrialization and the Birth of a Movement

During the 19th century, industrialization in the U.S. led to rapid immigration and migration to urban centers. Due to a combination of overcrowding and unchecked environmental contamination from city industries, people in this period were exposed to a wide range of environmental health issues [3]. Communicable disease spread and death rates in tenements were extremely high, with sanitation being almost impossible to maintain. This contributed to the public pressures that birthed both the progressive public health movement and urban planning.

Many cities responded to these threats and outbreaks by creating health boards to improve sanitation in cities. As cities instituted city-wide sewer systems for the first time, it became apparent that a more comprehensive approach was needed to ensure proper grading of roadbeds and housing drainage. These initial actions of sanitation reform merged concepts of the historical nuisance laws with late 19th-century police powers. There was also a growing, broader interest in more robust attempts at planning emanating from these public health reforms. The influential *Shattuck Report of the Sanitary Commission of the State of Massachusetts* of 1850 offered, among its many other public health recommendations, specific recommendations regarding how towns and villages could be laid out to reduce disease transmission [9]. Cities like Memphis, Tennessee, went further by pushing for "systematized" and "comprehensive" plans that would seek to improve health and sanitation within city boundaries [9].

> During the 19th century, industrialization in the U.S. led to rapid immigration and migration to urban centers.

Meanwhile, the unhygienic conditions of the late 19th-century help spur two competing visions of communities in the U.S. For some, the ills that were associated with cities, poor sanitation, overcrowding, and substandard housing, made urban living intrinsically unhealthful. Those in this camp believed the only way to substantially address these issues was to remove the population (or just those affluent enough to leave) from the unhealthful sectors of industry. Through ideas like the Garden Suburb and Garden City movement, reformers argued for the bucolic health benefits of country living. Specifically, they emphasized modest housing, small-scale agriculture, and local industry [10]. These movements helped affirm the principles that drove much of suburban development in the 20th century of (1) low population density, and (2) separation of industry from housing (3) physical separation of structures with green space [1].

Competing with the Garden City movement was the City Beautiful movement, which had a significant influence on urban planning. Stemming from a popular exhibition featured at the 1893 World's Columbian Exposition in Chicago, this movement popularized the aesthetics of cities and promoted social consciousness into urban life. Supporters believed that well-designed cities could improve health and inspire a sense of civic pride that would cure many modern ailments [10]. Specifically, it led to interest in: (1) planned civic centers, (2) formally arranged complexes, such as concert halls, and (3) open space systems and circulation systems [11]. In the years after the exhibition, there were several successful attempts to institute these ideals, under the goal of "designing" cities to promote health and bringing nature back into cities [3]. Daniel Burnham, one of the principal architects of the Chicago Exposition, was commissioned to develop a plan for the entire Chicago region. The Burnham Plan for Chicago, which was published in 1909, was viewed as the first metropolitan plan [12].

> The unhygienic conditions of the late 19th-century help spur two competing visions of U.S. communities.

The years during and after the City Beautiful movement led to a flurry of public movements in cities in the U.S. The policies produced in many of these cities later served as the foundation for urban planning. Hartford, Connecticut, was the first to establish an official planning commission in 1907. A year later, Wisconsin enacted legislation authorizing certain cities to create planning commissions. In 1913 Massachusetts required all cities with population exceeding 10,000 people to have planning commissions [13].

In New York City, led by the work of Lawrence Veiller, who produced a survey of poor housing conditions in the city, the right to regulate and enforce housing conditions was given to the public health departments, which allowed the group to clear slums. In 1916, a comprehensive zoning ordinance was also passed that regulated land use, height/bulk of buildings, and density of population. New York City's model zoning ordinance was quickly passed by many municipalities across the U.S. These early developments eventually evolved into more formal codifications of state authority to dictate the organization of communities. However, this and similar movements in the following years saw a rise in civic pride, where city residents considered it a sign of civic maturity that they had utilized planning [13].

At the same time, an equally influential development was occurring with the rise in automobiles. As cars became popular, city streets – which had always been dangerous – had become the epicenter of a public health crisis. Cities initially reacted against the small minority of the population that were car owners by restricting access and heavily limiting speed in cities. In 1923 the political power shifted. During that year, the city of Cincinnati responded to citizen petition to decrease all vehicle speed limits in the city to 25 miles per hour. This energized vehicle owners and manufacturers to mobilize pressure against the proposal, and were able to help defeat its passage. This mobilization led to the emergence of a powerful political organization group that pushed to make city streets more vehicle-friendly (and as a consequence less friendly to pedestrians) [14].

> As cars became popular, city streets – which had always been dangerous – had become the epicenter of a public health crisis.

6.3.2 U.S. Federal Government's Growth into Planning

In 1922 the U.S. Department of Commerce distributed a model law for local zoning programs called "A Standard State Zoning Enabling Act" (SZEA). It was promptly adopted by a majority of U.S. states and still serves as the institutional framework for modern local zoning purposes [15]. Responding to opponents who were pushing back at the rights of the government to dictate what was being done on private property, the U.S. Supreme Court affirmed the constitutional right of local governments to dictate planning ordinances. Under the 1926 ruling in the Village of Euclid, Ohio v. Amber Realty, Co., the ruling gave formal constitutional approval of zoning, emanating from the common law principles of public nuisance [16]. Finally, in 1929, the federal government expanded its recommendations for planning by creating model traffic laws that enacted strict pedestrian controls, such as the first rules against jaywalking. These types of federal interventions set the precedent, as all three branches of federal government took steps to (1) emphasize local control over the built environment, yet (2) influence how the nature of those decisions and principles were being emphasized [17].

> In 1926 the U.S. Supreme Court affirmed the constitutional right of local governments to dictate planning ordinances.

As the Great Depression began in 1929, the federal government slowly grew more involved in dictating built environment policy priorities through (a) low-income housing in city centers, (b) highway construction through cities, and (c) subsidizing development in suburban neighborhoods. Each of these developments saw a decreasing voice of public health in decision-making [18]. Factors that led to the birth of public health and urban planning – sanitation and infectious disease outbreaks – were no longer dominating the public dialogue because both fields had moved on to new, more pressing issues [9]. Public health officials shifted toward public infrastructure, food sanitation, prenatal care, and childhood vaccinations. Meanwhile, urban planning shifted gears toward such policies as "urban renewal" and mass suburbanization, which would have long-term effects on the built environment. As later detailed by researchers in the early 21st century, this separation allowed public policy to be made without knowledge of public health consequences.

The first low-rent housing program was created with the passage of the Public Housing Act of 1937 and signified a shift in the social responsibility of homelessness and poor housing. Even more so, with the creation of housing projects in city centers across the U.S. it affirmed two principles that shaped policy decision-making in urban planning: (1) while the federal government shouldered most of the financing of these buildings, almost all decision-making power was left to local housing authorities, and (2) the replacement of private slums with public-owned buildings gave the government greater ownership of city property, allowing elected officials greater leverage to shape the composition of neighborhoods. The public housing movements had a wide-ranging impact on several aspects of the built environment. In an effort to add housing stock

> The first low-rent housing program was created with the passage of the Public Housing Act of 1937 and signified a shift in the social responsibility of homelessness and poor housing.

it created the Federal Housing Authority (FHA) in 1934. This agency helped to improve financing of home purchases and in the long-run created incentives that favored suburban housing development over existing urban neighborhoods [9].

A second movement that both increased federal control over local planning and also had major public health consequences was the development of highways. During the Great Depression, several industry groups created the National Highway Users Conference to influence the U.S. public's role in creating roads, and in turn, pressure on how those policies would be drafted. They were successful, as Congress released a planning document in 1939 that (1) echoed industry's sentiments and (2) mapped out the idea of an interstate system [19]. By 1955 the U.S. Department of Commerce's "yellow book maps" furthered that idea by publishing a series of maps detailing the routes the interstate system would traverse through city centers. Notably, no public health or city planning professionals were a part of how these plans were designed [19]. As a result, the routes were not designed to promote mobility or in maintaining community structure, but rather to promote mobility through the city – which unsurprisingly led highways to cut through the core of every major city. The maps laid the foundation for the birth of the interstate system, which was implemented through the Federal Aid Highway Act of 1956. In exchange for 90% of the cost of the highways being paid by the federal government, U.S. states had to consent to the routes dictated by the yellow book [19]. Some cities (San Francisco, Cleveland, New York City) were able to move highway plans through particularly well-off, politically organized communities, yet every city eventually had neighborhoods become isolated due to road construction [15].

Finally, the third aspect that dramatically shifted the built environment was the emergence of the U.S. federal government's providing easy FHA mortgage money for some new suburban developments. The Housing Act of 1949 was enacted as part of President Truman's domestic agenda, the Fair Deal. It expanded federal power in issuing mortgage insurance and providing federal financing for "urban renewal" projects, which demolished low-income neighborhoods in urban centers. This legislation had two major legacies. It drastically expanded the exodus of White Americans to the suburbs, and displaced many African-American communities, where housing was replaced with either more expensive housing or non-residential public works [20]. Additionally, the legislation continued the pattern of federal housing policy, where cities are given federal funds to acquire and clear slums and make the land available for residential reuse. With the availability of incentives and the World War II post-war population boom, there was a run on suburban development. This led to poorly planned development of subdivisions with poor access to sewage, utilities, and public transportation. In cities, the creation of highways and the effects of the Housing Act of 1949 decimated huge swaths of many neighborhoods and redesigned most cities in order to prioritize highways systems and use of automobiles.

> A second movement that both increased federal control over local planning and also had major public health consequences was the development of highways.

The Housing Act of 1954 continued the trends of earlier U.S. federal policies. It established a mandatory requirement for comprehensive planning as a prerequisite to funding for urban renewal. It also put the federal government in the business of providing financial aid for planning. Any city wanting to undertake urban renewal was required to develop a "workable program for community improvement" (WPCI), including a comprehensive master plan. Under Section 701 of the Housing Act of 1954, the federal

> The Housing Act of 1954 established a mandatory requirement for comprehensive planning as a prerequisite to funding for urban renewal.

government mandated all state, area-wide, and local public agencies to use comprehensive planning to solve problems in urban areas [21]. Recognizing the need for planning and that the financial stress would be difficult, it granted funding for master plan studies through approved state agencies. There was 50/50 cost care for communities with less than 25,000 populations, and it dropped to a 25% local cost share with a larger population. Subsequent models in the 1960s used similar carrot/stick funding mechanisms to encourage local action to focus on the urban built environment (U.S. Department of Housing and Urban Development, Model Cities, New Towns, etc.).

6.3.3 New Federalism and the Reemergence of Public Health

With the election of Richard Nixon (R-CA) as U.S. President, the federal government's approach to local planning shifted with the Nixon administration's New Federalism Initiative. This strategy sought to redefine how power was shared between levels of

> President Nixon's New Federalism Initiative ceded more control to local authorities and created fewer stipulations for federal money.

government by ceding more control to local authorities and creating fewer stipulations for federal money. Specifically, one of the more recognized components of this policy was the movement away from categorical funding (those with specific functions) to block grants and revenue sharing.

The Housing and Community Development Act of 1974 consolidated various categorical grants into a single block grant. It merged separate programs like the Model Cities program into single "Community Development Act" funds. Communities were awarded federal money based on a formula designed to attempt to assure fair and equitable distribution. However, its structure also led to further eroding of urban centers. The program required no local financial contribution and no stipulation to focus on the most distressed neighborhoods. Additionally, reflective of the shift in national politics, the programs expanded access of recipient communities. No longer were cities the only eligible communities. Suburbs and rural areas – sometimes with high relative incomes – received federal dollars, which further shrank the pool for which city centers could rebuild [8].

By the late 20th century, the separation of public health from planning had never been more pronounced. As described by Perdue [8], the 1995 edition of *Urban Land Use Planning*, a standard text for planners, contained no mention of "health and safety." In 2000, however, CDC reasserted the public health impacts of the built environment by calling for increased focus on the way that sprawl (the expansions of developments away from city centers) impacts health [9]. To illustrate how out-of-the-ordinary this pronouncement was, the Southern California Building Industry Association reacted harshly to the report, calling it "a ludicrous sham" and argued that

CDC should stick to "fighting physical disease, not defending political ones." A representative from the National Association of Home Builders accused CDC of being overly focused on regulating lifestyle, arguing further that surveys show that people like sprawl. Nevertheless, this event represented a turning point in how public health pressures grew in urban planning [22].

> By the late 20th century, the separation of public health from planning had never been more pronounced.

This was further advanced through the Brookings Institute's (2007) Blueprint for American Prosperity initiative, which connected economic growth with the built environment that: (1) promoted healthful lifestyle and (2) was sustainable [22]. With the election of U.S. President Barack Obama (D-IL), this convergence was further cemented with the creation of the Sustainable Communities Initiative. This HUD/DOT/EPA Memorandum of Agreement (MOA) partnership consisted of two grant programs. The Sustainable Communities Regional Planning Grants supported local planning efforts to integrate planning with land use, economics, and workforce transportation. And the Community Challenge Planning Grants aimed to achieve communities that were both affordable and sustainable. Overall, they were driven by a call to "Work together to ensure that these housing and transportation goals are met while simultaneously protecting the environment, promoting equitable development, and helping to address challenges of climate change" [23].

Perspective: With the election of Donald Trump (R-NY) as U.S. President in 2016, the policy direction of the U.S. government shifted. The 2018 Presidential Budget called for a 13.2% decrease in the HUD budget. The cuts would target many low-income housing assistance efforts, and rental assistance programs. Related to the built environment, the Trump administration has proposed eliminating the Community Development Block Grant Program, which provided assistance to local governments to fund community development initiatives. The budget would reduce the Department of Energy's budget by 5.6%. The Weatherization Assistance Program would be eliminated, which helps retrofitting efforts on a household level. While the budget blueprint will likely be amended by the Congress, it signals a shift in federal policy.

6.4 IMPACTS OF THE BUILT ENVIRONMENT ON HUMAN HEALTH

The built environment can affect human health both directly and indirectly. Directly, the built environment can unwittingly expose populations to physical, radiological, and chemical agents in air, water, or building materials. For example, vehicle emissions are a major contributor to air pollution and have been linked to a range of respiratory and cardiovascular diseases and premature deaths annually, as discussed in Volume 1, Chapter 2.

Chemical exposure can also occur through proximity to industrial pollution. The Environmental Justice movement emerged in the 1980s in response to the unequal distribution of industrial emissions in communities of color and has brought attention to the number of facilities with high emissions disproportionately reside in majority non-White neighborhoods. In addition, proximity to high-traffic areas or industrial areas also exposes populations to noise pollution. This has been linked to chronic issues such as hypertension, high blood pressure, and heart disease, hearing impairment, and stress.

The Built Environment

The built environment has also been associated with three major environmental exposures that have caused widespread health issues. Lead is a deadly neurotoxin that was distributed through the built environment through paint, gasoline, and water pipes. Millions in the U.S. and abroad have likely been badly damaged due to these policies. Its pathway to the human population is still happening today in cities with older housing stock, which are more likely to have lead in paint, and water service lines, which can increase the risk of lead exposure, as was the situation in Flint, Michigan, when nonpotable municipal water supplies resulted in exposure of residents to lead (Volume 1, Chapter 3). Built-environment policies can decrease these types of exposure by developing and enforcing stricter building codes, or by setting standards for how rental units are inspected.

> The built environment has been associated with three major environmental exposures that have caused widespread health issues.

Relatedly, asbestos is an incredibly toxic mineral that due to built environment policies was put in close proximity to human populations for decades. Different from lead and asbestos, radon occurs naturally. However, built environment policies – how houses are built, and the choices of ventilation and insulation – can increase the risk of lung cancer due to exposure to radon. Ventilation and building codes also greatly influence the quality of the indoor air. Air exchange rates and standards of types of materials used in buildings can quickly raise the amount of volatile organic compounds in the air, as well as other aeropathogens and aeroallergens.

The built environment can also directly affect public health through injuries, as many community factors affect injury morbidity and mortality. Community violence rates have historically been higher in high-density communities (though some argue some of those differences are due to differences in lead exposure). However, built environment decisions, ranging from residential unit design to visibility for access to opportunities, have also been linked to changes in violence rates.

Most injuries and unintentional accidents result from motor vehicle crashes. This is heavily influenced by policies including the presence and quality of roads, enforcement of driving regulations, and misuse of alcohol [24]. Pedestrians and cyclists in the U.S. have a higher risk of being struck by cars than a similar person in Europe [25]. Other injury rates in communities may be affected by the built environment. Building conditions (e.g., how it is lighted, how it is designed, the materials it contains, the presence of smoke detectors), and the outdoor condition (e.g., the layout of roads and sidewalks) have their relationship to a population's risk of injuries [9].

Second, it is well-established that the built environment can influence human behavior, and a growing amount of research shows that certain built environment designs of where we live and work can influence our physical activity. It is known that a sedentary lifestyle combined with high-calorie, high-fat diets contribute to higher rates of obesity. Connections between design choices and the active lifestyles have been found, as well as obesity rates. Research has found an association between a "sprawl index," physical activity, and obesity. This

> Research has found an association between a "sprawl index," physical activity, and obesity.

"sprawl index" was also associated with other chronic medical conditions [26,27]. It was also found that people residing in communities with more interconnected street networks were more likely to have adequate physical activity per day barriers related to physical activity, which in turn influences the energy balance of individuals [28]. Principles associated with new urbanism (density, street connectivity, mixed-use) were found to be associated with increased physical mobility [16].

The built environment also indirectly affects health behavior through individual transportation choices and the environmental exposure that occurs from these design patterns. This can be seen in whether driving or walking is preferable and through the proximity of traffic congestion and food outlets. These traffic choices can affect public health by increasing pedestrian and cyclist safety. As people drive less, the risk of collision (whether as a cyclist or pedestrian) decreases. These health effects can also be decreased through modifications of the built environment to increasing walkability and better sightlines and bike lanes for cyclists.

At a regional level, the built environment also influences health by the decisions in where – and how quickly – communities grow. These decisions directly influence travel behavior, which can dictate how much time people in a community spend in a car versus in their neighborhood. It also determines the amount of sedentary time spent in cars. These decisions also, in turn, affect the direct environmental exposures to which populations are exposed. Sprawl into undeveloped regions can put populations closer to vector-borne diseases. It can also strain the local environmental health systems such as sewer or septic systems or water systems.

6.5 HEALTH BENEFITS OF URBAN NATURE

There exists a considerable body of published research on the ways that nature influences our physical, mental, and social lives. Over a period of three decades, a broad and diverse body of scientific literature emerged that describes the human health value of nature, confirming that trees, parks, gardens, and other natural settings are essential to livable and sustainable cities. This body of scientific literature was summarized in 2018 in a lengthy report by the U.S. Department of Agriculture's Forest Service [29]. Excerpts pertaining to public health are described herein. For instance, the reports notes that research studies show:

> Trees, parks, gardens, and other natural settings are essential to livable and sustainable cities.

- People living near parks and green space have less mental distress, are more physically active, and have extended life spans.
- Exposure to nature may impact human mortality from chronic disease.
- When people exercise outdoors in natural environments, they do so for longer periods of time and at greater intensities.
- There is strong evidence that time spent in nature can improve the attention capacity of children with attention deficit disorders.

Positive health effects are enhanced when green space includes the presence of water, or blue space. Similarly, some research shows that inner-city children who grow up

FIGURE 6.2 An urban forest in Malvern, Victoria, Australia. (City of Malvern, 2019, https://www.stonnington.vic.gov.au/Community/Parks-and-gardens/Malvern-Urban-Forest)

in public housing buildings with a view of nature have greater impulse control and are able to concentrate better and delay gratification longer.

Research confirms that the availability of parks, trails, and nature can positively affect attitudes toward being active and encourage physical activity. Illustrated in Figure 6.2 is an urban forest in Australia, an area that supports physical activity. Researchers are finding that time spent in nature provides a wealth of mental benefits, from increased cognitive performance and well-being (being at your best) to alleviation of mental health illnesses such as depression, attention deficit disorders, and Alzheimer's. For example, a population-level study of 2,479 individuals along a rural to urban gradient in Wisconsin compared mental health outcomes with a vegetation index and percentage of tree canopy coverage [30]. After controlling for a wide variety of socioeconomic factors, the authors identified a strong association between better mental health among both urban and rural residents in areas with more green space. Higher levels of neighborhood green space were associated with significantly lower levels of symptoms for depression, anxiety, and stress.

> Higher levels of neighborhood green space were associated with significantly lower levels of symptoms for depression, anxiety, and stress.

In another study, researchers tested for a variety of biomarkers of stress and heart disease risk in blood and urine samples from 408 patients at a cardiology clinic in Louisville, Kentucky [31]. They also used satellite data from the National Aeronautics and Space Administration (NASA) and the U.S. Geological Survey (USGS) to estimate the extent of greenery where each person lived. Compared to people in areas with the least amount of green space, residents of the greenest neighborhoods had lower urinary levels of the hormone epinephrine, indicating lower stress levels, the study found. They also had lower urinary levels a marker of oxidative stress known as F2-isoprostane. The connection between residential greenery and lower levels of certain markers of heart problems held up even after researchers accounted for other factors that can independently influence the risk of heart disease like age, sex, ethnicity, smoking status, patients' use of statins to control cholesterol, neighborhood poverty, and proximity to pollution from traffic fumes.

6.6 IMPACTS OF THE BUILT ENVIRONMENT ON ECOSYSTEM HEALTH

At its most basic function, a community is established by modifying the surrounding environment. As communities grew, so too did the fitting of the area to the needs of the population, its industry, and its energy demands. In the decades following World War II, the rise of the environmental movement can be linked to the visible degradation of the environment, for example, smog and acid rain. In response to these two pressures (increasingly industrialization and growing concern of environmental contamination) the sustainability movement emerged to try to ensure built environment policies ensured the long-term health of the ecosystem.

Much of the sustainability movement is rooted in the principle that city developments have short- and long-term effects on the surrounding environment. Every feature of the built environment requires resources from the natural environment, so the movement seeks to be efficient in how resources are used, and how to make those resources last longer.

The built environment also influences the ecosystem in how many areas become developed. As a result, the continuing sprawl of urban and suburban development into formerly undeveloped land has considerable impact on an ecosystem. Directly, sprawl spreads environmental health systems (sewage and drinking water) thin and can affect water supply and quality [1]. Indirectly, this development can influence other communities, especially when the sprawl develops upstream of the water sources of larger communities. Low-density development increases the regional dependence on motor vehicles and requires new infrastructure to be delivered to these remote locations. Sprawl also increases the increased prevalence of less permeable surfaces. Rain collects on these surfaces, picks up pollutants and waste in the community, and can drain into the nearest surface water. This not only affects the nearby ecosystem, but can feed downstream into larger communities that rely on that local water source [1,32].

> The sprawl of urban and suburban development into formerly undeveloped land has considerable impact on an ecosystem.

6.7 IMPACTS OF THE BUILT ENVIRONMENT ON THE SOCIAL ENVIRONMENT

It is increasingly recognized that certain communities with positive social indicators tend to also have certain built environment features that appear to facilitate those social interactions. The social environment is best described as the structure and characteristics of relationships among people in a community. Components include social networks, social capital, social support systems that interpersonal interaction provides [24].

The Built Environment

Much of the City Beautiful movement and other similar movements during the Progressive Era of 1890–1920 was that the "look" of a community could stimulate more social culture. In her groundbreaking book, *The Death and Life of Great American Cities*, the urban planner

> It is increasingly recognized that certain communities with positive social indicators tend to also have certain built environment features that appear to facilitate those social interactions.

Jane Jacobs reflected on how some neighborhoods were able to encourage community and safety through how it prioritized people's interactions with each other. Many of her suggestions became a centerpiece in the New Urbanism movement. And her view of how design elements and the social environment could reduce crime has become commonly accepted [33].

However, as planning moved away from these urbanism principles, and toward suburbanization, many of these principles (public/private demarcation, public focus on the streets, dense public transportation) were not prioritized in suburban development. Several decades later, in the equally seminal *Bowling Alone*, the sociologist Robert Putnam investigated why several metrics of social life and community participation had declined in the late 20th century. Putnam argued that the design of sprawling communities, where people lived and worked in different locations and spent much of their free time in cars, was a likely candidate for the loss of social capital. Specifically he stated that "each 10 additional minutes of daily commute time cuts involvement in community affairs by 10%" [34].

Putnam's hypothesis has been further supported by researchers in the early 21st century. Frank and Engelke [1,35] found that how communities are laid out can strongly influence the psychosocial health of their residents. This occurs by affecting how they see themselves within the community, their connection to others, their safety, and relationship with nature [35]. Other studies showed that higher collective efficacy – and higher political participations, social trust, and social engagement – was associated with residing in a community with mixed-use zoning (combination of businesses and residences), and high walkability scores [36].

This social environment is further hypnotized to affect health through a variety of pathways [24]. Norms believed to be associated in dense areas can promote healthful activities related to smoking, diet, exercise, and sexual behavior. Additionally, the relationships found in these communities are also believed to help connect at-risk individuals to the necessary services that can keep them healthy [37].

6.8 SOCIAL JUSTICE ISSUE

University of California Berkeley researchers reported in 2020 how redlining, a racist mortgage appraisal practice of the 1930s, which established and exacerbated racial residential segregation boundaries in the U.S., impacted communities' greenspace [38]. Researchers utilized investment risk grades assigned >80y ago through security maps from the Home Owners' Loan Corporation (HOLC). They assessed whether historical HOLC investment grades are associated with 2010 greenspace, a health-promoting neighborhood resource [38a]. Researchers noted that one resource thought to promote health and buffer stress is urban greenness or greenspace. Study findings, adjusted for historical characteristics, indicate that neighborhoods assigned worse HOLC grades in the 1930s are associated with reduced present-day greenspace. In other words, a racist mortgage policy was associated with lesser greenspace in communities, thus reducing health promotion of residents.

6.9 U.S. POLICY OVERVIEW

As with many other political/policy issues, prime authority rests with local municipalities and communities. However, the U.S. federal government's influences have built environment policy by both soft and hard influences. Soft influences involve federal publications, National Academies Press reports, and studies through the National Institutes of Health (NIH) and the Centers for Disease Control and Prevention (CDC), which affect the national discourse of positive and negative aspects of community design. Harder influences, for example, came from highway fund stipulations, which have traditionally been a popular mechanism to mandate particular policy ideas through the power of the purse.

As described by Perdue, there are five main policy tools that exist to modify the built environment [38]:

- Zoning ordinances that designate an area for special use and related development requirements
- Building and housing codes that set standards for structures
- Taxing policy that can encourage or discourage activities
- Government expenditures that directly provide resources for projects and programs related to the built environment
- Environmental regulations that set quality or emission standards

Two policy tools have emerged as powerful levers in emphasizing environmental and public health in built environment decision-making. One tool is the Health Impact Assessment (HIA). The other tool is the environmental impact assessment (EIA), which is mandated by the National Environmental Policy Act (NEPAct) of 1969. EIAs detail a process in which environmental effects are forecast based on the proposed project, policy, or public action. If those effects are determined to be severe enough, the EIA helps stakeholders find appropriate measures to mitigate these effects [39].

> Two policy tools, HIAs and EIAs, are powerful levers in emphasizing environmental and public health in built environment decision-making.

While EIAs are required to include considerations of human health effects, EIAs have rarely been used for those purposes. As a result, a specific policy tool was developed to focus solely on health impacts. Called HIA, this tool is used to evaluate the potential health effects of a project before it is built or a policy before it is implemented. It can provide recommendations to increase positive and minimize adverse health outcomes. A HIA follows a series of well-defined steps:

The Built Environment

- **Screening**: determines whether an HIA would be useful for the proposal under consideration
- **Scoping**: established a plan for conducting the HIA
- **Assessments**: Describes the baseline conditions of the group likely affected by the proposal, and how the proposal might affect their baseline conditions
- **Recommendation**: Based on the assessment, the HIA develops practical implications to improve the health
- **Reporting**: Engages stakeholders in discussing the results
- **Monitoring and evaluating**: Evaluates the HIA process and the impact they have.

While HIAs are largely done voluntarily, as of 2010 a total of 119 HIAs have been completed in the U.S. They have been completed by governmental and nongovernmental sectors. While a majority deal with a built environment, many deal with energy and education issues as well [40, 41]

6.10 CURRENT PRACTICES AND ISSUES

Several key practices can affect how built environment policies are developed. Some of these practices are described in the following sections.

6.10.1 Sprawl

As stated by the CDC, the growing sprawl of communities away from city centers has wide-ranging public health and environmental health consequences. Cities have continued to grow and expand over ever larger geographic distances, with undeveloped (forests, wetlands) or underdeveloped (farm land) being converted to residential use. This places additional strain on utilities, with increased demands on resources. Due to lower population density in these suburban communities, public transportation is not always financially or logistically feasible, leading to greater reliance on automobiles. The style that these communities represent (low density, low land-use mix, strict separation of work and home) can also lead to poor economic, health, and social outcomes [42].

6.10.2 Air Quality

While outdoor air is heavily regulated, as noted in Volume 1, Chapter 2, the influence of sprawl on transportation has been shown to have large effects on air pollution [3]. On the other hand, indoor air in the built environment is seldom regulated in the U.S. Due to issues of privacy and logistical complications, fewer guidelines exist in what constitutes clean indoor air. Radon – still mostly a regional problem – is a colorless odorless radioactive gas that can lead to an increased risk of cancer. Radon is most frequently addressed by reducing exposure in a building by sealing leaks in basements and constructing improved building foundations. EPA sets exposure standards above which radon levels can be considered a health hazard. However, no policy mechanisms exist to intervene or regulate. Volatile organic compounds and household chemicals can lead to numerous adverse health outcomes and in particular are connected with a built environment disorder called *Sick Building Syndrome*. These chemicals are addressed through policy by educating consumers and setting standards that address a building's ventilation and operations.

6.10.3 Housing and Health Issues

Several environmental health issues have emerged in the 20th century where populations are exposed through their interactions with the built environment. For most hazards, building codes regulate (or prohibit) their presence in new housing stock. For older housing stock, separate policies have been addressed to mandate removal in publicly-owned properties and to provide funding for the removal of hazards in private property. For instance, asbestos was seen as inexpensive, easily mined, easily manufactured, and, most importantly, a nonflammable form of insulation. However, with the discovery of a long range of serious health risks, policies were enacted to remediate older housing stock, to regulate how it was removed, and regulate how workers were informed of its presence in buildings. For mold, few governmental regulations exist for mold removal, but building codes often dictate that its presence is a violation of standards. Finally, with lead, its presence in paint required a similar approach to remediate, educate, and monitor. EPA and the Occupational Safety and Health Administration (OSHA) set lead exposure standards. Enforcement includes the adoption of strict laws that establish landlords' liability for ensuring that children are protected and the careful establishment of procedures that promote safe removal.

> Several environmental health issues emerged in the 20th century where populations are exposed through their interactions with the built environment.

6.10.4 Brownfields

Brownfields are sites that were the former location of abandoned industrial or commercial activities, and as a result of either real or perceived environmental risk, their future land use is limited (Chapter 2). Through the Comprehensive Environmental Response, Compensation, and Liability Act (CERCLAct), the U.S. federal government takes the responsibility to remediate brownfields, which are defined as "real property, the expansion, redevelopment, or reuse of which may be complicated by the presence or potential presence of a hazardous substance, pollutant, or contaminant" [43]. Typically, remediation is complicated by liability and costs. Determining liability for the clean-up can be complicated already, but fearing legal action, the current property owners might hide the true nature, location, and scope of the contamination. Also due to the assessment and remediation process, the investment (whether planned or potential) may be scared away. The Small Business Liability Relief and Brownfields Revitalization Act of 2002 authorized funding for site assessment and remediation and clarified liability and environmental insurance issues, which in turn helped to transfer liability, and stimulate investment [44,45].

6.10.5 U.S. Transportation Priorities

For much of the history of urban planning in the U.S., the automobile was the most important determinant in how street networks were constructed. Federal transportation spending underwent a significant reorganization in 1991 with the ISTEA [23]. It introduced new flexibility, public involvement, and accountability into the transportation planning process. One key change was to ensure that transportation projects would not lead to air pollution emissions that exceeded limits set in state air quality improvement plans. Since then, a growing built environment issue has been the battle over how alternative modes of transportation are encouraged.

Hazards of being a pedestrian in an automobile-focused environment are well-documented. As described by Pucher and Dijkstra [25], efforts to improve walkability often focus on six types of approaches: (1) better facilities for walking and cycling, (2) traffic calming lanes (lower speed limits) in residential neighborhoods, (3) urban design oriented to people, not cars, (4) restriction on motor vehicle use, (5) traffic education, (6) traffic regulations and enforcement. Policies for bikers have evolved to both: (1) decrease bike-related accidents and injuries, and (2) promote biking as an alternative mode of transportation. Bike lanes have emerged as one of the more popular products to deal with both goals, but whether they only affect people of higher socioeconomic status remains unknown [46].

Current issues of transportation in the U.S. are balancing several emerging principles in urban design. Mobility has historically been the main concern in planners, as their role was to connect various points in a regional area with the assumption that they would be made in an automobile. However, accessibility has emerged as an equally important principle. With this principle, the goal is not to drive, but to access goods and services. As a result, it leads to reducing the distance between two points (rather than just allowing cars to get there as quickly as possible). Transit-oriented development seeks to integrate land use and transportation by developing higher density housing, retail, and other uses within walking distance of a station.

Related to the issue of transportation is the problem of street design. This is also dictated by two principles. Streets can encourage proximity (where many destinations are concentrated in a small area) through mixing uses and increasing density. They can also encourage connectivity (where there are many linkages of transportation points that are direct). This is illustrated by a community with several street intersections (e.g., it has a grid network of streets). In this first case, the commuter can get between two points more directly. However, a community with few street intersections (e.g., it has many cul-de-sacs), would be unlikely for a commuter to get anywhere directly [1].

6.10.6 Sustainability

Finally, the issue of sustainability has grown to be one of the more central issues in 21st century built environment policy. Sustainable development emerged in the early 1970s as a realization that current projections of population, resource use, pollution, and economic growth were in a period of "overshoot," where needs were much greater than the planet could support for much longer. Early efforts aimed at finding how to reduce these demands in the long term. In 1987, the UN's World Commission on Environment and Development would define sustainable development as the "development that meets the needs of the present without compromising the ability of future generations to meet their own needs" [47].

Sustainability is also involved in a separate pursuit that was advanced by Ian McHarg, who postulated that humanity was part of a local ecosystem, not outside of it. Therefore, how humanity built and designed its cities could be improved if they acknowledge the interrelationship of people and their natural environment. The U.S. National Science Foundation initiated the long-term ecological research program in 1980 to better understand the processes in the environment and how they change in different regions and across time [48].

> The issue of sustainability has grown to be one of the more central issues in 21st century built environment policy.

Sustainability policies still exist as a patchwork of initiatives. The Partnership for Sustainable Communities is a collaboration among EPA, HUD, and the Department of Transportation that seeks to help families attain improved access to accessible and affordable transportation options. Along with that, the Smart Growth Network is a network of private, public, nongovernmental organization that seeks to improve development practices in the U.S.

6.10.7 CLIMATE CHANGE

Of the several key practices described that can affect how built environment policies are developed, none will be more important than climate change. And that assertion is not a matter of hyperbole. As discussed in Volume 1, Chapter 1, climate change has already in 2019 caused increased temperatures in both ambient air and bodies of water; increased acidification of the oceans, raised ocean levels due to melting polar ice, and contributed to more energetic natural disasters such as hurricanes and tornadoes. And without global resolve to implement policies and actions to mitigate the causes of climate change, the consequences of climate change will become more widespread and dire. What impact will climate change have on the built environment? Specific impacts will be numerous, but it is possible to forecast some of the broader impacts, as follows:

> Of the several key practices that can affect how built environment policies are developed, none will be more important than climate change.

- Rising ambient air temperatures will produce a greater demand for energy sources for use in residential and commercial air conditioning. Sources of renewable energy will continue to increase both in importance and availability. Solar panel will become commonplace as standard equipment on new residential and commercial buildings.
- Building materials used in the construction of residential and commercial properties will need to have better insulating properties for protection against rising ambient air temperatures. An example is a new wood developed by university researchers [51]. The material, if used on a building's

The Built Environment

exterior, could drop a building's temperature as much as 10°C and reduce cooling costs as much as 35%; this material by modifying natural wood using hydrogen peroxide.
- Buildings in tornado- and hurricane-prone areas of the U.S. and elsewhere will require greater strength to protect against more energetic natural disasters.
- Some buildings in coastal areas will need to be relocated as sea levels continue to rise. Some whole cities could require relocation and rebuilding. Other buildings in coastal areas might require modifications such as relocation of boat docks.
- Marine craft will need to be built or modified to protect against the increased acidity of ocean waters.
- Cities will add more trees and other green vegetation as contributions to climate change mitigation and for their contribution to lower ambient air temperatures. A study of 10 megacities found that urban forests can lower the "urban heat island effect" of cities, which are often several degrees warmer than nearby rural areas, worsening air, and water pollution, and making sweltering workers less productive [50]. Researchers noted that plants cool air around them through transpiration and their leaves block heat from the sun, as well as absorbing noise.
- During heatwaves, set to become more frequent as temperatures rise, trees can cool buildings, potentially reducing the cost of fans and air conditioning, the study said.
- One way to reduce greenhouse emissions is to make homes have "net zero" energy, such that they are able to produce as much energy as they consume through solar panels, heat pumps, a super-tight barrier between the inside and outside, above-code insulation, and energy-efficient windows, lighting, and appliances [50a, 51].

6.11 POLICIES

Due to the independent nature of local planning decisions, a laboratory effect takes places, as each city works to develop its own unique strategy to combat the issues of the built environment. Some of these policies grow regionally, nationally, and even internationally due to a bottom-up, or a top-down process. Policies move bottom-up as cities seek to replicate policies that appear to have worked in other locations. This leads to ideas quickly spreading around the globe. Top-down policies occur as organizations (international, national, or state or province) develop a new idea and offer incentives (whether financial or administrative support) for local planning governments to implement.

6.11.1 HEALTH IN ALL POLICIES

A major policy that evolved top-down is the Health in All Policies (HiAP). This is an approach that integrates health considerations across policy sectors. While it is a recent movement, it has its roots back in the WHO Declaration of Alma-Ata in 1978. It is based on the premise that health is influenced by a variety of determinants, and therefore, health should be integral to the development of a variety of policies. HiAPs differ from HIAs by expanding the HIA approach to use multiple strategies to systematize and integrate across agencies. In other words, HIA is one tool that can be used as a part of a larger array of tactics to address health through HiAP [52,53]. It took shape in 2006 through the EU to strengthen human protection in the creation of policies.

HiAP was first used in the U.S. through a partnership with HUD and EPA through the Partnership for Sustainable Communities, which prioritized six livability recommendations and targeted improvements in affordable housing, green buildings, transit, water management, and brownfield space. More recently, HiAP has been included in the Patient Protection and Affordable Care Act, which mandated the creation of the National Prevention Council, which plans and carries out programs that promote health and prevent disease [54]. In 2017 the Trump administration commenced efforts to repeal the Affordable Care Act, hence whether the National Prevention Council continues to exist is uncertain.

> HiAP is an approach that integrates health considerations across policy sectors.

6.11.2 COMPLETE STREETS

Complete Streets is another top-down policy approach that takes a similar holistic approach to improving transportation policy and city design. It establishes a standard of safe (and accessible) facilities for bikers and pedestrians for all transportation projects. This can be done by improving conditions and opportunities for walking and bicycling, integrating either into transportation systems and providing safe facilities for all modes. The policy has its roots in the 1970s but the term "complete streets" came about in the early 21st century through Smart Growth America [55]. The National Complete Streets Coalition was founded in 2005 and had a diverse array of backers such as the AARP, American Planning Association, and the National Association of Realtors. In 2010 the U.S. DOT issued a policy statement supporting the use of Complete Streets policy, and by 2016 several hundred jurisdictions, and dozens of U.S. states have enacted Complete Streets policies that mandate transportation plans to incorporate Complete Streets principles. An ideal Complete Streets policy:

- Includes a vision for how and why the community wants to complete its streets
- Specifies that "all users" includes pedestrians, bicyclists, and transit passengers of all ages and abilities, as well as trucks, buses, and automobiles
- Applies to both new and retrofit projects, including design, planning, maintenance, and operations, for the entire right of way
- Makes any exceptions specific and sets a clear procedure that requires high-level approval of exceptions
- Encourages street connectivity and aims to create a comprehensive, integrated, connected network for all modes
- Is adoptable by all agencies to cover all roads
- Directs the use of the latest and best design criteria and guidelines while recognizing the need for flexibility in balancing user needs

- Directs that Complete Streets solutions will complement the context of the community
- Establishes performance standards with measurable outcomes
- Includes specific next steps for implementation of the policy

An even broader attempt at addressing public and environmental health impacts of the built environment is WHO's Healthy Cities Initiative. It seeks to promote health in cities by measuring health in policies and making local agencies understand the health impacts of local planning. It is a notable policy, as it shows more global partnership in promoting the connection between health and city planning. Notably it follows the traditional role of the federal government in applying soft pressure by providing free research, analysis, and support to make switching to more health-friendly policies easier [23].

6.11.3 LEED

The U.S. Green Building Council's Leadership in Energy and Environmental Design (LEED) developed rating systems that establish different standards in how buildings are constructed at each stage. Buildings that meet high standards are given a high rating and credit as a green building. The Natural Resources Defense Council initiated the development of LEED in 1994 and deals with its own set of guidelines: sustainable sites, water efficiency, energy and atmosphere, materials and resources, indoor environmental quality, and innovation and design process. The LEED program also operates a LEED Neighborhood focus, which considers the broader picture of sustainability: street connectivity, access to public transportation, and storm water impact mitigation [55].

> LEED rating systems establish standards in how buildings are constructed at each stage.

6.11.4 GREEN ROOFS POLICIES

Residents of large cities have access to many amenities such as public transportation, access to merchants, entertainment venues, educational institutions, and such. But there can be challenges that reduce the quality of life in large cities. In particular, cities are islands of constructed building, sidewalks, street pavements, and ambient air polluting vehicles. As a consequence, cities can become "heat islands" and "domes of polluted air." And climate change and the continued proliferation of vehicles within city boundaries add to the problems of heat and air pollution. There are no simple solutions to these kinds of urban challenges, but some cities have taken an ecosystems approach by adopting policies that encourage greener architecture and landscape and by limiting vehicle traffic within downtown areas.

> Green roof is an eco-friendly design technique that sows plants above a roofline.

Green roof policies are an example of this ecosystems approach. Green roof is an eco-friendly design technique that sows plants above a roofline. Illustrated in Figure 6.3 is a green roof structure located on a roof in the UK. This structure was designed and constructed by a commercial green roof company.

FIGURE 6.3 Example of a green roof garden in the U.S. (USDA, 2021. https://www.green-roofs.com/projects/u-s-department-of-agriculture-usda-whitten-building/.)

Green roofs can reduce storm-water runoff, improve air quality, and help mitigate the urban heat island effect. For a building's tenants and owners, green roofs reduce the need for heating and cooling. They also can provide food and a recreational area for residents, depending on the design of a green roof [56]. As a matter of green policy, green roof legislation has been enacted in several countries. For instance, Germany's green roof industry has been legislated and supported by the government in various ways since the 1970s. In 2009, Toronto mandated green roofs on industrial and residential buildings. Cordoba became the first city in Argentina to require green roofs in July 2016 and France's legislation mandates at least partial coverage of green roof or solar technology on all new construction and went into effect in March 2017.

> Green roofs can reduce storm-water runoff, improve air quality, and help mitigate the urban heat island effect.

Green roof policies have emerged in the U.S. in some large cities. In October 2016, San Francisco became the first U.S. city to require that certain new buildings be built with a green roof. The law requires between 15% and 30% of roof space on most new office construction projects to incorporate solar, green roofs, or a blend of both [57]. The San Francisco ordinance builds on an earlier bill passed by the city's Board of Supervisors that requires new residential and commercial buildings ten stories or shorter to install solar panels or a solar heating system that covers 15% of the roof.

Of special policy interest, a voter-initiated green roof plebiscite was approved by 54% of Denver's voters in 2017 and then became a city ordinance effective January 2018. The initiative requires that larger new buildings constructed in the city – and some existing ones, when their roofs are replaced – include rooftop gardens, potentially in combination with solar panels. Denver's new rules, which allow the use of solar panels to offset some rooftop garden coverage requirements, will exceed those set out by green-roof pioneers San Francisco and Toronto. Those cities apply green-roof requirements to new buildings, but not to roof replacements on existing buildings that meet size thresholds, as Denver's law does [57].

Some green roofs have become "roof farms," where vegetables other food items are grown. Perhaps "the city of light," Paris, best exemplifies the policy of green roofs, including roof farms. By 2020, the French capital will host more than 100 hectares (0.39 square mile) of rooftop gardens and planted walls, the Paris City Hall said. Of this, one-third will be devoted to urban farming [58]. The first wave of projects will lead to the cultivation on 32 sites of 425 tons of fruits and vegetables, 24 tons of mushrooms, 30,000 flowers, the production of 8,000 liters of beer, and 95 kilograms of honey. The agricultural products will be marketed to Parisian customers. In the Paris neighborhood of Le Marais a major rooftop farm is sprouting, commencing in 2019. Twenty gardeners will tend to 30 different kinds of plants to produce organic, nutrient-rich vegetables for the community and adjacent food establishments. The six-story, 150,000 ft^2 garden is expected to yield a ton of organic produce every day for the French capital [59]. It is the world's largest urban farm.

Toronto was the first North American city to pass a green roof law. The 2009 law required new buildings or additions that are greater than 21,000 ft^2 to cover between 20% and 60% of their buildings with vegetation. Developers can opt out for a fee, but fewer than 10% choose to do so [60]. Since the law was enacted, about 640 green roofs, covering more than 5 million ft^2 collectively, have been constructed, making the city a leader in the green roof movement.

> Toronto was the first North American city to pass a green roof law.

6.11.5 Urban Agriculture

A development in urban built environments that has human and ecological health implications is called by names such as urban agriculture, city farming, and urban gardens. Urban and peri-urban agriculture (UPA) or urban and peri-urban agriculture and forestry (UPAF) is defined as the growing of trees, food, and other agricultural products (herbs, pot plants, fuel, fodder) and raising of livestock (and fisheries) within the built-up area or on the fringe of cities. UPAF includes production systems such as horticulture, livestock, (agro-) forestry and aquaculture and input supply, processing, and marketing activities.

The most striking feature of urban agriculture uses urban resources (land, labor, and urban organic wastes), grows to produce for urban citizens, is strongly influenced by urban conditions (urban policies and regulations, high competition for land, urban markets, prices, etc.) and impacts the urban system (having effects on urban food security and poverty, as well as on ecology and health).

6.11.6 Other Built Environment Policies

Beyond these larger programs, there are several smaller, developing policies that seek to address sustainability issues in the built environment. These include the following:

- **Urban greening**: Focuses on planting or replanting trees along streets. This aims to make the street network more walkable. It also addresses sustainability by mixing species, thus making the network of trees more resistant to disease [61].

- **Agrihoods™**: Agricultural neighborhoods are master-planned housing communities with working farms as their focus [62]. Often they have large swaths of green space, orchards, hoop houses and greenhouses, and some with barns, outdoor community kitchens, and environmentally sustainable homes decked with solar panels and composting.
- **Urban forests**: Climate scientists who calculated the value of urban forests to the world's great cities have now worked out how town planners can almost double their money. Just plant 20% more trees [63]. In 2017 researchers put a value on the contributions of the urban forest: $500 million to the average megacity in pollution absorbed, temperatures lowered, and moisture taken up [63]. A different evaluation, performed by the National Tree Benefit Calculator, estimated a single 36-inch diameter willow oak in a residential area in the Washington D.C. suburbs can provide nearly $330 in benefits per year [63a, 64].

  In 2015, the World Economic Forum's Global Agenda Council on the Future of Cities named increasing green canopy coverage as one of its top 10 urban initiatives. Municipalities from Barcelona, Spain; to Melbourne, Australia; to Chicago have put urban canopy coverage at the center of their long-term strategic plans [65]. As an example, the city of Milan, Italy, has announced their intent to plant 3 million trees throughout the city. The planting will contribute to Milan's plan to mitigate climate change. The plan to plant 3 million trees by 2030 will improve air quality, cool neighborhoods, and enhance public spaces [66]. No other city has announced such an ambition ecological plan.
- Planting a trillion trees worldwide for purpose of helping mitigate climate change was recommended by a research team in 2019 [67]. The study calculated that over the decades, those new trees could suck up nearly 750 billion tons of heat-trapping carbon dioxide from the atmosphere. That's about as much carbon pollution as humans have spewed in the past 25 years. Six nations with the most room for new trees are Russia, the U.S., Canada, Australia, Brazil, and China. The study team emphasized that planting trees is not a substitute for weaning the world off using fossil fuels.
- **Car-free city centers**: Several cities, e.g., Oslo and Copenhagen, have designated their central areas to be free of cars [68]. By the beginning of 2019, Oslo finished removing more than 700 parking spots – replacing them with bike lanes, plants, tiny parks, and benches – as a major step toward a vision of a car-free city center. In a new zoning plan, Oslo is taking its intentions further, giving pedestrians, cyclists, and public transportation greater priority than private cars, and planning a network of pedestrian zones that are fully car-free, as illustrated in Figure 6.4. These changes have generally been well received by local residents.
- **ReGen villages**: This policy is working to create a circle of homes in a closed-loop system that produces its own solar and biogas power, grows organic vegetables, farms fish and chickens, harvests water, and recycles waste into fertilizer [68].

The Built Environment

FIGURE 6.4 Downtown Oslo, Norway, illustrating no cars, bicycle rack, and solar panel. (City of Oslo Visitors Centre, 2019. https://www.visitoslo.com/en/tourist-information-centre/)

- **Zero net emission (ZNE) requirements**: The carbon footprint of commercial, residential, and municipal buildings will not consumes more energy for heating, hot water, lights, and appliances than they produce. In the U.S., the building sector is moving toward zero net emissions. The goal is to bring the carbon footprint of commercial, residential, and municipal buildings in leading cities to zero. New York, Boston, Washington D.C., Vancouver, San Francisco, Phoenix, and Austin are cities that have subscribed to this zero net emission policy [69]. In 2018 Canada opened its largest net zero energy building in December [70]. The Joyce Centre for Partnership and Innovation is located at Mohawk College, Hamilton, Ontario. The $54 million project is powered by solar energy. It was also made with 28 geothermal wells in order to tap into a clean renewable energy source.

Several policies seek to promote more density in city centers. Urban Growth Boundaries set a regional boundary that mandates high-density zoning on the urban side of the boundary and low-density zoning on the rural size. This has been incorporated in Portland, Oregon, and Lexington, Kentucky, and has been argued to reign in urban sprawl. Greenbelt policies go about this in another way. Through NGOs or governments, land on the peripheries of the urban center are purchased in order to keep it from being developed. This also nudges developers to build more dense properties in city centers, but can be very expensive. San Jose, California, has used this policy to buy several hundred thousand acres.

Another policy direction is to directly amend existing zoning/build code rules. Generally, zoning code reforms change codes so that they permit and encourage development that is more consistent with new ideas that promote sustainability and physical activity. Other strategies include density bonuses, which is an incentive that allows developers to build more units, taller buildings, or more floor space than otherwise would be allowed in exchange for their putting a specified number or percentage of units in their development that could be considered affordable. This has been argued to help with affordability and density. Upzoning is a similar policy that directly changes zoning in an area from pure residential to increase commercial use. This allows for greater density, but has been argued to increase congestion.

Several policies also exist to modify traffic patterns [71]. Along with the Complete Streets policy pursuit, the Traffic Choices Study is a federally-funded pilot program that tested various ways to reduce traffic congestion in cities and improve walkability. Road network tolling is often seen as a way to reduce congestion while raising revenue. Traffic calming policies try to deliberately slow down vehicles' speed so that the distance to cross streets is reduced and the sight lines for pedestrians are improved. This can be done by using corner "bulb-outs," making the sidewalk wider at the intersection so that the distance to cross the street is reduced, and raised pavements that can result in drivers instinctively reducing their speed.

Other policies include congestion charging (which imposes a toll on cars on entering into the city), and pedestrian zones (which eliminate automobile traffic in city centers to promote walkability). Finally, the road diet policy involves converting roads by reducing the number of lanes to allow for use of bike lanes, pedestrian refuge islands, transit stops, or parking. Some U.S. cities have adopted policies to reverse the effects that urban freeways have had on their city centers. For example, Rochester, New York, through a two-decade long effort with the help of a grant from the U.S. DOT, removed a 2/3-mile long stretch of highway called the Inner Loop. Cities that have removed freeways – sometimes by an act of nature – saw decreases in traffic congestions. In some of these cases, traffic congestion dropped as much as 50%. This has motivated a program through the DOT called the Every Place Counts Design Challenge, which aims to study situations where infrastructures divide communities.

6.12 GLOBAL PERSPECTIVE

Many of the built environment problems and potential solutions have been experienced in countries globally. Relatively speaking, the U.S. was late in the game, as Europe had been developing building ordinance and regulations of urban life as early as the 19th century and were standard by the 19th century. Zoning was formally established in Germany in 1870. This was partly driven by the fact that municipal governments often owned at least half of city land, which gave regulators greater power to shape cities.

6.13 SEMINAL ISSUE OF SELECTED GLOBAL BUILT POLICIES

Broadly, policies focused on the built environment (whether from China, the EU, India, or the U.S.) are mostly focused on the issue of reforming energy efficiency. As countries develop national strategies to address climate change, it is becoming clear that inefficiencies in the built environment offer an effective (and politically popular) way to reduce energy demand and improve built environment health. Four policy measures mostly constitute how countries have addressed energy efficiency.

- Financing schemes to incentives private property owners to upgrade to more efficient systems
- Acceleration of efficient products (e.g., air conditioners, laundry equipment)
- Smart billing that incentives private property owners to energy savings (or generated electricity from solar panels) back into the grid

6.14 HAZARD INTERVENTIONS

A multitude of hazards exists in the built environment, and communities have adapted many ways to mitigate those threats. Unfortunately, due to the complexity of human interactions, many of those solutions have created brand new problems. Regardless, the history of the built environment is a history of hazard interventions. Town fires were a major issue in early communities, and some of the first steps to modify the built environment was to standardize how houses should be built. For example, houses in Germany during the Middle Ages were required to build chimneys out of stone and roofs out of tile due to past disasters related to fire. The first sewer systems, which were built in response to cholera and other communicable disease outbreaks, were a more collective step to manage community health threats. After science uncovered the immense risk posed by the exposure to lead or asbestos, policies were enacted to stop their use in new developments and finance their removal from old housing stock. Additionally, after disasters exposed the weaknesses of buildings in the San Francisco earthquake, or the vulnerability of communities in New Orleans subsequent to Hurricane Katrina, policies were proposed and some implemented to modify the built environment to reduce future public risks.

While policy mechanisms exist to quickly address built environment hazards, there are a number of factors that make interventions unique. Due to the local control over planning and zoning, many cities can act quickly. However, due to that same local autonomy, it is hard to get cities in the U.S. to act uniformly. Likewise, modifications to how local planning boards incorporated health, such as through HIAs or HiAP, the changes came slowly through the soft influence of the federal government (e.g., providing research, support, and expertise).

A critical barrier to quick hazard intervention is the cost associated with built-environment developments. As discussed with the rush of suburban development, once property is developed according to the rules in place at that time, it can be very difficult to retrofit. Even after a disaster, it remains hard to uproot the community that calls a certain area home. After Hurricane Katrina and flooding struck New Orleans, there was a push to eliminate subdivisions and neighborhoods that were deemed to be too risky. However, through local pressure, these policy proposals were rejected. This further emphasizes the need for deliberate planning strategies that (1) incorporate all stakeholders – including those of lower socioeconomic status, and (2) consider all potential impacts of the proposed development, especially with the environmental and public health.

6.15 SUMMARY

As presented in this chapter, the broad area of concern over the modified environment in a community is often called the built environment. This encompasses the physical elements and structures where people spend their days and includes buildings, green spaces, sidewalks, vehicle and pedestrian traffic, watersheds, powerlines, and internal environment between members of a community. These built structures and their conditions of use can have beneficial as well as detrimental impacts on a community's well-being and its individuals. Central to how the built environment is designed and how it is now being used to improve human and ecological health are several policy tools: comprehensive plans, zoning codes, building codes, and subdivision regulations, which are available exclusively to local and regional government organizations. The U.S. federal government's gradual entry into planning of built environments became significant when public health representation was absent from planning forums. A seminal consideration in policies relevant to the built environment is how to make building more energy-efficient for purposes of lessening impacts on climate change. Another important consideration is how to build transportation systems that are less disruptive to social structures and more protective of environmental quality.

REFERENCES

1. Frank, L. D. and P. Engelke. 2005. Multiple impacts of the built environment on public health: Walkable places and the exposure to air pollution. *Int. Reg. Sci. Rev.* 28(2):193–216.
2. Renalds, A., T. H. Smith, and P. J. Hale. 2010. A systematic review of built environment and health. *Fam. Community Health* 33(1):68–78.
3. Lopez, R. 2012. *The built environment and public health*, vol. 16. New York: John Wiley & Sons.
4. Morehouse, T. A. 1969. The 1962 Highway Act: A study in artful interpretation. *J. Am. Inst. Planners* 35(3):160–8.
5. Young, E. 2012. *Understanding regionalism: What are COGs and MPOs?* Washington, DC: National Association of Regional Councils.
5a. Kranking, C. 2020. Regulators loosen California's groundbreaking rule to require residential rooftop solar. *Calmatters*, February 20.
6. Reynolds, L. 2003. Intergovernmental cooperation, metropolitan equity, and the new regionalism. *Wash. L. Rev.* 78:93.
7. Hommann, M. 1993. City planning in America: Between promise and despair. Santa Barbara, CA: Praeger Publishers.
8. Perdue, W. C., L. O. Gostin, and L. A. Stone. 2003. Public health and the built environment: Historical, empirical, and theoretical foundations for an expanded role. *J. Law Med. Ethics* 31(4):557–66.
9. Davies, K. 2013. *The rise of the US environmental health movement.* Lanham, MD: Rowman & Littlefield.
10. Wilson, W. H. 1994. *The city beautiful movement.* Baltimore, MD: JHU Press.
11. Neuman, M. and S. Smith. 2010. City planning and infrastructure: Once and future partners. *J. Plan. Hist.* 9(1):21–42.
12. Peterson, J. A. 2009. The birth of organized city planning in the United States, 1909–1910. *J. Am. Plan. Assoc.* 75(2):123–33.

13. Peterson, J. A. 2003. *The birth of city planning in the United States, 1840–1917.* Baltimore, MD: JHU Press.
14. Norton, P. D. 2011. Fighting traffic: The dawn of the motor age in the American city. Cambridge, MA: MIT Press.
15. Sullivan, E. J. and L. Kressel. 1975. Twenty years after–renewed significance of the comprehensive plan requirement. *Urb. L. Ann.* 9:33.
16. Schilling, J. and L. S. Linton. 2005. The public health roots of zoning: In search of active living's legal genealogy. *Am. J. Prev. Med.* 28(2):96–104.
17. Salkin, P. E. 2002. From Euclid to growing smart: The transformation of the American local land use ethic into local land use and environmental controls. *Pace Environ. L. Rev.* 20:109.
18. Stromberg, J. 2016. Highways gutted American cities. So why did they build them? *Vox*, May 11.
19. Wells, C. W. 2012. Fueling the boom: Gasoline taxes, invisibility, and the growth of the American highway infrastructure, 1919–1956. *J. Am. Hist.* 99(1):72–81.
20. Thomas, J. M. and M. Ritzdorf. 1997. *Urban planning and the African American community: In the shadows.* Thousand Oaks, CA: Sage Publications.
21. McGrath Jr, D. C. 1973. Appropriate relationships between comprehensive planning and transportation planning for the 1970s. *Transportation* 1(4):403–18.
22. Perdue, W. C. 2008. Obesity, poverty, and the built environment: Challenges and opportunities. *Geo. J. Poverty L. Policy* 15:821.
23. Katz, B. 2010. Obama's metro presidency. *City & Community* 9(1):23–31.
24. Frumkin, H. 2016. Environmental health: From global to local. New York: John Wiley & Sons.
25. Pucher, J. and L. Dijkstra. 2003. Promoting safe walking and cycling to improve public health: lessons from the Netherlands and Germany. *Am. J. Public Health* 93(9):1509–16.
26. Ewing, R., T. Schmid, R. Killingsworth, et al. 2003. Relationship between urban sprawl and physical activity, obesity, and morbidity. *Am. J. Health Promot.* 18(1):47–57.
27. Sturm, R. and D. A. Cohen. 2004. Suburban sprawl and physical and mental health. *Public Health* 118(7):488–96.
28. Feng J., T. A. Glass, F. C. Curriero, et al. 2010. The built environment and obesity: A systematic review of the epidemiologic evidence. *Health & Place* 16(2):175–90.
29. USDA. 2018. *Urban nature for human health and environment.* Washington, DC: U.S. Department of Agriculture, Forest Service.
30. Beyer, K. M. M., A. Kaltenbach, A. Szabo, et al. 2014. Exposure to neighborhood green space and mental health: Evidence from the survey of the health of Wisconsin. *Int. J. Environ. Res. Public Health* 11(3):3453–72.
31. Rapaport, L. 2018. Leafy green neighborhoods tied to better heart health. *Reuters*, December 28.
32. Brabec, E., S. Schulte, and P. L. Richards. 2002. Impervious surfaces and water quality: A review of current literature and its implications for watershed planning. *J. Plan. Lit.* 16(4):499–514.
33. Jacobs, J. 1961. *The death and life of great American cities.* New York: Knopf Doubleday Publishing Group, Vintage Books.
34. Putnam, R. D. 2001. *Bowling alone: The collapse and revival of American community.* New York: Simon and Schuster.
35. Aspinwall, L. S. and U. M. Staudinger. 2003. *The ecology of human strengths. A psychology of human strengths: Fundamental questions and future directions for a positive psychology*, 331–43. Washington, DC: American Psychological Association.
36. Cohen, D. A., T. A. Farley, and K. Mason. 2003. Why is poverty unhealthy? Social and physical mediators. *Soc. Sci. Med.* 57(9):1631–41.

37. Berkman, L. F. and T. Glass. 2000. Social integration, social networks, social support, and health. *Soc. Epidemiol.* 1:137–73.
38. Perdue, W. C., L. A. Stone, and L. O. Gostin. 2003. The built environment and its relationship to the public's health: The legal framework. *Am. J. Public Health* 93(9):1390–4.
38a. Nardone, A., K. E. Rudolph, R. Morello-Frosch, and J. A. Casey. 2021. Redlines and greenspace: The relationship between historical redlining and 2010 greenspace across the United States. *Environ. Health Perspect.* 129(1):017006.
39. Glasson, J., R. Therivel, and A. Chadwick. 2013. *Introduction to environmental impact assessment.* New York: Routledge.
40. Committee on Health Impact Assessment. 2011. *Improving health in the United States: The role of health impact assessment.* Washington, DC: National Academies Press.
41. Collins, J. and J. P. Koplan. 2009. Health impact assessment: A step toward health in all policies. *JAMA* 302(3):315–7.
42. Frumkin, H. 2002. Urban sprawl and public health. *Pub. Health Rep.* 117(3):201.
43. Mintz, J. A. 2002. New loopholes or minor adjustments: A summary and evaluation of the small business liability relief and Brownfields revitalization act. *Pace Environ. L. Rev.* 20:405.
44. Wedding, G. C. and D. Crawford-Brown. 2007. Measuring site-level success in brownfield redevelopments: A focus on sustainability and green building. *J. Environ. Manage.* 85(2):483–95.
45. Berman, L., C. A. De Sousa, T. Linder, and D. Misky. 2016. From blighted brownfields to healthy and sustainable communities: Tracking performance and measuring outcomes. In: *Reclaiming Brownfields: A comparative analysis of adaptive reuse of contaminated properties,* ed. R. C. Hula, L. A. Reese, C. Jackson-Elmoore, 311. Farnham: Ashgate Publishers.
46. Fraser, S. D. and K. Lock. 2011. Cycling for transport and public health: A systematic review of the effect of the environment on cycling. *Eur. J. Public Health* 21(6):738–43.
47. Brundtland, G. R. 1987. *Our common future: The world commission on environment and development.* Oxford: Oxford University Press.
48. Seltzer E. and A. Carbonell. 2011. *Regional planning in America: Practice and prospect.* Cambridge, MA: Lincoln Institute of Land Policy.
49. Sakharkar, A. 2019. A new wood could drop a building's temperature and reduce cooling costs. *Tech Explorist,* May 24.
50. Pujol-Mazzini, A. 2017. Cleaner air, cooler buildings: Urban trees save megacities millions. *Thomson Reuters Foundation News,* August 23.
50a. Huninilowycz, M. 2020. More and more homeowners are renovating existing homes to make them 'net zero' energy consumers. Here's how. *Ensia.com.* January 8.
51. Rudolph, L. and J. Caplan. 2013. *Health in all policies: A guide for state and local governments.* Washington, DC: American Public Health Association.
52. Bostic, R. W., R. L. Thornton, E. C. Rudd, and M. J. Sternthal. 2012, Health in all policies: The role of the US Department of Housing and Urban Development and present and future challenges. *Health Aff.* 31(9):2130–7.
53. Stoto, M. A. 2013. *Population health in the affordable care act era.* Kansas City: Children's Health Association.
54. Laplante, J. N. and B. McCann. 2008. Complete streets: We can get there from here. *ITE J.* 78(5):24–8.
55. U.S. Green Building Council. 2017. LEED for neighborhood development. A prescription for green healthy communities. http://www.greenhomeguide.org/livinggreen/led_for_neighborhood_development.
56. Snow, J. 2016. Green roofs take root around the world. *National Geographic,* October 27.
57. Murray, J. 2017. As nation's strictest green roof law takes effect in Denver, plenty of uncertainty swirls around voter-passed requirements. *The Denver Post,* December 31.

58. Staff. 2017. Post office workers grow vegetables, breed chickens on Paris rooftop "farm." *Reuters*, September 26.
59. Nielsen, D. 2019. The world's largest urban farm is opening next year in Paris. *Dwell*, August 16.
60. DiNardo, K. 2019. The green revolution spreading across our rooftops. *The New York Times*, October 9.
61. Bowler, D. E., L. T. M. Knight, and A. S. Pullin. 2010. Urban greening to cool towns and cities: A systematic review of the empirical evidence. *Landscape Urban Plan*. 97(3):147–55.
62. Adams, B. 2019. In Detroit, a new type of agricultural neighborhood has emerged. *Yes Magazine*, November 4.
63. Radford, T. 2018. Urban forests add to cities' health and wealth. *Climate News Network*, January 22.
63a. Biron, C. L. 2020. Money trees: U.S. cities find new ways of valuing urban forests. *Reuters*, February 20.
64. Hanes, S. 2019. How Baltimore is saving urban forests – And its city. *The Christian Science Monitor*, June 10.
65. Locker, M. 2019. Milan is embarking on a bold plan to plant 3 million trees. *Fast Company*, June 22.
66. Borenstein, S. 2019. Best way to fight climate change: Plant a trillion trees. *AP News*, July 4.
67. Peters, A. 2019. What happened when Oslo decided to make its downtown basically car-free? *Fast Company*, January 24.
68. Varinsky, D. 2016. The 'Tesla of eco-villages' is developing off-grid villages that grow their own food and generate their own power. *Business Insider*, September 28.
69. Partin, J. and M. Shank. 2016. Because national governments won't, cities are pushing zero-energy buildings. *Fast Company*, March 25.
70. Baldwin, E. 2018. Canada's largest net zero energy building opens in Ontario. *Arch Daily*, December 25.
71. Puget Sound Regional Council. 2008. *Traffic choices study–summary report. Prepared for the Value Pricing Pilot Program*. Washington, DC: Federal Highway Administration.

7 Transportation

7.1 INTRODUCTION

Policies that pertain to physical hazards in the environment are the focus of this book. Prior chapters have dealt with such hazards to human and ecological health as air pollution, water contamination, and climate change. Transportation of people and their products of commerce is a component of the physical environment globally and as such can present a hazard to human and ecological health. Therefore, the purpose of this chapter is to describe key government and private sector policies that affect transportation in the U.S. Of particular note will be a description of policies purposed for the protection of occupants in motor vehicles. Other transportation policies on motor vehicles' design, roadway safety features, and vehicle operators' infractions are presented. Emphasis is given in this chapter to policies and impacts that pertain to the use of automobiles, given their being the major source of transportation-related human fatalities. The adverse human and ecological health consequences of transportation are described, along with hazard interdictions. Illustrated in Figure 7.1 is

FIGURE 7.1 Modern urban transportation. (Aerial Photography of Cars on Road Intersection, Free Stock Photo (pexels.com).)

DOI: 10.1201/9781003212621-7

an example of a major transportation corridor in an urban setting. It is apparent from the number and density of vehicles that traffic mishaps will occur.

Living organisms move about either through self-effort or by some form of transport or conveyance, where transport is defined herein as "the movement of people or goods from one place to another" [1]. As humans and other organisms move about in the environment, they become part of the physical environment, occasionally presenting and experiencing a physical hazard in the environment, the focus of this book. The reasons for the need to move about are myriad. For many creatures, the search for food or for reproduction is a prime motivation for moving from one place to another. Human beings fit into this pattern, but with additional purposes for moving about including for purposes of security, commerce, recreation, and curiosity.

Humankind's earliest journeys were by walking, the most basic form of humans' transportation. Over the course of time, humans came to recognize and adopt other forms of movement, in particular, the use of other animals for conveyance. The domesticated horse became the favored animal for humans' use as a means of moving about and for transport of goods. Carts, wagons, and other wheeled instruments, pulled by domesticated animals such as horses, donkeys, camels, dogs, and goats became humans' principal means worldwide for transporting goods, products, household possessions, families, and individuals.

Our reliance on equines as a means for transportation remained well into the early 20th century, when mechanically powered vehicles came into popular use. The first of the mechanical devices was the steam engine, originally applied to manufacturing and farm equipment and later to steam-powered modes of transportation that included boats, trains, and personal vehicles. This mode of power was supplanted in the 20th century by internal combustion engines that required petroleum-based fuels such as gasoline and diesel mixtures. Various kinds of internal combustion engines were designed and placed into automobiles, trucks, construction equipment, trains, airplanes, and ships, each of which became a component of the physical environment experienced globally. And as with other physical elements in humankind's environment, modes and operation of transportation devices can present hazards to human and ecosystem health. As a matter of global public health, the World Health Organization (WHO) estimates that 3,400 people die each day globally from vehicle-related causes, which equates to more than 1.2 million premature deaths annually [2].

> As a matter of global public health, the WHO estimates that 3,400 people die each day globally from vehicle-related causes. This equates to more than 1.2 million premature deaths annually.

7.2 PRÉCIS HISTORY OF MOTOR-POWERED VEHICLES

A brief history of the evolution of motor-powered vehicles provides an interesting insight into how the creative minds of a relatively few inventors changed forever how humans transported themselves and their goods. As a digression, it is an interesting thought exercise to construct a short list of the inventions and discoveries that globally changed forever how human beings live. Shown in Table 7.1 is one person's short list of humankind's societal global-shaping events, one item of which is the invention of the automobile.

Transportation

TABLE 7.1
A Short List of Inventions and Discoveries that Forever Changed Human Existence

Year or Period	Event	Place	Societal Impact
Unknown	Evolution of human language	Global	Essential for cultural social development
7000 BCE	Discovery of metals and later, metallurgy	Anatolia (modern-day Turkey)	Metal tools could do old jobs better than ever and new jobs never done before. Metals also created new sources of wealth in their own right
2nd century BCE	Archimedes discovery of water's buoyancy	Greece	Knowledge of buoyancy facilitated building ships that could span the globe
4000 BCE	Discovery of lime mortar	Egypt?	Permitted the lasting construction of stone structures, e.g., city walls, cathedrals, pyramids
9th century	Invention of gunpowder	China	Gunpowder became the essential ingredient in devices used in war, law enforcement, hunting, and recreation
1796	Discovery of vaccines	England	Permitted humans to prevent some diseases that were formerly fatal.
19th century	Invention of internal combustion engines	Holland, France, Germany	Provided the power for vehicles that could travel great distances, enhancing humans' commercial and personal travel
19th century	Harnessing of electromagnetism	Europe	Understanding of electromagnetism led to electric power, electronic devices, and communication systems
1885	Invention of the automobile	Germany	Internal combustion vehicles became the primary means for personal travel and business endeavors
1930s	Invention of the digital computer	U.S.	Facilitated personal and societal communication, information sharing, and data processing

In many ways, the automobile revolutionized society and enabled people to explore the world around them with greater ease and freedom. The world would certainly not look the way it does today if the automobile had not come along when it did. The invention of the automobile cannot actually be attributed to one single individual. Instead, the automobile was invented in many stages and pieces by many people globally over the course of several decades. Experts estimate that more than 100,000 patents are related to the invention of the modern automobile [3].

The first theoretical plans for a motorized vehicle date back to Leonardo da Vinci and Isaac Newton. Their plans did not come to fruition in their lifetimes. The first self-propelled road vehicle was a steam-driven military tractor invented in 1769 by a French engineer and mechanic named Nicolas Joseph Cugnot [3]. It traveled at 2.5 miles per hour and was used by the French Army to haul artillery. The vehicle was a failure as a military weapon, owing to slow speed and large amounts of fuel required. Similarly, steam power proved to be less than ideal for automobile design because the engines were too heavy. The next step forward in automobile design came in the 1830s when Robert Anderson of Scotland developed the first electric automobiles, powered by rechargeable batteries. But, in addition to being slow and expensive, these early electric vehicles needed to stop frequently to recharge the batteries.

The automobile as we know it today came into existence when Gottlieb Daimler and Karl Benz, German mechanical engineers, invented the first practical gasoline-powered vehicles in the mid-1880s [3]. Many other inventors built upon the work of Daimler and Benz and the gasoline-powered car developed quickly at the end of the 19th century. By the early 1900s, gasoline-powered cars began to outsell all the other types of motor vehicles. Despite their great expense, there was soon a demand for them. Entrepreneurs quickly recognized the need for industrial production and more affordable alternatives.

One entrepreneur, Ransome Eli Olds, built the first mass-produced automobile in the U.S.: the 1901 Curved Dash Oldsmobile. Olds invented the basic idea of the assembly line and is also credited with launching the Detroit-area automotive industry. Henry Ford improved the assembly line system by installing the first conveyor belt-style assembly line in his Ford Motor Company plant in Highland Park, Michigan [3]. Ford's assembly process helped to reduce costs and make vehicles more affordable. His Ford Model T was a huge success, selling 15 million vehicles by 1927. Illustrated in Figure 7.2 is a 1926 Ford Model T automobile. The assembly line process remains as the principal manufacturing method used globally for producing motor vehicles and many other commercial products, although electromechanical robots have replaced human assembly workers in many instances.

> Gottlieb Daimler and Karl Benz, German mechanical engineers, invented the first practical gasoline-powered vehicles in the mid-1880s.

FIGURE 7.2 Ford Model T automobile. (The Henry Ford.)

The invention of internal combustion engines was the necessary companion for automobiles and other early forms of powered vehicles. A history source describes the evolution of the internal-combustion engine as follows [4].

> Ransome Eli Olds built the first mass-produced automobile in the U.S.: the 1901 Curved Dash Oldsmobile.

"The first person to experiment with an internal-combustion engine was the Dutch physicist Christian Huygens, about 1680. But no effective gasoline-powered engine was developed until 1859, when the French engineer Étienne Lenoir built a double-acting, spark-ignition engine that could be operated continuously. In 1862, Alphonse Beau de Rochas, a French scientist, patented but did not build a four-stroke engine; sixteen years later, when Nikolaus A. Otto built a successful four-stroke engine, it became known as the 'Otto cycle.' The first successful two-stroke engine was completed in the same year by Sir Dougald Clerk, in a form that (simplified somewhat by Joseph Day in 1891) remains in use today. In 1885, Gottlieb Daimler constructed what is generally recognized as the prototype of the modern gasoline engine: small and fast, with a vertical cylinder, it used gasoline injected through a carburetor. In 1889, Daimler introduced a four-stroke engine with mushroom-shaped valves and two cylinders arranged in a V, having a much higher power-to-weight ratio; with the exception of electric starting, which would not be introduced until 1924, most modern gasoline engines are descended from Daimler's engines" [4].

Perspective: These 19th-century inventions of various forms of internal combustion engines and their mechanical descendants evolved into power sources ranging from tiny garden equipment such as weed trimmers to gigantic machines used for moving vast quantities of raw materials such as coal and various ores. One can assert

> The invention of internal combustion engines was the necessary companion for automobiles and other early forms of powered vehicles.

with confidence that few people on planet Earth have not been impacted by the use of some kind of internal combustion-powered device. But one can also assert that emissions from internal combustion engines have become a major contributor to air pollution and climate change and the ensuring adverse effects on public health, as detailed in Chapters 6 and 8, respectively.

7.3 IMPACTS OF TRANSPORTATION ON HUMAN HEALTH

How humans choose to move about brings the risk of adverse outcomes. This has probably always been the case. Some of our ancestors fell from horses; some drowned when canoes sank; others were accosted when walking. In modern times, human fatalities related to motor vehicles have become a costly concern to public and private health agencies as well as to members of the general public. As a matter of global public health, the World Bank's Global Road Safety Facility (GRSF) reports "There is growing awareness worldwide that road traffic crashes and injuries represent an unacceptable, and underfunded, public health crisis. Ninety percent of deaths on roads occur in low- and middle-income countries. Road traffic injuries are already the leading cause of death among youths and young adults [5]."

Another source, the WHO's Global Status Report on Road Safety 2015, which reflects information from 180 countries, indicates that worldwide the total number of road traffic deaths has plateaued at 1.25 million per year, with the highest road traffic fatality rates in low-income countries

> WHO's Global Status Report on Road Safety 2015 indicates that worldwide the total number of road traffic deaths has plateaued at 1.25 million per year.

[6]. The WHO further observed in their report that while there has been progress toward improving road safety legislation and in making vehicles safer, urgent action is needed to achieve the ambitious target for road safety reflected in WHO's newly adopted 2030 Agenda for Sustainable Development: halving the global number of deaths and injuries from road traffic crashes by 2020.

Concerning the U.S. data on motor vehicle fatalities, the U.S. Department of Transportation (DOT) began in 1975 an annual census of motor vehicle deaths, recording information on crash type, vehicle type, road type, driver characteristics, and a variety of other factors. DOT researchers analyze these data each year to quantify the public health problem of motor vehicle deaths. According to DOT data, a total of 35,092 people died in motor vehicle crashes in the U.S. in 2015 [7]. These deaths occurred in 32,166 crashes involving 48,923 motor vehicles. This was a 7% increase in deaths compared with 2014 and was the highest number of traffic deaths since 2008. Contributing to the death toll were alcohol, speeding, lack of safety belt use, and other problematic driver behaviors. Moreover, death rates varied by vehicle type, driver age and gender, and other factors. In 2016, the NTSHA reported that motor vehicle-related crashes on U.S. highways claimed 37,461 lives [8]. The agency estimated that 94% of serious crashes were due to dangerous choices or errors people make while operating a motor vehicle.

The CDC reports that motor vehicle crashes are a leading cause of premature death among those aged 1–54 in the U.S. [9]. Most crash-related deaths in the U.S. occur to passenger vehicle occupants. For adults

> According to DOT data, a total of 35,092 people died in motor vehicle crashes in the U.S. in 2015.

and older children (who are large enough for seat belts to fit properly), seat belt use is one of the most effective ways to save lives and reduce injuries in crashes. The CDC notes that millions do not buckle up on every trip, contributing to the following impact on public health in the U.S.: A total of 22,441passenger vehicle occupants died in motor vehicle traffic crashes in 2015. More than half (range: 52%–59%) of teens (13–19 years) and adults aged 20–44 years who died in crashes in the U.S. in 2015 were unrestrained at the time of the crash. Illustrated in Figure 7.3 is a motor vehicle collision in which two vehicles received considerable damage.

The CDC further notes that more than 2.6 million drivers and passengers were treated in emergency departments in the U.S. as a result of being injured in motor vehicle traffic crashes in 2016 [10]. Young adult drivers and passengers (18–24) have the highest crash-related nonfatal injury rates of all adults. Nonfatal crash injuries to drivers and passengers resulted in more than $48 billion in lifetime medical and work

Transportation

FIGURE 7.3 Automobile collision. (https://www.pexels.com/photo/red-and-yellow-hatchback-axa-crash-tests-163016/.)

loss costs in 2010. For comparison, the 2017 discretionary funding for the National Institutes of Health was $34.1 billion [11].

An estimated 6,227 pedestrians were killed in traffic in 2018 according to a study from the Governors Highway Safety Association [12]. The number approaches a 30-year high. The report cited alcohol use, speeding, unsafe infrastructure, and the prevalence of SUVs as some of the biggest problems contributing to the fatalities. It also suggested that the increased use of smartphones might contribute to such deaths.

Perspective: Morbidity and mortality associated with motor vehicle crashes are a public health problem both domestically and globally. In the U.S., more than 35,000 premature deaths occurred as a consequence of motor vehicle crashes. Globally, more than 1.25 million lives are lost annually to vehicle crashes. Nonfatal injuries caused in vehicle crashes number in the millions both domestically and globally. Public health departments and practitioners can assist in reducing this social burden of death and injuries through proactive campaigns for driver education and advocacy of safety measures while operating motor vehicles and riding as passengers. An emphasis should be given to seat belt use and abstinence of distracting behaviors such as cell phone texting and the use of intoxicating drugs. Protection of children as passengers should be given the highest priority.

> CDC notes that more than 2.6 million drivers and passengers were treated in emergency departments in the U.S. as a result of being injured in motor vehicle traffic crashes in 2016.

7.4 IMPACTS OF TRANSPORTATION ON ECOSYSTEM HEALTH

Data are lacking on any association between modes of transportation impacting ecological health. However, one can make some general observations as follows:

- Highways and streets are constructed over land and some bodies of water, e.g., a highway bridge. This removes the land from the use by feral animals and other life that could have inhabited there.
- Similarly, the construction of airports on land will result in loss of habitat for some feral animals and other creatures.
- Disposal of used motor vehicles presents challenges in proper waste disposal, e.g., batteries when improperly disposed can leach metals into soil and unprotected streams of water.
- Aircraft in flight can be a hazard to migrating birds and vice versa.
- Cargo can be damaged during loading or unloading a ship, sometimes causing water pollution problems.

7.5 U.S. FEDERAL TRANSPORTATION POLICIES

With the commercial adoption of internal combustion engines, linked to mass production of automobiles and subsequent other kinds of motor vehicles, came the need for policies to control the means, application, and consequences of transportation on public roads in the U.S. and elsewhere. As the number and variety of motor vehicles increased, so did the associated morbidity and mortality caused by vehicle crashes. This led to policies in U.S. states and local governments to enact policies for reducing this public health toll. Over time, the U.S. federal government began to assume a larger role in enacting policies intended for the same purpose of reducing public health consequences of transportation, with a focus on surface transportation of automobiles and other motor vehicles. The following sections will present the primary federal policies and statutes bearing on motor vehicle surface transportation. Shown in Table 7.2 are the statutes to be profiled in this section.

As background, the U.S. has no national transportation policy. Rather, the federal government through Congressional actions has implemented various policies on motor vehicles' design and use, road construction and maintenance grants, safety standards, air transportation, and policies bearing on air traffic control and operation of passenger and freight trains. Additionally, U.S. states have individually promulgated policies that bear on the modes of transportation and their use, with emphasis on policies that pertain to automobiles. These various federal and state policies will be subsequently described, but suffice it to say here that the public's safety is the central concern in all transportation policies. Following is a short list of federal statutes that have implications for safe surface transportation and therefore are relevant for public health purposes. Commentary for each statute is provided with an emphasis on health implications.

7.5.1 FEDERAL-AID HIGHWAY ACT, 1956

The concept of an interstate highway system as we know it was first described in a 1939 report to Congress called *Toll Roads and Free Roads* [13]. The report rejected

TABLE 7.2
The Principal U.S. Federal Surface Transportation Statutes

Statute	Year of Enactment	Purpose
Federal-Aid Highway Act	1956	Authorized the construction of a 41,000-mile network of interstate highways that would span the nation
Urban Mass Transportation Act	1964	Provided $375 million for large-scale urban public or private rail projects in the form of matching funds to cities and states
National Traffic and Motor Vehicle Safety Act	1966	Empower the federal government to set and administer new safety standards for motor vehicles and road traffic safety
National Minimum Drinking Age Act	1984	Required all states to enforce a minimum legal drinking age of 21 or else risk losing 10% of all federal highway construction funds
Intermodal Surface Transportation Efficiency Act	1991	Directed a major change to transportation planning and policy, as the first U.S. federal legislation in the post-Interstate Highway System era
Motor Carrier Safety Improvement Act	1999	Established the Federal Motor Carrier Safety Administration
Fixing America's Surface Transportation Act	2015	Provided long-term funding certainty for surface transportation infrastructure planning and investment

the toll superhighway network Congress had suggested; revenue from tolls on most segments would not support the bonds issued for their construction. However, the report added that the country needed a toll-free express highway network. Thomas H. MacDonald, Chief of the U.S. Bureau of Public Roads, and Herbert S. Fairbank, Chief of the agency's Division of Information, prepared the report. The ideas expressed in the "free roads" portion of the report evolved through further study and experience before approval of the Federal-Aid Highway Act of 1956, but the vision of an U.S. interstate system of highways began with MacDonald and Fairbank.

However, it remained for others to actually accomplish the construction of an interstate system. Among these was the man who would become U.S. President, Army General Dwight D. Eisenhower [13]. Eisenhower was supreme commander during WWII of the Allied Expeditionary Force and given the responsibility of spearheading the planned Allied invasion of Nazi-occupied Europe. During World War II, Eisenhower had been stationed in Germany, where he was impressed by the network of high-speed roads known as the Reichsautobahnen. After he became president in 1953, Eisenhower (R-KS) was determined to build the highways that lawmakers had been talking about for years. In 1954, he informed the nation's governors of his plan to build a new network of coast-to-coast highways, a project that soon came to be called

> The Federal-Aid Highway Act of 1956 created a 41,000-mile "National System of Interstate and Defense Highways."

the Interstate Highway System. He asserted that for its future ground transportation needs, the U.S. would be making a major investment in new roadways. To achieve his plan would require support and action by the U.S. Congress, perhaps a task as difficult as actually constructing the highways.

> In 1954 President Dwight Eisenhower informed the nation's governors of his plan to build a new network of coast-to-coast highways, the Interstate Highway System.

For instance, the Federal-Aid Highway Act of 1944 had authorized the construction of a 40,000-mile "National System of Interstate Highways" through and between the nation's cities, but offered no way to fund the project. It took several years of wrangling, but a new Federal-Aid Highway Act passed in June 1956. On June 29, 1956, President Eisenhower signed the Federal-Aid Highway Act of 1956. The bill created a 41,000-mile "National System of Interstate and Defense Highways" that would, according to Eisenhower, eliminate unsafe roads, inefficient routes, traffic jams and all of the other things that got in the way of "speedy, safe transcontinental travel." At the same time, highway advocates argued, "in case of atomic attack on our key cities, the road net [would] permit quick evacuation of target areas." For all of these reasons, the 1956 law declared that the construction of an elaborate expressway system was "essential to the national interest." The law also allocated $26 billion to pay for the new system of highways. Under the terms of the law, the federal government would pay 90% of the cost of expressway construction, with U.S. states funding the remainder. The money came from an increased federal gasoline tax that went into a nondivertible Highway Trust Fund [13]. The federal gas tax applies to both gasoline and diesel fuel, as well as to a number of less common fuels like liquefied natural gas and methanol. Since 1993, drivers have been paying 18.4 cents in tax per gallon of gasoline and 24.4 cents per gallon of diesel fuel.

> The money to construct the Interstate Highway System came from an increased federal gasoline tax that went into a non-divertible Highway Trust Fund.

The new interstate highways were controlled-access expressways with no at-grade crossings–that is, they had overpasses and underpasses instead of intersections. They were at least four lanes wide and were designed for high-speed driving. They were intended to serve several purposes: eliminate traffic congestion; replace what one highway advocate called "undesirable slum areas" with pristine ribbons of concrete; make coast-to-coast transportation more efficient; and make it easy to get out of big cities in case of an atomic attack [13]. It is difficult today to imagine the impact of the absence of the interstate system of highways in the U.S. The impact on the growth of the U.S. economy alone would have been enormous, since the delivery of goods between U.S. states would have been severely hampered.

7.5.2 URBAN MASS TRANSPORTATION ACT, 1964

The Urban Mass Transportation Act of 1964 (UMT Act) represents a seminal transportation policy for the U.S. As preface to the UMT Act, a bill was introduced in 1960 in the Senate that would have provided federal assistance for mass transportation. While it actually passed the Senate, it never emerged from committee in the House of Representatives. In 1961, Eisenhower's successor President John F. Kennedy (D-MA) supported a Senate proposal to establish federal assistance for mass transportation, but as part of a larger urban housing bill, and it was enacted into law [14]. President Kennedy signed the Omnibus Housing Act on June 30, 1961, and said that mass transportation is "…a distinctly urban problem and one of the key factors in shaping community development."

The 1961 act did not initiate broad-scale federal assistance for mass transportation, rather, the act provided $50 million for loans and $25 million – taken out of urban renewal funds – in outright grants for demonstration pilot projects in mass transportation. In 1962, President Kennedy sent a major transportation message to Congress. It called for the establishment of a program of federal capital assistance for mass transportation, with a core provision of good urban transportation. Sadly, President Kennedy's assassination on November 22, 1963, foreclosed his ability to see the bill enacted into law. However, Congress acted in 1964 on Kennedy's proposed program and on July 9, 1964, President Lyndon Johnson (D-TX) signed the UMTAct into law [14]. The new measure provided $375 million in capital assistance over 3 years. This law of the 1960s remains today as the principal federal statute for providing assistance for mass transportation in U.S. states, territories, and Native American lands.

> The Urban Mass Transportation Act of 1964 remains today as the principal federal statute for providing assistance for mass transportation in U.S. states, territories, and tribal lands.

7.5.3 DEPARTMENT OF TRANSPORTATION ACT, 1966

U.S. President Lyndon Johnson (D-TX) signed into law the Department of Transportation Act on October 15, 1966 [15]. The act resulted in producing the most sweeping reorganization of the Federal Government since the National Security Act of 1947. With the stroke of his pen, President Johnson created the fourth largest Federal agency and brought approximately 95,000 employees into the new organization. The Department of Transportation Act brought 31 previously scattered Federal elements under the wing of one Cabinet Department. The legislation provided for five initial major operating elements within the Department.

Four of these organizations were headed by an Administrator: the Federal Aviation Administration (FAA), which was previously the independent Federal Aviation Agency; the Federal Highway Administration; the Federal Railroad Administration; and the Saint Lawrence Seaway Development Corporation. The new Department also contained the U.S. Coast Guard, which was headed by a Commandant and had previously been part of the Treasury Department. The Department of Transportation (DOT) Act also created, within the new Department, a five-member National Transportation Safety Board (NTSB).

7.5.4 NATIONAL TRAFFIC AND MOTOR VEHICLE SAFETY ACT, 1966

President Lyndon Johnson (D-TX) and Congressional allies foresaw the need for additional federal legislation on motor vehicle safety beyond what was implied in the UMTAct [16]. On September 9, 1966, Johnson signed the National Traffic and Motor Vehicle Safety Act (NTMVSAct) into law. Immediately afterward, he signed the Highway Safety Act (HSAct). The two bills made the federal government responsible for setting and enforcing safety standards for cars and roads. Unsafe highways, Johnson argued, were a menace to public health: "In this century," Johnson said before he signed the bills, "more than 1,500,000 of our fellow citizens have died on our streets and highways; nearly three times as many Americans as we have lost in all our wars." It was a genuine crisis, and one that the automakers had proven themselves unwilling or unable to resolve. "Safety is no luxury item," the President declared, "no optional extra; it must be a normal cost of doing business [16]."

> The National Traffic and Motor Vehicle Safety Act and the Highway Safety Act made the federal government responsible for setting and enforcing safety standards for cars and U.S. roads.

By the mid-1960s, many reformers and safety experts were beginning to argue that there were actions the federal government could do to make the roads less dangerous. Automakers, these safety advocates argued, had the technology and the know-how to build stronger, sturdier cars and trucks that could keep people safe in case of an accident – and could even help prevent accidents altogether.

Although vehicle manufacturers' companies' lobbyists managed to weaken the safety standards in the final bill, the NTMVS Act did result in sturdier cars: it required "seat belts for every passenger, impact-absorbing steering wheels, rupture-resistant fuel tanks, door latches that stayed latched in crashes, side-view mirrors, shatter-resistant windshields and windshield defrosters, lights on the sides of cars as well as the front and back, and the padding and softening of interior surfaces and protrusions [16]." Additionally, the HS Act required that roadbuilders install guardrails, better streetlights, and stronger barriers between opposing lanes of traffic. These two federal statutes of the 1960s committed the U.S. federal government as a supporter of actions intended to protect the safety of the U.S. motoring public.

7.5.5 Motor Vehicle Information and Cost Savings Act. 1972

In 1972, the Motor Vehicle Information and Cost Savings Act, enacted October 20, 1972, expanded NHTSA's scope to include consumer information programs.

7.5.6 National Minimum Drinking Age Act, 1984

The National Minimum Drinking Age Act of 1984 (NMDAAct) was an important federal statement on reducing the toll of drivers' unwise consumption of alcoholic beverages on their operation of motor vehicles. This statute has important public health implications that resonate today, given the linkage between alcohol consumption and automobile fatalities. The NMDAAct essentially sets a national standard for the minimum drinking age

> The National Minimum Drinking Age Act of 1984 required all states to raise their minimum purchase and public possession of alcohol age to 21 years.

Transportation

of alcoholic beverages, delegating to the states the responsibility for enforcement of the age standard [17].

The principal provisions of the NMDA Act with implications for public health are excerpted from the Act as follows [17]:

- Required all states to raise their minimum purchase and public possession of alcohol age to 21 years. U.S. states that did not comply faced a reduction in highway funds under the Federal Highway Aid Act.
- Requires states to prohibit purchase and public possession of alcoholic beverages. It does not require the prohibition of persons under 21 (also called youth or minors) from drinking alcoholic beverages. The term "public possession" is strictly defined and does not apply to possession for the following four exceptions:
  1. "An established religious purpose, when accompanied by a parent, spouse, or legal guardian age 21 or older;
  2. Medical purposes when prescribed or administered by a licensed physician, pharmacist, dentist, nurse, hospital, or medical institution;
  3. In private clubs or establishments; and
  4. In the course of lawful employment by a duly-licensed manufacturer, wholesaler or retailer" [17].

Perspective: The NMDA Act is grounded on the public health concern of alcohol consumption as a key element in vehicle crashes. The statute is also interesting in regard to how it shares responsibility between federal policy and U.S. state policies on minimum age for alcohol consumption.

7.5.7 Intermodal Surface Transportation Efficiency Act, 1991

On December 18, 1991, President George H.W. Bush (R-TX) signed the Intermodal Surface Transportation Efficiency Act of 1991. The purpose of the Act was clearly enunciated in its statement of policy: "to develop a National Intermodal Transportation System that is economically efficient, environmentally sound, provides the foundation for the Nation to compete in the global economy and will move people and goods in an energy efficient manner [18]." The Act provided authorizations and appropriations for highways, highway safety, and mass transit for the next 6 years. Total funding of about $155 billion was available over fiscal years 1992–1997.

The provisions of the Act reflect these important policy goals. Some of the major features included:

- A National Highway System (NHS), consisting primarily of existing interstate routes and a portion of the Primary System, was established to focus Federal resources on roads that are the most important to interstate travel and national defense, roads that connect with other modes of transportation, and are essential for international commerce.
- State and local governments were given more flexibility in determining transportation solutions, whether transit or highways, and the tools of enhanced planning and management systems to guide them in making the best choices.

- New technologies, such as intelligent vehicle highway systems (now known as Intelligent Transportation Systems or ITS) and prototype magnetic levitation systems, were funded to push the U.S. forward into thinking of new approaches in providing 21st-century transportation.
- The private sector was tapped as a source for funding transportation improvements. Restrictions on the use of Federal funds for toll roads were relaxed and private entities were entitled to own such facilities.
- The Act continued discretionary and formula funds for mass transit.
- Highway funds were available for activities that enhance the environment, such as wetland banking, mitigation of damage to wildlife habitat, historic site, activities that contribute to meeting air quality standards, a wide range of bicycle and pedestrian projects, and highway beautification.

> The Intermodal Surface Act made highway funds available for environment activities, such as wetland banking.

- Highway safety was further enhanced by a new program to encourage the use of safety belts and motorcycle helmets. The Act mandated that passenger automobiles and light trucks built after September 1, 1998, to have airbags installed as standard equipment for the driver and the right front passenger.
- State uniformity in vehicle registration and fuel tax reporting was required. This was meant to ease the recordkeeping and reporting burden on businesses and contribute substantially to increased productivity of the truck and bus industry.

7.5.8 Motor Carrier Safety Improvement Act, 1999

U.S. President Bill Clinton (D-AR) signed the Motor Carrier Safety Improvement Act of 1999 (MCSI Act) into law on December 9, 1999. The MCSI Act amended federal transportation law to establish a Motor Carrier Safety Administration within the DOT, with a separate motor coach division, superseding the functions performed by the Office of Motor Carrier and Highway Safety of the Federal Highway Administration. The sections of the MCSI Act that have implications for public health are the following excerpts [19]:

> The Motor Carrier Safety Improvement Act of 1999 transformed policies to guide the growth and development of the U.S. transportation infrastructure.

> (§ 5) Amends federal transportation law to prohibit a U.S. state from issuing a special license or permit to an individual who holds a commercial driver's license that permits the individual to drive a commercial motor vehicle during a period in which: (1) the individual is disqualified from operating a commercial vehicle; or (2) the individual's driver's license is revoked, suspended, or canceled.
>
> (§ 5) Requires a U.S. state to maintain, as part of its driver information system, a record of each violation by, or conviction under, a U.S. state or local motor vehicle traffic control law while operating a motor vehicle (except a parking violation) for each individual who holds a commercial driver's license.

(§ 5) Directs the Secretary to initiate rulemaking to (1) require a federal medical qualification certificate as part of U.S. state-issued commercial drivers' licenses and (2) establish a national registry of preferred medical providers.

(§ 6) Directs the Secretary, through the National Highway Traffic Safety Administration in cooperation with the Motor Carrier Safety Administration, to carry out a program with the U.S. states to improve the collection and analysis of data on crashes, including crash causation, involving commercial motor vehicles.

(§ 6) Applies specified safety regulations to eight-passenger vehicles.

7.5.9 FIXING AMERICA'S SURFACE TRANSPORTATION ACT, 2015

Two bipartisan statutes joined together by the Congress and the U.S. President Barack Obama (D-IL) administration resulted in transportation policies pertaining to surface transportation in the U.S. Specifically, on July 6, 2012, President Barack Obama (D-IL) signed into law the Moving Ahead for Progress in the 21st Century Act (MAP-21) [20]. The act funded surface transportation programs at more than $105 billion for fiscal years (FY) 2013 and 2014. Of note, MAP-21 was the first long-term highway authorization enacted since 2005. Of note, the Act transformed the policy and programmatic framework for investments and guides the growth and development of the U.S. transportation infrastructure. MAP-21 created a streamlined, performance-based, and multimodal program to address the many challenges facing the U.S. transportation system. These challenges included improving safety, maintaining infrastructure condition, reducing traffic congestion, improving the efficiency of the system and freight movement, protecting the environment, and reducing delays in project delivery. MAP-21 was built on and refined many of the highway, transit, bike, and pedestrian programs and policies established in 1991.

Success with MAP-21 led to a second bipartisan policy statement from Congress. On December 4, 2015, President Obama signed into law the Fixing America's Surface Transportation Act (FAST Act) [21]. The FAST Act funds surface transportation programs, including, but not limited to, federal-aid highways at more than $305 billion for fiscal years 2016 through 2020. It was the first long-term surface transportation authorization enacted in a decade that provides long-term funding certainty for surface transportation in the U.S. The Federal Highway Administration (FHWA) has the major authorities under the FAST Act. As noted by the FHWA, the FAST Act built on the changes made by MAP-21 and set a course for transportation investment in U.S. highways. Specifically, the FAST Act:

- "Improves mobility on America's highways: The FAST Act establishes and funds new programs to support critical transportation projects to ease congestion and facilitate the movement of freight on the Interstate System and other major roads. Examples include developing a new National Multimodal Freight Policy, apportioning funding through a new National Highway Freight Program, and authorizing a new discretionary grant program for Nationally Significant Freight and Highway Projects.

- Creates jobs and supports economic growth: The FAST Act authorizes $226.3 billion in federal funding for FY 2016 through 2020 for road, bridge, bicycling, and walking improvements. In addition, the FAST Act includes a number of provisions designed to improve freight movement in support of national goals.
- Accelerates project delivery and promotes innovation: Building on the reforms of MAP-21 and FHWA's Every Day Counts initiative, the FAST Act incorporates changes aimed at ensuring the timely delivery of transportation projects. These changes will improve innovation and efficiency in the development of projects, through the planning and environmental review process, to project delivery [21]."

7.6 U.S. STATES' TRANSPORTATION POLICIES WITH HEALTH IMPLICATIONS

The federal transportation statutes and policies discussed in the previous sections are generally directed to providing resources and standards to states for road construction and implementation of standards such as drivers' age of alcohol purchases. But it is the U.S. states where literally "the rubber hits the road" as concerns enforcement of traffic and vehicle safety laws and policies. Prior to a description of three salient state laws that relate to public health, it is important to distinguish between primary and secondary laws. A primary law means that a law enforcement officer can pull you over for the offense without having to witness some other violation. For instance, an officer sees a driver texting and issues a citation. A secondary law refers to the fact that an officer can pull you over and issue a ticket only if the officer has witnessed some other violation – for example, the driver ran a stop light while texting.

> A *primary law* means a law enforcement officer can cite you for an offense without having to witness some other violation. A *secondary law* refers means an officer can cite you only if the officer has witnessed some other violation.

This section will describe three of the more important U.S. state enforcement policies that are relevant for public health purposes. These are establishment and enforcement of U.S. state statutes on vehicle speed limits, driving while under the influence of alcohol or drugs, and use of cell phones while driving. Each of these three transportation policies will be described in summary detail.

7.6.1 MOTOR VEHICLE SPEED LIMITS

As a matter of public health, limits on the speed of motor vehicles driven on public roads were found to be necessary for the purpose of reducing the potential for vehicle crashes and attendant harm to drivers, passengers, and pedestrians. The policy of establishing vehicle speed limits is in essence an acknowledgment of the physics of moving objects. The higher velocity of an object brought to a stop (e.g., crash), the greater the force exerted on the object. For motor vehicles, this translates into greater force on a vehicle's occupants as speed increases. As a consequence of the physics of moving objects, speed limits were established on motor vehicles that use public thoroughfares.

Setting speed limits on vehicles traveling on public roads is a political/policy process as well as an engineering endeavor. Speed limits are designed to provide a balance between the safety of passengers and convenience, i.e., time spent in travel [22]. Because speed limits are enforceable laws, they are set by elected officials. Speed limits are established through legislative actions, resulting in statutes and regulations enacted by U.S. state legislatures. State governments, e.g., Georgia, establish speed limits, which can be modified by county or municipal governments [23].

As to engineering, traffic engineers analyze roads for which speed limits are to be established [22]. This is often done through a speed study. A speed study consists of a review of the function, design, and actual use of the road under study. For instance, is the road intended for heavy commuter traffic or light local traffic? How wide are the lanes? How wide are the shoulders? Is it straight or curvy? Are there a lot of intersections? A primary principle of traffic engineering is to provide safe and quick-as-possible travel between destinations. The engineer has to balance those two objectives. The end product of a traffic engineering study is a recommendation for an appropriate speed limit for specific modes of traffic, e.g., recommended speed limit for personal use vehicles and large commercial trucks. The actual establishment of a road's speed limit is the responsibility of local government authorities, e.g., a city council.

> The policy of establishing vehicle speed limits is in essence an acknowledgment of the physics of moving objects.

7.6.2 Driving under the Influence of Drugs or Alcohol

Impaired performance by persons operating motor vehicles is a major public health problem. In particular, driving under the influence (DUI), also known as driving while intoxicated, drunk driving, or impaired driving is the crime of driving a motor vehicle while impaired by alcohol or drugs, including those prescribed by physicians [24]. The use of illicit drugs that refers to the use of illegal drugs (including marijuana according to federal law) or misuse of prescription drugs can make driving a car unsafe – just like driving after drinking alcohol. Drugged driving puts the driver, passengers, and others who share the road at risk. In 2015, the National Council on Alcoholism and Drug Dependency cited the following facts [25]:

> An estimated 32% of fatal car crashes in the U.S. in 2015 involved an intoxicated driver or pedestrian.

- "FACT: An estimated 32% of fatal car crashes involve an intoxicated driver or pedestrian.
- FACT: 3,952 fatally injured drivers tested positive for drug involvement.
- FACT: More than 1.2 million drivers were arrested in 2011 for driving under the influence of alcohol or narcotics.

- FACT: Car crashes are the leading cause of death for teens, and about a quarter of those crashes involve an underage drinking driver.
- FACT: On average, two in three people will be involved in a drunk-driving crash in their lifetime" [25].

Other troubling data include according to the National Survey on Drug Use and Health (NSDUH), in 2014, 27.7 million people in the U.S. aged 16 years or older had driven a vehicle while under the influence of alcohol in the past year and 10.1 million had driven under the influence of illicit drugs [26]. Other findings from the survey show that men are more likely than women to drive under the influence of drugs or alcohol. And a higher percentage of young adults aged 18–25 drives after taking drugs or drinking alcohol than do adults 26 or older.

> In 2014, 27.7 million people in the U.S. aged 16 years or older had driven a vehicle while under the influence of alcohol in the past year.

With alcohol, a drunk driver's level of intoxication is typically determined by a measurement of blood alcohol content (BAC). A BAC measurement in excess of a specific threshold level, such as 0.05% or 0.08%, defines a criminal offense with no need to prove impairment. In some jurisdictions, there is an aggravated category of the offense at a higher BAC level, such as 0.12%. As an example, the Georgia DUI law has three BAC conditions [27]:

- 0.08% or higher, if they are 21 years old or older operating regular passenger vehicles.
- 0.04% or higher, if they are operating commercial vehicles.
- 0.02% or higher, if they are younger than 21 years old.

The Georgia law also prescribes penalties when DUI infractions occur. The nature of DUI penalties depends on a person's age, license type, and previous DUI convictions. As examples, for drivers 21 years and older, the penalties for the first DUI are suspended license for up to 1 year, a $300–$1,000 fine, mandatory 40 hours of community service, possible imprisonment for 1 year, and possible limited driving permit. All penalties are subject to a court's final decision. For drivers age 16–20 years, a third DUI offense within 5 years will result in the following penalties: driver's license suspension for 5 years, $410 fee, DUI alcohol or drug risk reduction program and all associated costs, clinical evaluation and possible treatment, and ignition interlock device and habitual violator probationary license with court permission. While the nature and severity of DUI penalties vary across the U.S. states, it can be stated that DUI is a criminal offense with consequential penalties and significant public health implications.

> On average, two in three people will be involved in a drunk-driving crash in their lifetime.

All 50 U.S. states, the District of Columbia, and U.S. territories have laws that specifically target drugged drivers. Almost one-third of states have adopted the *per se* standard that forbids any presence of a prohibited substance or drug in a driver's body while in control of the vehicle, without any other evidence of impairment.

Others have established specific limits for the presence of intoxicating drugs, while still others follow a zero-tolerance rule with regard to the presence of intoxicating drugs in a person's system.

In addition to alcohol, according to the National Institute on Drug Abuse (NIDA) in 2016, after alcohol, marijuana is the drug most often found in the blood of drivers involved in crashes [28]. Tests for detecting marijuana in drivers measure the level of delta-9-tetrahydrocannabinol (THC), marijuana's mind-altering ingredient, in the blood. But the role that marijuana plays in crashes is often unclear. THC can be detected in body fluids for days or even weeks after use, and it is often combined with alcohol. The risk associated with marijuana in combination with alcohol, cocaine, or benzodiazepines appears to be greater than that for the drug by itself.

Several studies have shown that drivers with THC in their blood were about twice as likely to be responsible for a deadly crash or be killed as drivers who had not used drugs or alcohol. However, a large National Highway Traffic Safety Administration (NHTSA) study found no significant increased crash risk traceable to marijuana after controlling for drivers' age, gender, race, and presence of alcohol [28]. More research is needed, particularly, since several U.S. states have approved the use of marijuana for medical and recreational uses, thereby likely increasing the number of motor vehicle operators with THC in their blood.

Along with marijuana, prescription drugs are also commonly linked to drugged driving crashes. A nationwide study in 2010 of deadly crashes found that about 47% of drivers who tested positive for drugs had used a prescription drug, compared to 37% of those who had used marijuana and about 10% of those who had used cocaine [28]. The most common prescription drugs found were pain relievers. However, the study did not distinguish between medically supervised and illicit use of prescription drugs.

> Several studies have shown that drivers with THC in their blood were about twice as likely to be responsible for a deadly crash or be killed as drivers who had not used drugs or alcohol.

7.6.3 Use of Cell Phones while Driving a Vehicle

Cell phones have proliferated globally, commencing in the mid-first decade of the 21st century. The devices have had many positive benefits that include enhanced personal communications, commerce, recreation, and response to emergencies. But with the positive have come some adverse outcomes, including the public health problem of cell phones usage associated with motor vehicle crashes. Many localities have enacted restrictions or bans on cell phones or text messaging. In some but not all states, local jurisdictions need specific statutory authority to do so. In addition, most school bus drivers are banned from texting and using handheld cell phones by state code, regulation, or school district policy. One source notes that as of September 2017, vehicle drivers talking on a handheld cell phone while driving is banned in 15 U.S. states and the District of Columbia [29].

> The use of all cell phones by novice drivers is restricted in 38 states and the District of Columbia. Text messaging is banned for all drivers in 47 states and the District of Columbia.

The same source observes that the use of all cell phones by novice drivers is restricted in 38 states and the District of Columbia. Text messaging is banned for all drivers in 47 states and the District of Columbia. In addition, novice drivers are banned from texting in one state (Missouri).

An example of one state's cell phone law is provided by the state of Georgia. The Georgia cell phone law states the following [30]:

- "Except for bus drivers and novice drivers, there is no prohibition on cell phone use while driving in Georgia. Georgia prohibits all drivers from texting while driving.
- **Cell phone use**: There is no handheld cell phone prohibition for drivers. Novice drivers in Georgia – drivers younger than 18 – are prohibited from all cell phone use (handheld and hands-free) while driving.
- **Texting laws**: All drivers are prohibited from texting while driving. The law prohibits drivers from using a cell phone, text messaging device, personal digital assistant, computer, or similar wireless device to write, send, or read text data while driving. The ban applies to text messages, instant messages (IM), email, and internet data. The law creates exceptions for emergency personnel, drivers responding to emergencies, and drivers who are fully parked.
- **Bus drivers**: Bus drivers are prohibited from handheld and hands-free cell phone use while driving.
- How is it enforced? Offenders are subject to a $150 fine and one point against their Georgia driving record. Georgia's cell phone and texting laws are considered "primary" laws."

In 2018 Georgia stiffened the aforementioned law against driving and texting on cell phones. Called the Hands Free Law, the following is a brief description of what the law states [31]: "A driver cannot have a phone in their hand or use any part of their body to support their phone. Drivers can only use their phones to make or receive phone calls by using speakerphone, earpiece, wireless headphone, phone is connected to vehicle or an electronic watch. GPS navigation devices are allowed. ... A driver may not send or read any text-based communication unless using voice-based communication that automatically converts message to a written text or is being used for navigation or GPS. A driver may not write, send or read any text messages, e-mails, social media or internet data content. ... Exceptions are as follows: (1) Reporting a traffic crash, medical emergency, fire, criminal activity, or hazardous road conditions. (2) An employee or contractor of a utility service provider acting within the scope of their employment while responding to a utility emergency. (3) A first responder (law enforcement, fire, EMS) during the performance of their official duties. (4). When in a lawfully parked vehicle – this does not include vehicles stopped for traffic signals and stop signs on the public roadway." Commercial Motor Vehicle Operators can only use one button to begin or end a phone call... "The driver of a school bus cannot use a wireless telecommunication device or two-way radio while loading or unloading passengers. The driver of a school bus can only use a wireless telecommunication device while the bus is in motion as a two-way radio to allow live communications between the driver and school and public safety officials." First conviction: $50, one point on a license; second conviction: $100, two points on a license; third and subsequent convictions: $150, three points on a license.

Perspective: It is interesting in a public health policy context to note that the Georgia law contains special provisions on cell phone use while driving that are specific to bus drivers and novice drivers. It is also interesting that the Georgia legislature, concerned about the number of fatalities associated with cell phone texting while operating a motor vehicle, enacted a tough law call the Hands Free Law. Public health data supported the enactment of this law.

7.7 SAFETY TECHNOLOGY AND HEALTH IMPLICATIONS

Complementing federal and associated U.S. state policies purposed to make transportation in the U.S. safer for travelers are advancements in technologies that improve motor vehicle safety. To be described in this section are technological contributions to safer transportation including seat belts and other restraints, airbags, vehicle designs, and road designs. Each new technology is grounded in the purpose of improving the safety of persons experiencing vehicular transportation, a purpose that comports with good public health practice.

7.7.1 Seat Belts

As will be described in this section, persons seated in motor vehicles are made safer by wearing seat belts. This dictum came by way of both observational data and research findings. Observational data accrued as persons who responded to vehicle crashes began to realize that persons who remained inside the vehicles in which they were riding were less often fatally injured than the passengers who were thrown from the vehicle. In other words, a vehicle's shell and frame provided a degree of protection to passengers. Further, airline passengers in the U.S. had long been required to wear a seat belt during take-off and landings of aircraft. Subsequent to the observational data, safety analysis investigations were conducted that supported the observational data.

Following is a short history of seat belts as a safety component in automobiles and other motor vehicles, followed by a summary of the use of seat belts by operators and passengers in motor vehicles, all of which have importance for public health. Further, for policy context, it is important to distinguish between

> Persons seated in motor vehicles are made safer by wearing seat belts. This dictum came by way of both observational data and research findings.

primary and secondary enforcement of seat belt policies. Primary enforcement [of seat] belt use laws permit seat belt use law violators to be stopped and cited independently of any other traffic behavior. Secondary enforcement laws allow violators to be cited only after they first have been stopped for some other traffic violation, e.g., disregarding a stop sign [32].

According to Defensive Driving, a safe driving NGO, "the seat belt was invented by George Cayley, an English engineer who in the late 1800s created seat belts to help keep pilots inside their gliders [33]. However, the first patented seat belt was created by American Edward J. Claghorn on February 10, 1885, in order to keep tourists safer in taxis in New York City. Over time, the seat belt slowly started showing up in manufactured cars to help passengers and drivers stay put inside their car seats.

However, it was not until the mid-1930s when several U.S. physicians began testing lap belts, saw their impact on passengers' safety, and began urging manufacturers to provide seat belts in all cars. In 1954, the Sports Car Club of America required competing drivers to wear lap belts during competitions and in the following year, the Society of Automotive Engineers appointed a Motor Vehicle Seat Belt Committee. Race car drivers were the first to really wear seat belts to help protect themselves against serious internal injuries [33]."

The real breakthrough with modern seat belts came in 1958 when Swedish engineer Nils Bohlin invented the three-point seatbelt. Up until this event, seat belts in cars were two-point lap belts, which strapped across the body, with the buckle placed over the abdomen. The Volvo Car Corporation (Volvo) hired Bohlin in 1958, and he designed the three-point seat belt that better protects the driver and passenger in a vehicle crash. The three-point design was created to help secure both the upper and lower body. When Bohlin died in 2002, the Volvo Company had estimated that the three-point seat belt had saved more than 1 million lives in the four decades since it was introduced [33].

> Volvo estimated in 2002 that the three-point seat belt invented by Nils Bohlin had saved more than 1 million lives.

"Once the idea of safety benefits of seat belts caught on with the U.S. public, sales of seat belts skyrocketed. Auto companies offered seat belts as optional equipment and were even sold at local gas stations. Since 1966, American vehicles are required to have seat belts in their cars. As such, by 1975, most first-world countries had a seat belt requirement in their cars. Once they became more common in cars, laws soon followed afterwards. By 1970, the world's first seat belt law was created in Victoria, Australia, which required passengers to wear their seat belts at all times. In the U.S., seat belt laws came around the time of the creation of The National Traffic and Motor Vehicle Safety Act of 1966" [33].

All new passenger cars in the U.S. had some form of seat belts beginning in 1964 [34], shoulder belts in 1968, and integrated lap and shoulder belts in 1974 [35]. Initially, few occupants wore the belts: surveys in various locations recorded belt use of about 10%. The first widespread survey, taken in 19 cities in 1982, observed 11% belt use for drivers and front-seat passengers. New York enacted the first belt use law in 1984. Other U.S. states soon followed. In a typical U.S. state, belt use rose quickly to about 50% shortly after a state's belt law went into effect. By 1996, every state, with the exception of New Hampshire, had a mandatory seat belt use law covering drivers and front-seat occupants [35].

> New York enacted the first belt use law in 1984. Other U.S. states soon followed.

Studies in U.S. states that changed their law from secondary to primary enforcement show that belt use increased across a broad range of drivers and passengers. In some states, belt use increased more for low-belt-use groups, including

Transportation

Hispanics, African-Americans, and drinking drivers, than for all occupants [35]. In 2016, seat belt use in the U.S. ranged from 70.2% in New Hampshire, to 97.2% in Georgia, with a national average of 89.7% according to the NHTSA [36].

7.7.2 Child Safety Seat or Booster Seat

As observed and reported by the CDC, motor vehicle injuries are a leading cause of death among children in the U.S. Further, the CDC opines that many of these deaths can be prevented. In the U.S., 663 children ages 12 years and younger died as occupants in motor vehicle crashes during 2015, and more than 121,350 were injured in 2014 [37]. Further, the CDC study found that more than 618,000 children ages 0–12 years old rode in vehicles without the use of a child safety seat or booster seat or a seat belt at least some of the time. Of the children ages 12 years and younger who died in a crash in 2015 (for which restraint use was known), 35% were not buckled up [37]. All U.S. states have enacted statutes and attendant regulations that address the presence of young children riding in motor vehicles. As an example, according to Georgia's law [38]:

> Of the children ages 12 years and younger who died in a vehicle crash in 2015, 35% were not buckled up according to the CDC's injury data.

- "All children under the age of 8 whose height is less than 57 inches must ride in the backseat of a car. A child is safer in the back and farthest away from the force of an airbag, an important consideration given that the force of a deploying airbag can cause injuries to a child.
- Children less than the age of 8 years old are required to be in either a car seat or a booster seat suitable for their age and height.
- If there is not a back seat in the vehicle (e.g., a truck) or if other restrained children are in the back seat, Georgia law permits a child under the age of 8 years old to sit in front if restrained in the proper car seat or booster and the child weighs at least 40 pounds.
- Georgia's Primary Safety Belt Law allows law enforcement officers to issue a citation if they observe a seat belt offense. They do not need to stop the driver for another traffic violation first, as in some other states.
- Violating these laws can result in a fine of up to $50 and one point against your license per improperly restrained child. A second incident may double the fines and points" [32].

7.7.3 Airbags

Airbags are a type of automobile safety restraint like seat belts. An airbag is a safety feature designed to protect passengers in a head-on or side-impact collision. They are gas-inflatable cushions built into the steering wheel, dashboard, door, roof, or seat of a car. A crash sensor triggers a rapid expansion of an airbag to protect occupants from the impact of a vehicle crash. Most cars today have driver's side airbags and many have one on the passenger side as well. Located in the steering wheel assembly on the driver's side and in the dashboard on the passenger side, the airbag device responds within milliseconds of a crash. After the initial impact, a folded nylon bag becomes rapidly inflated with nitrogen gas. This acts as a cushion for passengers, preventing them from being thrown forward into the steering column or dashboard, sparing them the trauma of striking an object.

The airbag was first conceived by John W. Hetrick in 1952 [39]. He created the idea as a result of an event that had occurred in the spring of 1952. Hetrick applied an event he had observed while in the U.S. Navy to the design of the airbag. He was repairing a torpedo that had a canvas covering.

> The Intermodal Surface Transportation Efficiency Act of 1991 required both driver's side and passenger airbags to be installed in all new vehicles by 1998.

When the compressed air that was in the torpedo was released, it quickly inflated the canvas shooting it to the ceiling. With this knowledge, he developed his design until he was able to obtain a patent on an airbag device on August 5, 1952. Automakers were interested in the airbag after Hetrick patented the idea. Both Ford and General Motors (GM) began to experiment with airbag designs in the late 1950s.

With the success of their tests, Ford decided that it would use the safety device in its 1971 line of full-size Ford and Mercury automobiles. However, this plan was quickly abandoned by Ford's engineers during prototype testing due to the difficulty of manufacturing reliable airbag systems. At the same time, GM was working on their version of an airbag, eventually allowing them to develop an effective airbag system. GM installed this system in 1,000 Chevrolet Impalas in 1973. This was the first time that airbag systems were offered to the public. However, the airbag did not sell well because they were offered as extra-expense options. During this period, only 10,321 airbag-equipped cars were sold, leading GM to set aside airbag its program.

> The airbag was reborn after Mercedes-Benz offered it as an option in its 1984 models.

The airbag was reborn after Mercedes-Benz offered it as an option in its 1984 models. Within 2 years, airbags had become standard equipment on all of Mercedes-Benz's automobiles. The airbag crusade gained further momentum when in 1991 Congress enacted the Intermodal Surface Transportation Efficiency Act. This ordered the National Highway Traffic Safety Administration (NTHSA) to require both driver's side and passenger airbags in all new vehicles by 1998 [39]. And as with the incorporation of seat belts as safety devices in motor vehicles, airbags have proven to be a public health success, preventing thousands of premature death in drivers and passengers.

7.7.4 Vehicle Designs

Motor vehicles are designed by vehicle manufacturers for appeal to potential customers and to meet specific applications in commercial affairs. Appeal to customers includes design features such as vehicle shape, color, cost to purchase and operate, engine efficiency, serviceability, and available special technologies such as communication devices. Vehicles designed for specific commercial applications include buses, trucks of various sizes and engine efficiency, vehicles for law enforcement, military equipment, and farming implements. Automobiles, in particular, are designed for sales appeal and customer utility.

In the U.S. motor vehicles are subject to adherence to safety standards established and enforced by the NHTSA, a component of the DOT. Federal Motor Vehicle Safety Standards (FMVSS) are U.S. federal regulations specifying design, construction, performance, and durability requirements for motor vehicles and regulated U.S. safety-related components, systems, and design features [40,41]. FMVSS are developed and enforced by NHTSA pursuant to statutory authorization in the National Traffic and Motor Vehicle Safety Act of 1966. Examples of some FMVSS regulations that relate to safety and therefore have relevance to public health are cited in the following subsection.

7.7.4.1 Vehicle Crash Design

Several FMVSS standards pertain to motor vehicle crash designs. The following excerpted examples portray the wide breadth of these kinds of design. The scope and purpose are illustrated. The actual details of each standard's specifications are not shown, but are available as NHTSA resources [40,41].

Standard No. 208; Occupant crash protection.

- S1. Scope. This standard specifies performance requirements for the protection of vehicle occupants in crashes.
- S2. Purpose. The purpose of this standard is to reduce the number of deaths of vehicle occupants, and the severity of injuries, by specifying vehicle crashworthiness requirements in terms of forces and accelerations measured on anthropomorphic dummies in test crashes, and by specifying equipment requirements for active and passive restraint systems.

Standard No. 224; Rear impact protection.

- S1. Scope. This standard establishes requirements for the installation of rear impact guards on trailers and semitrailers with a gross vehicle weight rating (GVWR) of 4,536 kg or more.
- S2. Purpose. The purpose of this standard is to reduce the number of deaths and serious injuries occurring when light-duty vehicles impact the rear of trailers and semitrailers with a GVWR of 4,536 kg or more.

Standard No. 218; Motorcycle helmets.

- S1. Scope. This standard establishes minimum performance requirements for helmets designed for use by motorcyclists and other motor vehicle users.
- S2. Purpose. The purpose of this standard is to reduce deaths and injuries to motorcyclists and other motor vehicle users resulting from head impacts.

Standard No. 222; School bus passenger seating and crash protection.

> S1. Scope. This standard establishes occupant protection requirements for school bus passenger seating and restraining barriers.
>
> S2. Purpose. The purpose of this standard is to reduce the number of deaths and the severity of injuries that result from the impact of school bus occupants against structures within the vehicle during crashes and sudden driving maneuvers.

The preceding standards must be utilized by motor vehicle and equipment manufacturers and apply to all vehicles manufactured or imported for sale in the U.S.

7.7.4.2 Crash Avoidance Technology

Crash avoidance features are rapidly making their way into the vehicle fleet. Shown in Table 7.3 are NHTSA data that illustrate the evolution of safety technology in motor vehicles [42]. Noteworthy is the evolution of more sophisticated technologies with the passage of time. As of 2017 six of the most common new technologies are forward collision warning, autobrake, lane departure warning, lane departure prevention, adaptive headlights, and blind spot detection. Information on the availability of features comes from the manufacturers and as of 2017 none of these safety features were mandated by U.S. federal regulations.

> 20 automakers representing more than 99% of the U.S. auto market committed to make automatic emergency braking (AEB) a standard feature by 2022.

However, the NHTSA and the Insurance Institute for Highway Safety announced on March 17, 2016, a historic commitment by 20 automakers representing more than 99% of the U.S. auto market to make automatic emergency braking (AEB) a standard feature on virtually all new vehicles no later than NHTSA's 2022 reporting year, which begins Sept. 1, 2022 [43]. AEB systems help prevent crashes or reduce their severity by applying the brakes for the driver. These systems use on-vehicle sensors such as radar, cameras, or lasers to detect an imminent crash, warn the driver, and apply the brakes if the driver does not take sufficient action quickly enough.

Automakers making the commitment are Audi, BMW, FCA US, Ford, General Motors, Honda, Hyundai, Jaguar Land Rover, Kia, Maserati, Mazda, Mercedes-Benz, Mitsubishi Motors, Nissan, Porsche, Subaru, Tesla Motors, Toyota, Volkswagen, and Volvo Car USA. According to the NHTSA, "the unprecedented commitment means that this important safety technology will be available to more consumers more quickly than would be possible through the regulatory process [43]." NHTSA estimates that the agreement will make AEB standard on new cars 3 years faster than could be achieved through the formal regulatory process. During those 3 years, according to Insurance Institute for Highway Safety estimates, the commitment will prevent 28,000 crashes and 12,000 injuries, a significant contribution to public health.

7.7.4.3 Autonomous Vehicles

An autonomous car (also known as a driverless car, self-driving car, robotic car) is a vehicle that is capable of sensing its environment and navigating without

TABLE 7.3
The Evolution of Automated Safety Technologies in Motor Vehicles [41]

	Safety Technology				
Period	Safety/Convenience Features	Advanced Safety Features	Advanced Driver Assistance Features	Partially Automated Safety Features	Fully Automated Safety Features
1950–2000	Cruise control, Seat belts, Antilock brakes				
2000–2010		Electronic stability control, Blind spot detection, Forward collision warning, Land departure warning			
2010–2016			Rearview video systems, Automatic emergency braking, Pedestrian automatic emergency braking, Rear cross traffic alert, Lane centering assist		
2016–2025				Lane keeping assist, Adaptive cruise control, Traffic jam assist, Self-park	
2025+					Highway autopilot

human input [44]. Autonomous vehicles date back to the 1939 World's Fair in New York where the General Motors exhibit predicted the development of driverless, radio-controlled electric cars.

> An autonomous car is a vehicle that is capable of sensing its environment and navigating without human input.

As TVs and modern appliances emerged in the U.S. in the 1950s, more images of driverless cars debuted. In the 1980s, experiments that detected the painted lines in the road were performed in the U.S. and Europe, and in 2011, Nevada became the first state in the U.S. to legalize their use.

Driverless cars use lasers that scan the environment more than a million times per second. Combined with digital GPS maps of the area, driverless cars detect white and yellow lines on the road as well as every stationary and moving object in their perimeter. Autonomous vehicles can drive themselves as long as a human driver can take control immediately when necessary.

Accident avoidance is the major incentive because the car can respond faster than a human driver. In addition, people can arrive more relaxed after a long trip. Vehicles can travel closer together on the road, and computers can operate them more economically than people. The ultimate manifestation is the reduction of vehicles. For example, driverless taxis could replace a family's second car that sits idle most of the time. Of course, fewer cars overall has other implications.

Two U.S. cities have given approval to companies using driverless cars. Over the last few years, GM's Cruise self-driving division has been operating a fleet of autonomous Chevy Bolt EVs in San Francisco. The test program is regulated by the California Division of Motor Vehicles and in October 2020 the agency gave them permission to operate vehicles without a backup driver. GM asserted they will be the first to test a truly driverless system in a major U.S. city. A second autonomous car company, Waymo, launched its own driverless ride-hailing service to the public in Phoenix [44a]. It remains to be determined whether the public will accept this form of city transportation.

Perspective: If thousands of lives can be saved each year, driverless cars will be a huge benefit. However, there are situations that are not so straightforward. For example, drivers on their daily commute in winter months know when steep hills are coming and might slow down considerably when temperatures fall below freezing. In addition, how will an automatic vehicle analyze hand signals from a police officer or road worker when a traffic incident has occurred or when repairs are taking place? It will take time and validated engineering to iron out the many exceptions to routine driving.

7.7.5 ROAD DESIGN

One of the primary reasons for building the Interstate System was to improve the safety of highway users: drivers, passengers, and pedestrians. Over the past 50 years, the Interstate System has done much to make highway travel safer and more efficient. Relative safety is measured by the "fatality rate" (fatalities per 100 million vehicle miles traveled, a measure used so data can be compared as traffic volumes change) [45]. The Interstate System is the safest road system in the U.S., with a fatality rate of 0.8 – compared with 1.46 for all U.S. roads in 2004.

When the Interstate Construction Program began in 1956, the national fatality rate was 6.05. This improvement in safety has been the result of many factors working together: the shifting of traffic onto the safer interstate highways and technological advances in safety, such as wider shoulders; slid-resistant pavements; better guardrail, signs, and markings; clearer sight distances; and breakaway sign posts and utility poles [45]. In addition, many other factors have contributed to improved safety on the U.S. highway system, including new vehicle safety features, such as shatter proof glass, padded interiors, safety belts, and airbags; programs to reduce impaired driving; and the combined, coordinated efforts of many private organizations and public agencies working together to make U.S. highways ever safer.

7.8 U.S. FEDERAL AGENCIES WITH TRANSPORTATION RESPONSIBILITIES

All cabinet-level U.S. federal departments are involved with transportation issues and programs to some degree, given the need for departments and agencies to transport personnel or cargo from place to place. For example, the Department of Defense must move military material and troops to and from locations around the world. And the Department of Homeland Security contains the Federal Emergency Management Agency and the U.S. Coast Guard, both of which must transport personnel, equipment, and supplies to U.S. locations where emergencies require their presence. However, for the purposes of this chapter, the U.S. Department of Transportation (DOT) deals most directly with issues and programs of greatest relevance to public and ecological health. Therefore this section and subsections that follows will describe the authorities and programs of the DOT.

7.8.1 U.S. Department of Transportation

Prior to the establishment of the DOT, the Under Secretary of Commerce for Transportation administered the functions now associated with the DOT. In 1965, the administrator of the Federal Aviation Agency – the future Federal Aviation Administration (FAA) – suggested to President Lyndon B. Johnson (D-TX) that transportation should be elevated to a cabinet-level post, and that the FAA be folded into the DOT [46]. President Johnson accepted the proposition that a single federal department was needed to develop and conduct comprehensive transportation policies and programs across all transportation modes. In 1966, Congress authorized the creation of a cabinet department that would combine major federal transportation responsibilities. This new department, the DOT, began full operations on April 1, 1967. The DOT comprises several organizational components of relevance to motor vehicle safety and passengers' well-being, matters of public health, as noted in the following sections.

7.8.1.1 Federal Highway Administration

The Federal Highway Administration (FHWA) was established in 1966 as the successor to a series of earlier federal organizations dating back to 1893 that had worked to improve U.S. public roads. The agency states its mission as "to improve mobility on the Nation's highways through national leadership, innovation, and program delivery" [47]. The FHWA supports U.S. state and local governments in the design, construction, and maintenance of the U.S. highway system (Federal Aid Highway Program) and various federally- and tribal-owned lands (Federal Lands Highway Program). The strategic priorities of the FHWA are stated to be "national leadership in transportation policy and innovation, effective delivery of the federal highway programs, improved safety and performance of the U.S. highway systems, and enhancement of the FHWA's corporate capacity to achieve its mission" [47]. The FHWA is headquartered in Washington, D.C., with satellite offices in each U.S. state. The following four program goals of the FHWA have implications for public and private health agencies.

> The number of highway-related fatalities in the U.S. has decreased by about 18% between 2006 and 2015.

- **"Making roads safer**: The number of highway-related (emphasis added) fatalities in the U.S. has decreased by about 18% between 2006 and 2015. The decrease coincides with, and is at least partly attributable to, the establishment and continuation of the Highway Safety Improvement Program (HSIP) as a core Federal-Aid Highway program with the requirement for States to have "Strategic Highway Safety Plans"; and the integration of both with other safety programs across the DOT.
- **Improving highway conditions and performance**: The share of National Highway System (NHS) travel occurring on pavements with good ride quality rose significantly from 48% in 2001 to 59% in 2014, despite an increase in NHS mileage of more than 50,000 miles due to the Moving Ahead for Progress in the 21st Century Act (MAP-21). By bringing pavements up to a condition of good repair, system users are benefited by decreasing wear and tear on vehicles and resulting repair costs, reducing traveler delays, and lowering crash rates.
- **Improving bridge condition and performance**: Over the decade commencing in 2006, even as the total number of bridges in the Nation's inventory increased from 597,561 to 611,845, the percentage of bridges classified as structurally deficient dropped from 12.6% in 2006 to 9.6% in 2015. Similarly, the percentage of the deck area (considering bridge size) on bridges classified as structurally deficient decreased from 9.6% in 2006 to 6.7% in 2015.
- **Spurring innovation**: The Every Day Counts (EDC) program seeks to increase innovation at every stage of the highway project lifecycle. Launched in 2010 as a partnership with State and local agencies, EDC has initiated its fourth 2-year cycle focusing on "efficiency through collaboration and technology." These innovations continue the goals of accelerating project delivery, enhancing roadway safety, reducing congestion, improving infrastructure performance, and improving environmental sustainability [48].

7.8.1.2 Federal Motor Carrier Safety Administration

The Federal Motor Carrier Safety Administration (FMCSA) was established within the DOT on January 1, 2000, pursuant to the Motor Carrier Safety Improvement Act of 1999. Formerly a part of the Federal Highway Administration, the FMCSA's primary mission is stated to be "to prevent commercial motor vehicle-related fatalities and injuries" [49]. The FMCSA asserts that activities of the FMCSA contribute to ensuring safety in motor carrier operations "through strong enforcement of safety regulations; targeting high-risk carriers and commercial motor vehicle drivers; improving safety information systems and commercial motor vehicle technologies; strengthening commercial motor vehicle equipment and operating standards; and increasing safety awareness [49]."

To accomplish these activities, the FMCSA works with federal, state, and local enforcement agencies, the motor carrier industry, labor and safety interest groups, and others. In regard to states, the FMCSA provides financial assistance for roadside inspections and other commercial motor vehicle safety programs, and it promotes motor vehicle and motor carrier safety. Other FMCSA activities include: (1) the development of unified motor carrier safety requirements and procedures throughout North America. (2) enforcement of regulations ensuring safe highway transportation of hazardous materials, (3) develops standards to test and license commercial motor vehicle drivers, (4) collects and disseminates data on motor carrier safety and directs resources to improve motor carrier safety, (5) operates a program to improve safety performance and remove high-risk carriers U.S. highways. The FMCSA is headquartered in Washington, D.C., with field offices in each U.S. state.

7.8.1.3 Federal Railroad Administration

The Federal Railroad Administration (FRA) was created by the Department of Transportation Act of 1966. The FRA's mission is "to enable the safe, reliable, and efficient movement of people and goods [50]." The FRA's functions include railroad safety and customer training (including state safety inspectors), accident and employee fatality investigations and reporting, partnerships between labor, management, and the agency that address systemic initiatives, development and implementation of safety rules and standards [50]. The FRA's Office of Railroad Safety promotes and regulates safety throughout the U.S. railroad industry. The staff includes 400 Federal safety inspectors who operate out of eight regional offices. Each region's personnel include grade crossing safety managers and safety inspectors for five of the safety disciplines, focusing on compliance and enforcement in: hazardous materials, motive power and equipment, operating practices, signal and train, control, track. The FRA is headquartered in Washington, D.C., with seven regional offices that span the continuous U.S. states and Alaska, with no office in Hawaii.

> The Federal Railroad Administration's Office of Railroad Safety promotes and regulates safety throughout the U.S. railroad industry.

7.8.1.4 Federal Aviation Administration

The Federal Aviation Agency became on April 1, 1967, one of several organizations within the DOT and received a new name, the Federal Aviation Administration (FAA). The FAA's mission is stated as, "to provide the safest, most efficient aerospace system in the world" [51]. Two of the FAA components with important responsibilities that relate to public health are the Air Traffic Organization (ATO) and the Aviation Safety (AS) office. The ATO is the operational arm of the FAA. It is responsible for providing safe and efficient air navigation services to 30.2 million square miles of U.S. airspace. This represents more than 17% of the world's airspace and includes all of the U.S. and large portions of the Atlantic and Pacific Oceans and the Gulf of Mexico.

Separate from the ATO, the AS is responsible for the certification, production approval, and continued airworthiness of aircraft; and certification of pilots, mechanics, and others in safety-related positions. The AS is also responsible for: "certification of all operational and maintenance enterprises in domestic civil aviation, certification and safety oversight of approximately 7,300 U.S. commercial airlines and air operators, and civil flight operations, and for developing regulations" [52]. The FAA is headquartered in Washington, D.C., with nine regional offices.

7.8.1.5 Federal Transit Administration

The Federal Transit Administration (FTA) provides financial and technical assistance to local public transit systems, including buses, subways, light rail, commuter rail, trolleys and ferries. Its mission is "to enhance citizens' mobility, accessibility, and economic well-being through the development and management of public transport services that are comprehensive, affordable, efficient, reliable, safe, and environmentally sound" [53]. The FTA also oversees safety measures and helps develop next-generation technology research. The FTA notes that transit ridership in the U.S. has reached its highest level since 1957, with more than 10.5 billion individual transit trips taken in 2015. To increase and improve the quality of public transportation, the FTA provides grant funding, technical assistance, safety oversight, policy development, planning support, and innovative technology research, to enhance mobility and access for people nationwide.

> Transit ridership in the U.S. reached more than 10.5 billion individual transit trips taken in 2015.

The FTA states that it has partnered since 1964 with U.S. state and local governments to create and enhance public transportation systems, investing more than $12 billion annually to support and expand public transit rail, bus, trolley, ferry, and other services [53]. The FTA is headquartered in Washington, D.C., with 10 regional offices.

7.8.1.6 Maritime Administration

The Maritime Administration (MARAD) is the agency within the DOT dealing with waterborne transportation. Established in 1950 under the auspices of U.S. President Harry Truman's (D-MO), the MARAD traces its origins to the Shipping Act of 1916, which established the U.S. Shipping Board, the first federal agency tasked with promoting a U.S. merchant marine and regulating U.S. commercial shipping. Congress enacted the 1916 law in part because of the severe disruptions in shipping caused by World War I. Six years later Congress passed the Merchant Marine Act of 1936, creating the U.S. Maritime Commission, which assumed the duties, functions, and property of the Shipping Board Bureau [54].

The MARAD works in many areas involving ships and shipping, shipbuilding, port operations, vessel operations, national security, environment, and safety. In 1981, the MARAD was transferred to the DOT [54]. The MARAD is headquartered in Washington, D.C., with 10 gateway port offices, located in the largest ports on the West, East, and Gulf Coasts, the Great Lakes, and the inland river system.

7.8.1.7 National Highway Traffic Safety Administration

The National Highway Traffic Safety Administration (NHTSA) was established by the Highway Safety Act of 1970 as the successor to the National Highway Safety Bureau, to carry out safety programs under the National Traffic and Motor Vehicle Safety Act of 1966 and the Highway Safety Act of 1966 [55]. The NHTSA also carries out consumer programs established by the Motor Vehicle Information and Cost Savings Act of 1972. On December 4, 2015, the Fixing America's Surface Transportation (FAST) Act was enacted into law, providing long-term funding certainty for highway and motor vehicle safety. Several of NHTSA's programs bear on issues of public health and safe driving. These include the following [55]:

- "NHTSA's Office of Emergency Medical Services (EMS) offered in 1971 the first national guidelines for training emergency medical technicians. NHTSA continues to advance a national vision for EMS through projects and research; fosters collaboration among Federal agencies involved in EMS planning; measures the health of the Nation's EMS systems; and delivers the data EMS leaders need to help advance their systems.
- The first child passenger safety law was enacted in Tennessee in 1978. All U.S. states enacted laws by 1985.
- In 2002, NHTSA launched its first nationwide Click It or Ticket seat belt campaign. By 2015 seat belt use across the U.S. reached an all-time high rate of 87%.
- The 21-year-old minimum drinking age laws were achieved nationwide in 1988 and nationwide zero-tolerance drunk driving laws for drivers less than 21 were achieved in 1998.
- In 2003, the first national media campaign to fight drunk driving was launched. In 2005, the 0.08 BAC drunk driving laws were enacted nationwide. In the following decade, as a result of the laws, high-visibility enforcement, and media efforts, drunk driving fell 23%.
- The New Car Assessment Program (NCAP) was established to provide crashworthiness ratings for new vehicles. Five-Star Safety Ratings were offered in 1993 to help consumers compare ratings and safety features of new cars and trucks. Research and Rulemaking efforts are underway to update the NCAP program to accommodate emerging vehicle safety technologies.
- Driver and passenger side air bags were required in cars and light trucks by model year 1998. In 2013, a NHTSA study estimated that 43,000 lives were saved by air bags" [55].

7.8.1.8 National Transportation Safety Board

The National Transportation Safety Board (NTSB) is an independent U.S. federal agency charged by Congress with investigating every civil aviation accident in the U.S. and significant accidents in other modes of transportation – railroad, highway, marine, and pipeline. The NTSB states its mission to be "Making transportation safer by conducting independent accident investigations, advocating safety improvements, and deciding pilots' and mariners' certification appeals" [56].

> The NTSB investigates every civil aviation accident in the U.S. and significant accidents in other modes of transportation.

The NTSB originated in the Air Commerce Act of 1926, in which the U.S. Congress charged the U.S. Department of Commerce with investigating the causes of aircraft accidents. Later, that responsibility was given to the Civil Aeronautics Board's Bureau of Aviation Safety, when it was created in 1940. In 1974, Congress reestablished the NTSB as a completely separate entity, outside the DOT, reasoning that "...No federal agency can properly perform such (investigatory) functions unless it is totally separate and independent from any other ... agency of the United States." The NTSB has no authority to regulate, fund, or be directly involved in the operation of any mode of transportation. Rather, it conducts investigations and makes recommendations from an independent viewpoint [57].

The agency states that since its inception, it has investigated more than 132,000 aviation accidents and thousands of surface transportation accidents and has issued more than 13,000 safety recommendations to more than 2,500 recipients [57]. The NTSB's headquarters are in Washington, D.C., with five regional field offices [56].

An example of a NTSB investigation is the fatal 2016 crash of an autonomous vehicle in Florida [58]. The agency's crash report of 2017 found that the vehicle's autopilot had functioned properly but lacked safeguards to prevent drivers from using it improperly. The fatality occurred when the vehicle's driver ignored warning from the autopilot, which resulted in a crash with a large truck. The vehicle's manufacturer revised its autopilot as a consequence of the NTSB report.

Perspective: The NHTSA's authorities and programs relate directly to the public health goals of prevention of premature death and disabilities. As enumerated in the preceding text, the agency's efforts in campaigns to promote the use of seat belts, installation of airbags in motor vehicles, and advocacy of emerging vehicle safety technologies all comport with public health benefits.

7.8.2 OTHER FEDERAL AGENCIES WITH TRANSPORTATION POLICIES

In addition to the DOT, other U.S. federal departments and agencies also have policies that pertain to transportation of personnel and cargo. For example, the Department of Defense (DOD) has large transportation policies, equipment, and services that are concerned with transporting military personnel, and equipment to many global locations. Similarly, the DHS has policies, equipment, and services to transport to domestic locations in instances of natural disasters, border security (Coast Guard), and homeland security activities. However, these departments and agencies do not have statutory authority for transportation issues, as does the DOT and are therefore not elaborated herein. A similar parallel exists in U.S. state governments, where states have departments of transportation, which has primary authority on matters of roads, safety, and driving regulations.

7.9 INTERNATIONAL GOVERNMENT ORGANIZATIONS WITH TRANSPORTATION POLICIES

Global transportation became a reality in the 20th century. The invention and adoption of airplanes as a common means to transport passengers and cargo across large expanses of the globe became commonplace, as did large container ships for the movement of vast quantities of goods across oceans and other bodies of water. Within national borders, motor vehicles and railways gradually replaced animal-drawn carts and similar conveyances. International travel and transportation necessitated the establishment of agencies that could interact with domestic government and private sector entities for the purpose of coordination, regulation, and interaction on areas of specialty need, e.g., coordination of air travel. The following United Nations organizations play important roles in international transportation policies and actions.

7.9.1 INTERNATIONAL CIVIL AVIATION ORGANIZATION

The International Civil Aviation Organization (ICAO) is a UN specialized agency, established by its Member States in 1944 to manage the administration and governance of the Convention on International Civil Aviation (Chicago Convention) [59]. The ICAO works with the Convention's 191 Member States and industry groups to reach consensus on international civil aviation Standards and Recommended Practices (SARPs) and policies in support of a safe, efficient, secure, economically sustainable and environmentally responsible civil aviation sector. These SARPs and policies are used by ICAO Member States to ensure that their local civil aviation operations and regulations conform to global norms, which in turn permits more than 100,000 daily flights in aviation's global network to operate safely and reliably in every region of the world.

> The International Civil Aviation Organization is a UN specialized agency to manage the administration and governance of the Convention on International Civil Aviation.

In addition to its core work resolving consensus-driven international SARPs and policies among its Member States and industry, and among many other priorities and programs, The ICAO states that it also coordinates assistance and capacity building for Member States in support of numerous aviation development objectives; produces global plans to coordinate multilateral strategic progress for safety and air navigation; monitors and reports on numerous air transport sector performance metrics; and audits States' civil aviation oversight capabilities in the areas of safety and security [59]. The ICAO is headquartered in Montreal, Canada.

7.9.2 INTERNATIONAL MARITIME ORGANIZATION

The International Maritime Organization (IMO) is a specialized UN agency responsible for regulating shipping. The IMO was established in Geneva, Switzerland in 1948 and came into force 10 years later, meeting for the first time in 1959 [60]. The IMO has 172 Member States and three Associate Members. The main mission and responsibility of the IMO is to "develop and preserve a comprehensive framework of regulations and policies for the shipping industry and its activities like maritime security, safety, technical cooperation, environmental concerns, and legal matters" [60].

> The International Maritime Organization is a specialized UN agency responsible for regulating shipping.

The major areas of concern that the IMO has been able to bring under regulation have been "prevention of accidents, setting up safety standards for ships and other vessels (including design and materials) for the member states to abide by, maintaining adherence to the established treaties of safety and security, prevention of pollution and other avoidable human disasters [60]." The IMO also facilitates technical co-operation among member states, setting up an audition and monitoring scheme for these rules, standards and finally monitoring liabilities and compensation in case of breach of any of these regulations. The IMO is headquartered in London, UK.

7.9.3 WORLD HEALTH ORGANIZATION

The World Health Organization (WHO) is the UN's principal agency and authority on global health. Among its many responsibilities are transportation-related health issues. In particular, in April 2004, the UN General Assembly resolution A/RES58/289 on "Improving global road safety" invited the WHO, working in close cooperation with the UN regional commissions, to act as coordinator on road safety issues across the UN system [61]. The World Health Assembly accepted this invitation in May 2004 and the WHO subsequently established the UN Road Safety Collaboration (UNRSC), which holds biannual meetings to discuss global road safety issues.

The Collaboration is an informal consultative mechanism whose members are committed to road safety efforts and in particular to the implementation of the recommendations of the *World Report on Road Traffic Injury Prevention*. The goal of the Collaboration is to facilitate international cooperation and to strengthen global and regional coordination among UN agencies and other international partners to implement UN General Assembly resolutions and the recommendations of the world report thereby supporting country programs.

> In 2015 the Secretary-General of the UN noted that each year some 1.3 million people are fatalities and up to 50 million people are injured on the world's roads.

Of note, UN Secretary-General Ban Ki-moon announced on April 29, 2015, the appointment of a Special Envoy for Road Safety. On this occasion, the Secretary-General noted that each year some 1.3 million people are fatalities and up to 50 million people are injured on the world's roads [62]. Half of all road traffic deaths are among vulnerable road users such as pedestrians, cyclists, and motorcyclists. Road traffic deaths are also the leading cause of death for young people aged 15–29 years old, and road traffic injuries are the eighth leading cause of death globally.

7.10 NGOs CONCERNED WITH TRAFFIC AND VEHICLE SAFETY

In addition to government organizations and agencies that are involved with issues of transportation safety, there exist several non-government organizations (NGOs) with similar aims and programs. Some of these NGOs are described herein.

7.10.1 GLOBAL ROAD SAFETY FACILITY

The GRSF is a global partnership program administered by the World Bank. It was established in 2006 with a mission to help address the growing crisis of road traffic deaths and injuries in low- and middle-income countries (LMICs). According to the World Bank, the GRSF provides funding, knowledge, and technical assistance designed to scale up the efforts of the LMICs to build their scientific, technological, and managerial capacities [63]. Since its inception, the GRSF has operated as a hybrid grant-making global program, allowing it to distribute funding externally for global, regional, and country activities, and internally through World Bank-executed grants, which enhance the work of the World Bank's transport global practice and leverage road safety investments in transport operations in client countries. The World Bank is headquartered in Washington, D.C., and operates out of 135 offices worldwide.

7.10.2 INSURANCE INSTITUTE FOR HIGHWAY SAFETY

The Insurance Institute for Highway Safety (IIHS) is an independent, nonprofit scientific and educational organization dedicated to reducing deaths, injuries, and property damage from motor vehicle crashes. The IIHS was founded in 1959 by three major insurance associations representing 80% of the U.S. auto insurance market [64]. The Institute's initial purpose was to support highway safety efforts by others organizations. A decade later, the IIHS was reinvented as an independent research organization, leading the transformation of the highway safety field from one focused solely on crash prevention to one using a scientific approach to identify a full range of options for reducing crash losses. In particular [64]:

> The Insurance Institute for Highway Safety is an independent, nonprofit scientific and educational organization dedicated to reducing motor vehicle crashes.

- "Human factors research addresses problems associated with teenage drivers, alcohol-impaired driving, truck driver fatigue and safety belt use, to name a few.

- Vehicle research focuses on both crash avoidance and crashworthiness. Crash tests are central to crashworthiness research, and IIHS testing expanded with the opening of the Vehicle Research Center (VRC).
- Research about the physical environment includes, for example, assessment of roadway designs to reduce run-off-the-road crashes and eliminate roadside hazards" [64].

The IIHS is headquartered in Arlington, Virginia.

7.10.3 Air Transport Action Group

The ATAG avers that it is the only global industry-wide body to bring together all aviation industry players to promote aviation's sustainable growth [65]. The ATAG is funded by its members, which include airlines, airframe and engine manufacturers, air navigation service providers, airline pilot and air traffic controller unions, chambers of commerce, tourism and trade partners, and ground transportation and communications providers. The ATAG observes that its diversity of membership adds to its credibility and high level of influence with international decision-makers. The ATAG is headquartered in Geneva, Switzerland.

7.10.4 The International Freight Forwarders Association

The FIATA represents the freight forwarding industry. They assert that their members cover approximately 40,000 forwarding and logistics firms and employ approximately 8–10 million people in 150 countries [66]. The FIATA notes that it has created standard documents and their electronic equivalents for use by freight forwarders worldwide. The FIATA is headquartered in Glattbrugg, Switzerland.

7.10.5 The International Air Transport Association

The IATA is an NGO representing the airline industry. They cite that their members represent some 240 airlines comprising 84% of total air traffic [67]. The IATA provides a standard approach for cargo facilitation to comply with government regulations requiring the provision of cargo information. The IATA is headquartered in Montreal, Canada, with Executive Headquarters in Geneva, Switzerland.

7.10.6 International Chamber of Shipping and International Shipping Federation

These two NGOs are the principal international trade association and employers' organization, respectively, for merchant ship operators. They state that the two represent all sectors and trades and about 80% of the world's merchant fleet [68]. They represent the industry on trade facilitation issues, such as maritime safety and shipbuilding standards. These two NGOs are headquartered in London, England.

7.10.7 THE INTERNATIONAL ROAD TRANSPORT UNION

The IRU asserts that it is the world's road transport NGO. They state that they represent the interests of truck operators (as well as the interests of bus, coach, and taxi operators) for the mobility of people and goods by road [69]. The IRU is active in trade facilitation for road transport and attempts to harmonize legislation governing road transport. The IRU's Secretariat is headquartered in Geneva, Switzerland.

7.10.8 THE INTERNATIONAL UNION OF RAILWAYS

The International Union of Railways (UIC, French: Union internationale des chemins de fer) is an international rail transport industry body, an NGO representing the railway industry [70]. The UIC states that they set and publish standards for railway sectors, such as for wagons, railway equipment, and railway stations. They have also developed standards for the exchange of information between railway companies and railway infrastructure operators. The UIC is headquartered in Paris, France.

7.10.9 THE SMDG

The SMDG (actual name) is an NGO user group for shipping lines, container terminals, and port authorities [71]. The organization says they have developed standards for the maritime container industry for the exchange of information of stowage plans and of individual movements of sea containers to, within, and from ports. The SMDG is headquartered in Rots, France.

Perspective: In a policy context, NGOs such as those detailed here offer several advantages over government-based organizations. In particular, NGOs can act more quickly on matters of commerce and safety when circumstances dictate, since government organizations often need to consolidate resources and obtain coordination and assistance from other government agencies and affected private sector entities. An example would be the establishment of recommended safety standards, where NGOs could act without public input and review usually required of government organizations. And, of course, one could also argue that lack of public input and review represents a discredit of NGOs.

7.11 MOTOR VEHICLES AND CLIMATE CHANGE

Motor vehicles powered by carbonaceous forms of fossil fuel (gasoline, diesel, jet) contribute to climate change because of emissions of combusted fuel. As will be described in the next subsection, emissions from motor vehicles make a substantial contribution to climate change. However, another source of transportation, commercial aircraft, also contributes to emissions that contribute to climate change. This will be discussed in an ensuing subsection. And although marine craft such as cargo ships use fossil fuels, their overall contribution to greenhouse gas emissions is thought to be low and hence is not discussed here. Also to be discussed are emerging efforts to reduce the impacts of motor vehicle and aircraft emissions as contributors to climate change.

7.11.1 MOTOR VEHICLE EMISSIONS AND CLIMATE CHANGE

The Union of Concerned Scientists observed in 2017 that "Our personal vehicles are a major cause of global warming. Collectively, cars and trucks account for nearly one-fifth of all U.S. emissions, emitting around 24 pounds of CO_2 and other global-warming gases for every gallon of gas. About 5 pounds comes from the extraction, production, and delivery of the fuel, while the great bulk of heat-trapping emissions – more than 19 pounds per gallon – come right out of a car's tailpipe. In total, the U.S. transportation sector – which includes cars, trucks, planes, trains, ships, and freight – produces nearly 30% of all U.S. global warming emissions, more than almost any other sector [72]."

> In total, the U.S. transportation sector produces nearly 30% of all U.S. global warming emissions, more than almost any other sector.

The same source recommends several methods to reduce climate-harmful emissions from motor vehicles. In particular, three of their recommendations are summarized as follows:

- "Fuel-efficient vehicles use less gas to travel the same distance as their less efficient counterparts. When less fuel is combusted, fewer emissions are generated. When emissions decrease, the pace of global warming slows.
- Cleaner fuels produce fewer emissions when combusted. Some fuels – such as those made from cellulosic biofuels – can reduce emissions by 80% compared to gasoline. And better regulations would help prevent the gasoline we do use from getting any dirtier.
- Electric cars that use batteries or fuel cells for power use electricity as fuel, producing fewer harmful emissions than their conventional counterparts. When the electricity comes from renewable sources, all-electric vehicles produce zero harmful emissions when driven [72]."

7.11.1.1 Corporate Average Fuel Economy Standards, 1975

Automobile emissions of air pollutants are correlated with the amount of fuel combusted in internal combustion engines. Quite simply, the less fuel combusted, the lesser amount of pollution emitted. Fuel economy is therefore relevant for air pollution policies. In the U.S. the Corporate Average Fuel Economy (CAFE) standards are regulations, first enacted by Congress in 1975, after the 1973–1974 Arab Oil Embargo, to improve the average fuel economy of cars and light trucks produced for sale in the U.S. [73]. CAFE standards are established by the National Highway Transportation Safety Administration (NHTSA) and regulate how far vehicles in the U.S. must travel on a gallon of fuel.

> Corporate Average Fuel Economy (CAFÉ) standards are regulations to improve the average fuel economy of cars and light trucks produced for sale in the U.S.

The NHTSA sets CAFE standards for passenger cars and for light trucks (collectively, light-duty vehicles), and separately sets fuel consumption standards for medium- and heavy-duty trucks and engines. NHTSA also regulates the fuel-economy window stickers on new vehicles. This site contains information about many aspects of these programs, and we encourage you to check back as new information is posted. The EPA enforces the fuel standards through its vehicles' emissions regulations.

The fuel economy standards were set by the U.S. President Barack Obama (D-IL) administration in 2012 with the approval of automakers [74]. In particular, there was an agreement between the Obama administration and the automobile industry that fuel economy standards would gradually increase the average miles per gallon requirements for cars to 54.5 mpg by 2025. The increase in vehicle fuel efficiency was a component of the Obama administration's climate change program, with a focus on using less fossil fuel (coal, oil) in order to lessen the production of corresponding greenhouse gases. However, that agreement with automobile manufacturers was set aside by the Trump administration in 2018 [74].

7.11.1.2 The Safer Affordable Fuel-Efficient Vehicles Rule, 2018

In 2018 the Trump administration set aside the Obama administration's CAFÉ standards for vehicle fuel efficiency. Their Safer Affordable Fuel-Efficient (SAFE) Vehicles Rule would amend existing Corporate Average Fuel Economy (CAFE) and tailpipe CO_2 emissions standards for passenger cars and light trucks and establish new standards covering model years 2021 through 2026. The Trump administration announced plans to roll back the Obama-era standard to about 37 miles per gallon. The proposal would retain the model year 2020 standards for both programs through model year 2026. The NHSTA asserts that if the new rule is adopted, the proposed rule's preferred alternative would save more than $500 billion in societal costs and reduce highway fatalities by 12,700 lives (over the lifetimes of vehicles through Model Year 2029) [73]. The SAFE rule is undergoing judicial review in 2019 in response to litigation by environmental organizations and the matter of CAFE standards remains a subject of political and judicial debate.

However, California and 23 other states vowed to keep enforcing the stricter rules (i.e., the Obama rules), potentially splitting the U.S. auto market in two. In reaction to the political debate at the federal government level concerning CAFE standards, one U.S. state, California, has acted independently of federal fuel efficiency standards. In 2019, with car companies facing the prospect of having to build two separate lineups of vehicles, they opened secretive talks with California regulators in which the automakers – BMW, Ford Motor Company, Honda, and Volkswagen of America – won rules that are slightly less restrictive than the Obama standards and that they can apply to vehicles sold nationwide Under the agreement, the four automakers, which together make up about 30% of the U.S. auto market, would face slightly looser standard than the original Obama rule: Instead of reaching an average 54.5 miles per gallon by 2025, they would be required to hit about 51 miles per gallon by 2026 [75]. The agreements were finalized by the California Air Resources Board (CARB) on August 17, 2020, giving the four automakers until 2026 to produce fleets averaging 51 mpg [76].

Perspective: The brouhaha between the Trump administration and some U.S. states in regard to vehicle fuel efficiency standards will become a matter of serious litigation, but the ultimate decision about vehicles' fuel standards will be settled in the commercial market of vehicle sales. Specifically, if customers of automotive vehicles demand fuel-efficient products, automobile manufacturers will produce them irrespective of what federal and state governments stipulate. One has only to look at the trend in replacement of fossil-fueled vehicles with electric vehicles to appreciate the policy impact of climate change concern as a stimulus of changes in vehicle marketplaces.

7.11.2 THE FUTURE OF TRANSPORTATION IS ELECTRIC VEHICLES

The old cliché "The handwriting is on the wall" certainly applies to the future of internal combustion-powered vehicles. Put simply, there is no future for internal combustion engines. Rather, the future is with electric vehicles (EVs), which will be powered by batteries, fuel cells, or perhaps solar systems. These assertions are based on policies stated by some national governments and by some vehicle manufacturers.

An analysis by the International Energy Agency in 2019 found that carmakers plan more than 350 electric models by 2025, mostly small-to-medium variants [77]. Plans from the top 20 car manufacturers suggest a tenfold increase in annual electric car sales, to 20 million vehicles a year by 2030, from 2 million in 2018. Starting from a low base, less than 0.5% of the total car stock, this growth in electric vehicles means that nearly 7% of the car fleet will be electric by 2030. An indication of the trend toward electric cars is the sale of more than 500,000 full (i.e., not hybrid) electric cars sold in Europe in 2020 [78]. Tighter vehicle emissions by EU countries have led to sales of more electric cars. Examples of the trend toward electric vehicles are provided by announcements by General Motors and Volvo. The former will sell only zero-emission vehicles by 2035 and Volvo will cease selling gasoline-powered cars by 2030 [79,80]. Both manufacturers will produce battery-powered vehicles.

7.11.2.1 Emerging Government Policies

Some governments have developed policies that favor EVs. For example, France and Britain committed in July 2017 to ban the sales of all gasoline- and diesel-powered cars by 2040, motivated largely by health concerns about air pollution [81]. Then China, the world's largest auto market, announced in September 2017 that it will set a deadline for automakers to stop selling internal combustion engine vehicles.

> China, the world's largest auto market, announced in September 2017 that it will set a deadline for automakers to stop selling internal combustion engine vehicles.

and correspondingly set emission targets for automakers [82]. In a different approach, the Germany's government and car industry agreed in 2019 to increase joint subsidies for electric car buyers and extend the program to 2025, a policy intended to increase EVs in Germany [83].

Of special note is the creation by 10 large corporations of EV100, a coalition of industrial groups that will advocate for the adoption of EVs throughout their industries [82]. The 10 corporations include utilities and an international delivery company.

In a similar action, the mayors of Auckland, Barcelona, Cape Town, Copenhagen, London, Los Angeles, Mexico City, Milan, Paris, Quito, Seattle, and Vancouver constitute a group of 12 urban leaders who have vowed to buy only zero-emission buses starting in 2025 and make major portions of their metros fossil fuel-free by promoting walking, cycling, and public transportation and limiting the use of private vehicles [84]. The 12 cities have a combined population in 2019 of about 80 million people. In the agreement, the leaders took aim at localized air pollution

> Mayors of 12 major cities have vowed to buy only zero-emission buses starting in 2025.

as well as climate change. These policy decisions by three major nations (Britain, China, France) and 12 cities are reflective of concerns about air pollution and climate change, as noted in previous chapters of this book. In California, the California Energy Commission awarded $70 million in 2019 to state schools to replace more than 200 diesel school buses with all-electric school buses [85]. Given the concern for children's health, replacement of diesel buses with electric vehicles will become more commonplace generally. As another example of a city transitioning to electric buses, Toronto is operating in 2020 the largest battery-powered electric bus fleet in North America. The city's 59 electric buses replaced diesel-powered buses for purposes of lower ambient air emissions.

One U.S. state, California, has announced its intention to phase out the sale of all gasoline-powered vehicles by 2035, stating their intention to lead the U.S. in reducing greenhouse gas emissions by encouraging the state's drivers to switch to electric cars [86]. In response, the EPA stated its belief that California would need to request a waiver from the agency and implied that the state's electricity infrastructure is insufficient for a shift toward electric vehicles [87]. It is likely that the issue will reside with a federal court decision. As a matter of public health, it is remarkable that EPA would oppose a U.S. state policy that would decrease air pollution from motor vehicles.

These policies to forego internal combustion vehicles have prompted major automakers to announce plans for electric vehicle fleets and the phase out of internal combustion cars. For example, the General Motors Corporation has announced that it will sell only electric vehicles by 2035 [88]. Similarly, Jaguar will be EV-only from 2025, while Land Rover vehicles will be 60% battery EV. While the direction toward EVs has gained policy support by some national governments, the change from internal combustion to EVs will not occur quickly, given challenges to the design and manufacture of EVs. Particularly challenging will be the design of longer-lasting batteries and the need to establish recharging stations for batteries and fuel cell refueling stations that supply hydrogen, the key fuel of fuel cells.

7.11.2.2 Electric Trucks

Elsewhere, of special relevance for the purpose of climate change implications is the increase globally in the number of electric trucks. One climate change source observes that "Electric trucks are driving out of factories and into service, and multiple vehicle companies are gearing up to make them. The result could be a significant reduction in greenhouse gas emissions – especially if the deliveries turn out to be cheaper than old-fashioned diesel engines. In the U.S., more than 6% of greenhouse gas emissions emitted in 2015 were from medium- and heavy-duty trucking. Transportation, including trucking, was responsible for 14% of emissions globally in 2010."

Because trucks need so much hauling power, they have eluded electrification until recently; a battery that could pull significant weight would itself be too hefty and too expensive. But now, improvements in battery technology have occurred, bringing down both battery size and cost. According to a recent report by a market source, the number of hybrid-electric and electric trucks is set to grow almost 25% annually, from 1% of the market in 2017 to 7% in 2027, a jump from about 40,000 electric trucks worldwide this year to 371,000 [89]." An example of this future of electric trucks is the 2020 order of 10,000 electric delivery trucks by the U.S. company United Parcel Service (UPS) [90]. The 10,000 trucks will be delivered to UPS in the second half of 2020, continuing through the year 2025 by EV maker Arrival. Another vehicle manufacturer, General Motors, has made a major commitment to the production of EVs. In October 2020 the company unveiled the 2022 GMC Hummer EV sport utility truck, or SUT.

7.11.2.3 Electric Trains

In addition to the advance in prevalence of electric trucks, one nation is committing to the development of battery-powered trains [91]. Ireland has announced plans to modernize its trains, harkening back to a technology it relied on in the 1930s and 1940s. National carrier Iarnród Éireann launched its request for proposals in May 2019 to purchase 600 new train carriages by 2040. This number approximates the total number of 2019 train carriages in Ireland. All those new trains would run on electricity, and up to half could be powered by batteries. That would represent a remarkable transformation for a network that today runs largely on diesel.

Battery-powered trains have been operated in Japan since 2014, while the U.K. started some services in 2015. Austria also tested its first battery-powered model in May 2019, while in 2019 Scotland's ScotRail ordered 70 new battery-powered trains [91]. This trend in battery-powered train carriages and equipment will phase out diesel-powered equipment, which is a public health benefit, owing to the removal of fine particulates emitted by diesel engines.

> Battery-powered trains have been operated in Japan since 2014; the U.K. started some services in 2015; Austria tested its first battery-powered model in May 2019.

Perspective: Motor vehicles have become a way of life for people around the globe. They provide a freedom of movement that enhances commerce, recreation, and personal welfare. But that comes at a cost to the environment, given the emissions released from vehicles powered by internal combustion engines. Climate change is occurring and internal combustion vehicles are a substantial contributor to the carbon load that forms greenhouse gases. But awareness of this relationship between such engines and climate change is leading to global policies that will in time see a phasing out of vehicles with internal combustion engines, being replaced by electric vehicles. This portent is a seminal example of science driving policy that in turn is driving a major societal outcome, the phasing out of internal combustion engines.

7.11.3 AIRCRAFT EMISSIONS AND CLIMATE CHANGE

The preceding sections have discussed the role of motor vehicles used in surface transportation and their contribution to climate change. Additionally, aircraft also contribute their emissions to climate change conditions. As previously cited, the Air

Transport Action Group cites "Worldwide, flights produced 781 million tons of CO_2 in 2015. Globally, humans produced more than 36 billion tons of CO_2." [..] "Aviation is responsible for 12% of CO_2 emissions from all transports' emissions compared to 74% from road transport" [65]. Therefore, although important as a contributor to climate change, the emphasis in this chapter has been motor vehicles.

In recognition of the contribution of aircraft to climate change, efforts to replace jet fuel with less-polluting fuel are underway. Some commercial aircraft across Europe have begun to run on renewable fuel made from vegetable oil and animal fat in an effort to reduce their carbon emissions [92]. The new fuel, which is produced by a Finnish company, is mixed with traditional kerosene at the Geneva, Switzerland, international airport commencing in 2018. The new fuel was tested by Lufthansa on more than 1,000 flights in 2011 with a 50–50 blend of renewables to kerosene. Carbon emissions were reduced by 47%.

> Aviation is responsible for 12% of CO_2 emissions from all transports' emissions compared to 74% from road transport.

The same source observes that sustainable fuels have only made a small dent in the airline industry, but their impact is growing. For example, several airlines have begun to use jet fuels made from waste. United Airlines agreed in 2016 to purchase 15 million gallons of biofuel from a U.S. firm, while British Airways has signed a contract with a British firm to supply jet fuel made from some of the 15 million tons of waste U.K. households send to landfill each year [92].

7.11.4 URBAN TRAFFIC CONTROL AND AIR POLLUTION

In order to reduce levels of air pollution in urban areas, some cities have developed traffic control policies that limit vehicle access to downtown areas. For example, in 2019, Paris instituted a policy that will prevent nearly 3 million cars from being driven in France's capital during weekdays [93]. The move comes as part of a greater effort to reduce air pollution in light of a report that showed Paris has the worst air particulate pollution of any EU capital. The policy bans diesel cars and vans made before January 1, 2006, from the capital between 8 a.m. and 8 p.m. on weekdays. Motorbikes and three-wheeled vehicles produced before June 30, 2004, are banned from Paris's roads Monday through Friday. According to the policy, cars barred from Paris on the weekdays will still be permitted in Paris's surrounding suburbs. But the residents risk fines once they enter the capital.

> To reduce levels of air pollution in urban areas, some cities have developed traffic control policies that limit vehicle access to downtown areas.

In a second example, Oregon senators passed a bill in 2019 regulating diesel trucks in the Portland, Oregon, metro area [94]. The bill requires truck owners to replace older diesel engines with newer models by 2025. The goal is to reduce toxic diesel pollution by requiring 2010 model year engines or newer. The newer engines filter out almost all of the diesel particulate that is known to cause cancer and other respiratory diseases. The bill would help pay for engine upgrades using about $50 million in settlement money from the Volkswagen diesel emissions cheating scandal, as noted elsewhere in Volume 1, Chapter 2.

7.12 GLOBAL IMPLICATIONS OF TRANSPORTATION

Transportation occurs globally, though the modes of transport can differ according to global location. For instance, the mode of transportation in Polar Regions must be radically different from that in metropolitan cities. Similarly, modes of transportation in rural areas will differ from those in highly urbanized areas, where few farm implements are needed and correspondingly few examples of mass transportation are supportable in rural regions [95]. As can be observed from the following descriptions of one region's and three countries' transportation issues, policies differ across national borders due to such factors as cultural traditions, natural and fiscal resources, commerce, and availability of vehicles to purchase and operate.

Of particular note in a public health sense is the incidence of vehicle-related fatalities in Africa, where like the rest of the world is witnessing a road safety crisis. According to the World Bank, Africa has the highest per capita rate of road fatalities in the world, where road deaths in sub-Saharan Africa are projected to more than double from some 243,000 deaths projected for 2015 to 514,000 by 2030 [96].

> Road deaths in sub-Saharan Africa are projected to reach 514,000 by 2030.

The export of used vehicles to low- to middle-income countries contributes to host countries' problems with air pollution and climate change [97]. According to the UN Environment Program (UNEP), millions of used cars, vans, and minibuses exported from Europe, the U.S., and Japan to the developing world are of poor quality, contributing significantly to air pollution and hindering efforts to mitigate the effects of climate change. UNEP estimates that between 2015 and 2018, 14 million used light-duty vehicles were exported worldwide. Some 80% went to low- and middle-income countries, with more than half going to Africa. While the used vehicles can contribute to host countries' commerce, the adverse effects on public health are seldom considered.

7.13 HAZARD INTERVENTIONS

As described in this chapter, how human beings transport themselves and commercial cargo can become a hazard to health and well-being, as illustrated by the global number of fatalities associated with the operation of motor vehicles. Several hazard interventions can reduce the toll of transport-related morbidity and mortality, as suggested by the following interventions:

- Operate vehicles that are equipped with the latest safety devices such as automatic braking and lane detection.
- Maintain all transportation equipment such as automobiles and aircraft in road and air worthy conditions through regular maintenance schedules.
- Use correctly installed and safety-certified seats and restraints for children when they are transported.
- Support policies that construct and maintain roads and other thoroughfares that meet traffic safety standards.

- Advocate for policies that reduce vehicle crashes due to adverse influences such as texting and DUI while operating motor vehicles.
- Inform yourself on the road, air, train, or marine worthiness of any form of chartered or rented transportation.
- Where possible, advocate that local traffic authorities implement controlled traffic or aircraft densities that promote fewer incidents of morbidity or mortality.

7.14 SUMMARY

As discussed in this chapter, how people move about and transport themselves and their goods constitutes part of our physical environment. And like other elements of humans' physical environment, transportation can bring public health consequences. Motor vehicles, in particular, can bring public health challenges due to crashes between vehicles. Domestic and international governments have developed policies such as motor vehicle standards and roadway construction programs that are intended to promote safer travel. However, as discussed in this chapter, vehicle crashes due to vehicle operators' lax judgment can result in injuries or death to passengers and operators. Such injuries and fatalities constitute a public health burden in terms of personal loss and economic costs.

REFERENCES

1. Cambridge Dictionary. 2018. Transport definition. https://dictionary.cambridge.org/dictionary/english/transport.
2. WHO (World Health Organization). 2017. *Fourth UN Global Road Safety Week kicks off worldwide*. Geneva: Media Centre.
3. NCFL (National Center for Family Learning). 2018. Wonder of the Day #1375. Who invented the automobile? https://wonderopolis.org/wonder/Who-Invented-the-Automobile.
4. The Columbia Encyclopedia, 6th ed. 2016. Internal-combustion engine. https://www.encyclopedia.com/science-and-technology/technology/technology-terms-and-concepts/internal-combustion-engine.
5. The World Bank. 2017. World Bank's Global Road Safety Facility (GRSF) receives a royal award for innovations in reducing traffic fatalities. Press release December 12. Washington, D.C.
6. WHO (World Health Organization). 2015. Global status report on road safety 2015. Geneva, Switzerland: Media Centre.
7. HLDI (Highway Loss Data Institute). 2015. *General statistics*. Arlington, TX: Insurance Institute for Highway Safety.
8. NHTSA (National Highway Traffic Safety Administration). 2017. *USDOT releases 2016 fatal traffic crash data*. Washington, D.C.: U.S. Department of Transportation.
9. CDC (Centers for Disease Control and Prevention). 2015. *Seat belts: Get the facts*. Atlanta, GA: National Center for Injury Prevention and Control.
10. CDC (Centers for Disease Control and Prevention). 2017. How big is the problem of crash-related injuries and deaths to drivers and passengers? https://www.cdc.gov/motorvehiclesafety/seatbelts/facts.html.
11. NIH (National Institutes of Health). 2016. *FY 2016 budget*. Washington, D.C.: U.S. Department of Health and Human Services.

12. Caron, C. and N. Chokshi. 2019. Pedestrian deaths in U.S. approach highest number in nearly 30 years, study shows. *The New York Times*, March 7.
13. FHWA (Federal Highway Administration). 2018. *Interstate highway system: Interstate frequently asked questions*. Washington, D.C.: U.S. Department of Transportation.
14. FTA (Federal Transit Administration). 2017. *A brief history of mass transit*. Washington, D.C.: U.S. Department of Transportation.
15. DOT (U.S. Department of Transportation). 2019. Mission, history, goals. https://checkthebox.dot.gov/transition/mission-history-goals.
16. A&E Networks. 2018. This day in history: 1966 President Johnson signs the National Traffic and Motor Vehicle Safety Act. https://www.history.com/this-day-in-history/president-johnson-signs-the-national-traffic-and-motor-vehicle-safety-act.
17. NHTSA (National Highway Traffic Safety Administration). 1999. *Fact sheet: Minimum drinking age laws*. Washington, D.C.: U.S. Department of Transportation.
18. FHWA (Federal Highway Administration). 2018. *Intermodal Surface Transportation Efficiency Act of 1991*. Washington, D.C.: U.S. Department of Transportation.
19. U.S. House of Representatives. 2017.H.R. 3419 (106 th): Motor Carrier Safety Improvement Act of 1999. Washington, D.C.: 106th Congress.
20. FHWA (Federal Highway Administration). 2012. *Moving Ahead for Progress in the 21st Century Act (MAP-21)*. A summary of highway provisions. Washington, D.C.: U.S. Department of Transportation.
21. FHWA (Federal Highway Administration). 2016. *Fixing America's Surface Transportation Act or "FAST Act."* Washington, D.C.: U.S. Department of Transportation.
22. Dahl, D. 2016. Who sets speed limits and how are they decided? *The Bellingham Herald*, May 15.
23. Justica. 2017. 2010 Georgia Code Title 40 – Motor Vehicles and Traffic. Chapter 6 – Uniform Rules of the Road. Article 9 – Speed Restrictions. Sec. 40-6-183 – Alteration of speed limits by local authorities. https://law.justia.com/codes/georgia/2010/title-40/chapter-6/article-9/40-6-183/.
24. Anonymous. 2019. Driving while intoxicated (DWI) law and legal definition. *US Legal*. https://definitions.uslegal.com/d/driving-while-intoxicated/.
25. NCADD (National Council on Alcoholism and Drug Dependence). 2015. *Driving while impaired – alcohol and drugs*. New York: NCADD.
26. Lipari, R, A. Hughes, and J. Bose. 2016. Driving under the influence of alcohol and illicit drugs. The CBHSQ Report, December 17. Washington, D.C.: Substance Abuse and Mental Health Services Administration.
27. DMV.com. 2018. DUI & DWI in Georgia. https://www.dmv.org/ga-georgia/automotive-law/dui.php.
28. NIDA (National Institute on Drug Abuse). 2016. *Drugged driving*. Bethesda, MD: National Institutes of Health.
29. HLDI (Highway Loss Data Institute). 2016. *Distracted driving*. Insurance Institute for Highway Safety, Arlington, VA.
30. Stim, R. 2016. Georgia text messaging and cell phone laws. http://www.drivinglaws.org/georgia.php.
31. GOHSG (Governor's Office of Highway Safety). 2018. *House Bill 673 – "Hands Free Law."* Atlanta, GA: Governor's Office.
32. GHSA (Governors Highway Safety Association). *Seat belts*. Washington, D.C.: GHSA.
33. Defensive Driving. 2016. A history of seat belts. https://www.defensivedriving.com/blog/a-history-of-seat-belts/.
34. Harris, D. 1996. 1964 brought seat belts for every buyer. *Automotive News*, June 26.
35. CDC (Centers for Disease Control and Prevention). 2015. *Primary enforcement of seat belt laws*. Atlanta, GA: National Center for Injury Prevention and Control.

36. NHTSA (National Highway Traffic Safety Administration). 2018. *Seat belt use in 2017 – use rates in the states and territories.* Washington, D.C.: U.S. Department of Transportation.
37. CDC (Centers for Disease Control and Prevention). 2017. *Child passenger safety: Get the facts.* Atlanta, GA: National Center for Injury Prevention and Control.
38. Consumer Protection Unit. 2018. *Child car seats.* Atlanta, GA: Georgia Department of Law.
39. Dirksen, S. 2018. History of American technology. What are air bags. http://web.bryant.edu/~ehu/h364proj/sprg_97/dirksen/airbags.html.
40. NHTSA (National Highway Traffic Safety Administration). 2018. *Federal motor vehicle safety standards and regulations.* Washington, D.C.: U.S. Department of Transportation.
41. NHTSA (National Highway Transportation Safety Administration). 2018. Federal Motor Vehicle Safety Standards. 49 C.F.R. 571. Washington, D.C.
42. NHTSA (National Highway Traffic Safety Administration). 2017. *Automated vehicles for safety.* Washington, D.C.: U.S. Department of Transportation.
43. HLDI (Highway Loss Data Institute). 2016. *U.S. DOT and IIHS announce historic commitment of 20 automakers to make automatic emergency braking standard on new vehicles.* Arlington, TX: Insurance Institute for Highway Safety.
44. Encyclopedia. 2018. Definition of autonomous vehicle. *PC Magazine*, July 14.
44a. Lambert, F. 2020. GM Cruise gets greenlight to operate true driverless Chevy Bolt EVs. *Electrek*, October 15.
45. FHWA (Federal Highway Administration). 2017. *Interstate frequently asked questions.* Washington, D.C.: U.S. Department of Transportation.
46. Anonymous. 2018. United States Department of Transportation. *Wikipedia.* https://en.wikipedia.org/wiki/United_States_Department_of_Transportation.
47. FHWA (Federal Highway Administration). 2012. *About U.S. Department of Transportation.* Washington, D.C.: FHWA.
48. FHWA (Federal Highway Administration). 2017. *Understanding Federal Highway Administration.* Washington, D.C.: U.S. Department of Transportation.
49. FMCSA (Federal Motor Carrier Safety Administration). 2018. *About us.* Washington, D.C.: U.S. Department of Transportation.
50. FRA (Federal Railroad Administration). 2016. *Railroad safety.* Washington, D.C.: U.S. Department of Transportation.
51. FAA (Federal Aviation Administration). 2013. *A brief history of the FAA.* Washington, D.C.: U.S. Department of Transportation.
52. FAA (Federal Aviation Administration). 2013. *Air Traffic Organization.* Washington, D.C.: U.S. Department of Transportation.
53. FTA (Federal Transit Administration). 2016. *Mission statement.* Washington, D.C.: U.S. Department of Transportation.
54. MARAD (Marine Administration). 2018. *About us.* Washington, D.C.: U.S. Department of Transportation.
55. NHTSA (National Highway Traffic Safety Administration). 2017. *Understanding the National Highway Traffic Safety Administration.* Washington, D.C.: U.S. Department of Transportation.
56. NTSB (National Transportation Safety Board). 2018. About the National Transportation Safety Board. https://www.ntsb.gov/about/Pages/default.aspx.
57. NTSB (National Transportation Safety Board). 2018. History of the National Transportation Safety Board. https://www.ntsb.gov/about/history/Pages/default.aspx.
58. Boudette N. and B. Vlasic. 2017. Tesla self-driving system faulted by safety agency in crash. *The New York Times*, September 12.
59. ICAO (International Civil Aviation Organization). 2018. *About ICAO.* Montreal, Canada.

60. IMO (International Maritime Organization). 2016. What is the International Maritime Organization (IMO)? https://www.marineinsight.com/maritime-law/what-is-international-maritime-organization-imo/.
61. UN (United Nations). 2010. The United Nations and road safety. http://www.un.org/en/roadsafety/background.shtml.
62. UN (United Nations). 2015. *Secretary-General appoints Jean Todt of France as special envoy for road safety*. New York: Office of the General-Secretary.
63. The World Bank. 2014. *Global Road Safety Facility (GRSF) strategic plan 2013–2020 (English)*. Washington, D.C.: The World Bank.
64. HLDI (Highway Loss Data Institute). 2018. *About the institutes*. Arlington, TX: Insurance Institute for Highway Safety.
65. ATAG (Air Transport Acton Group). 2018. Facts & figures. https://www.atag.org/facts-figures.html.
66. FIATA (International Federation of Freight Forwarders Associations). 2018. *Who is FIATA*. Glattbrugg, Switzerland: FIATA.
67. IATA (International Air Transport Association). 2018. About us. https://www.iata.org/about/Pages/index.aspx.
68. ICS (International Chamber of Shipping). 2018. About us. http://www.ics-shipping.org/about-ics/about-ics.
69. IRU (International Road Transport Union). 2018. Mission. https://www.iru.org/who-we-are/about-iru/mission.
70. UIC (International Union of Railways). 2018. UIC mission, objectives, challenges. https://uic.org/about.
71. SMDG. 2018. About us. http://www.smdg.org/index.php/about-us/.
72. Union of Concerned Scientists. 2018. *Car emissions & global warming*. Cambridge, MA.
73. NHSTA (National Highway Safety Transportation Administration). 2018. Corporate average fuel economy. https://www.nhtsa.gov/laws-regulations/corporate-average-fuel-economy.
74. Guess, M. 2016. EPA reaffirms 54.5mpg target fuel economy by 2025; automakers turn to Trump. *Ars Technica*, December 1.
75. Davenport, C. and H. Tabuchi. 2019. Automakers, rejecting Trump pollution rule, strike a deal with California. *The New York Times*, July 25.
76. Beitsch, R. and R. Frazin. 2020. California finalizes fuel efficiency deal with five automakers, undercutting Trump. *The Hill*, August 17.
77. Cozzi, L. and A. Petropoulos. 2019. *Commentary: Growing preference for SUVs challenges emissions reductions in passenger car market*. Paris: International Energy Agency.
78. Jolly, J. 2020. More than 500,000 full electric cars sold so far this year in Europe. *The Guardian*, December 3.
79. Boudette, N. E. and C. Davenport. 2021. G.M will sell only zero-emission vehicles by 2035. *The New York Times*, January 28.
80. Ewing, J. 2021. Volvo says it will stop selling gasoline-powered cars by 2030. *The New York Times*, March 2.
81. Gies, E. 2017. Are electric vehicles pushing oil demand over a cliff? *Inside Climate News*, October 10.
82. Fairley, P. 2017. 10 top companies commit to electric vehicles, sending auto industry a message. *Inside Climate News*, September 19.
83. Staff. 2019. German government expands subsidies for electric cars. *DW News*, May 11.
84. Dovey, R. 2017. 12 cities plan for emissions-free neighborhoods. *Next City*, October 23.
85. Dzikly, P. 2019. California replacing 200 polluting diesel school buses with all-electric buses. *Electrek*, July 17.
86. Sommer, L. and S. Newman. 2020. California governor signs order banning sales of new gasoline cars by 2035. *NPR*, September 23.

87. Frazin, R. 2020. EPA questions legality of California's attempt to phase out sales of gas-powered cars. *The Hill*, September 29.
88. Boudette, N. and C. Davenport. 2021. G.M. will sell only zero-emission vehicles by 2035. *The New York Times*, January 28.
89. Gies, E. 2017. Electric trucks begin reporting for duty, quietly and without all the fumes. *Inside Climate News*, December 18.
90. Yamanouchi, K. 2020. UPS orders 10,000 electric delivery trucks, plans test of self-driving vans. *Atlanta Journal-Constitution*, January 20.
91. O'Sullivan, F. 2019. A pioneer of battery-powered trains now wants a nationwide fleet of them. *Citylab*, June 5.
92. Chapman, B. 2017. Planes across Europe to start running on vegetable oil and animal fat in bid to tackle climate change and toxic air. *The Independent*, October 30.
93. Wise, J. 2019. Paris bans nearly 3 million vehicles during weekdays in effort to crack down on pollution. *The Hill*, July 1.
94. Profita, C. 2019. Lawmakers approve new diesel truck regulations for Portland. *Oregon Public Broadcasting*, June 30.
95. O'Neill, P. 2011. The problem with rural transport is that it is rural, the solution is in branding. Transport for development blog November 7. Washington, D.C.: The World Bank.
96. The World Bank. 2014. Tacking the road safety crisis in Africa. Feature story June 6. Washington, D.C.
97. UNEP (United Nations Environment Programme). 2020. New UN report details environmental impacts of export of used vehicles to developing world 26 October. Nairobi.

Lessons Learned and Authors' Reflections

The seven chapters in this volume are considered by the authors as addressing the emerging environmental hazards affecting human and ecological health. This assertion is based on the public health data, bodies of basic science, and observational signs that humans have amassed on specific environmental hazards. However, the authors are cognizant that some persons might not consider some or all of these seven environmental hazards as matters of public health concern. We respond by observing that five of the seven hazards are causal factors in premature loss of life. And the remaining two, noise and light pollution and the built environment are causal factors in producing adverse health morbidity. Programs of public health can reduce or prevent premature mortality and morbidity via efforts in public education, surveillance activity, and education outreach.

Covered in this volume were discussions of environment-related infectious diseases, solid and hazardous waste, drugs and alcohol, firearms violence, noise and light pollution, the built environment, and transportation. Each of these subjects the authors assert presents adverse health effects that result in actions to prevent their impact. And prevention of adverse effects is the gist of public health. Described in each of the seven chapters are mitigation policies and actions to prevent the described adverse health effects. Reflections on these seven environmental hazards gave rise to the following observations.

- The seven environmental hazards are global in their adverse effects. We humans and other creatures with which we share the planet will all experience adverse health effects, depending on the nature, quality, geography, and extent of exposure.
- Environment-related infectious diseases are the environmental hazard that has the greatest potential for the most serious pandemic adverse health effect.
- Mitigation policies such as government statutes and accompanying regulations, while generally well purposed to protect public health, are only as effective as their enforcement.
- In regard to environmental hazards, the health of children should be a principal component of mitigation strategies and actions.
- Three emerging environmental hazards, drugs and alcohol, firearms violence, and transportation involve issues of personal decision, i.e., persons choose to expose themselves to the hazard through the acquisition of the hazard and its employment. Deleterious personal choices should be respected but subject to education mitigations.

- Two emerging environmental hazards, noise and light pollution and the built environment bring societal issues of community development and subsequent impact. Public health impacts should be included in zoning and development review and planning.
- Government agencies that have statutory and ethical responsibilities for the mitigation of environmental hazards should not become subjects of political interference lacking substantive sociopolitical debate.
- Individuals and other private sector entities have responsibilities for the prevention of adverse consequences of environmental hazards.
- Youth education about environmental hazards and their health effects should be included in school curricula.

Environmental Policy and Public Health
Emerging Health Hazards and Mitigation, Volume 2 Workbook

Barry L. Johnson and Maureen Y. Lichtveld

Barry L. Johnson, MSc, PhD, FCR RADM (Ret.), U.S. Public Health Service Adjunct Professor Rollins School of Public Health Emory University Atlanta, Georgia, 30322	**Maureen Y. Lichtveld, MD, MPH** Dean, Graduate School of Public Health Professor, Environmental and Occupational Health Jonas Salk Chair in Population Health University of Pittsburgh Pittsburgh, Pennsylvania 15261

2022
All rights are reserved. No part of this material may be reproduced in any form or by any means without prior written permission from the publisher, CRC Press, Boca Raton, Florida, USA.
Address all requests and comments to Barry L. Johnson, bljradm@gmail.com, and/or Maureen Y. Lichtveld, mlichtve@pitt.edu.

RESEARCH PROJECTS

1. Described in this volume are various policies that are intended to mitigate the effects of emerging environmental hazards such as infectious diseases and firearms violence. U.S. federal government environmental statutes were discussed specifically. Research how federal laws are developed and promulgated. Be specific by identifying by name the various committees in Congress that will be involved in enacting an environmental statute.
2. The World Health Organization (WHO) was mentioned in this volume as the global leader on issues of epidemic and pandemic infections due to such environmental hazards as novel viruses and air pollution. Research the origins of the WHO and discuss and compare the sociopolitical aspects of the WHO, for instance, funding mechanism and governance. In your analysis of the work of the WHO, cite in some detail how the WHO coordinated the global eradication of smallpox.
3. The U.S. Centers for Disease Control and Prevention (CDC) was mentioned in this volume as the U.S. leader on issues of epidemic and pandemic infections due to such environmental hazards as novel viruses. Research the origins of the CDC and discuss and compare the sociopolitical aspects of the CDC, for instance, funding mechanism and governance. In your analysis of the work of the CDC, cited in some detail the agency's involvement, both positively and negatively, in responding to the COVID-19 pandemic.
4. This volume identified and discussed six emerging environmental health hazards. Each hazard was discussed for mitigation purposes in terms of various U.S. federal government policies such as statutes. At the core of such statutes was the authority given to regulatory agencies for their development and enforcement of environmental regulations such as control of emissions of toxic materials from electric power plants. Research and describe in detail the legal policies that must be followed in the development and promulgation of federal government environmental regulations.

CASE STUDIES

1. The COVID-19 pandemic commenced in February 2020. As a case study, research the public health and sociopolitical actions taken by the U.S. federal government in response to the pandemic's spread domestically. In particular, contrast the actions taken by U.S. state governments for intended control of COVID-19. State your opinion as to how any future pandemics should be managed in the U.S.
2. Described in this volume is the global ecological threat to marine life caused by dumping of plastic waste into oceans and other bodies of water. As a case study, investigate and report on global actions taken to lessen the volume of plastic waste being generated and dumped into bodies of water. Discuss the success and failure of international actions intended to reduce plastic waste.
3. As a thought exercise, assume gunpower had never been invented. Discuss how humans would have found ways to settle issues of national and international disputes such as property and wealth aspirations.

4. Choose as a case study at any school shooting in the U.S. Describe the shooting and its implications for gun control in the U.S. Emphasis should be given to a discussion of causal factors inherent in the shooter's actions.
5. Choose as a case study the construction of any rooftop garden in any building in a metropolitan area of the U.S. Ascertain the sociopolitical policies inherent in its construction and utilization. Discuss the history of the selected garden and its longevity. Speculate on whether the selected garden could have been constructed in your area of residence. Would you have supported amendments to local zoning laws and regulations that could encourage the establishment of rooftop gardens? Discuss your reasons for or against zoning policy.
6. Conduct a case study of the U.S. Supreme Court's decisions over the previous 60 years on cases concerning the Second Amendment of the U.S. Constitution. Be sure to cite key language in each majority opinion of the court. Opine on whether the court's decisions contributed to the prevalence of firearms in the U.S. population.
7. A major opioid epidemic occurred in the U.S. in 2020, as described in Chapter 3. Conduct a case study of the epidemic, discussing causal factors for the epidemic and attendant legal outcomes on behalf of persons and organizations that contributed to the epidemic.
8. Using WHO and CDC data on the incidence of alcohol-related disease annually in the U.S., conduct a case study of socioeconomic impact of alcohol consumption. Organize your study around a specific form of consumed alcohol, not specifying a brand name.

VOLUME 2: CHAPTER POLICY QUESTIONS

CHAPTER 1: ENVIRONMENT-RELATED INFECTIOUS DISEASES

1. As presented in this chapter, some varieties of mosquitoes can transmit viruses and/or parasites that can cause human infectious diseases. In your opinion, should all forms of mosquito be eradicated from planet Earth? Present your opinion in an essay of appropriate depth and include pros and cons for your opinion.
2. Humankind has eradicated smallpox as a threat to humanity and is nearing the eradication of polio and Guinea worm disease. Will humankind ever rid itself of malaria? Justify your opinion by providing an analysis of the challenges facing eradication.
3. Discuss how climate change could affect the prevalence of tick-borne disease in the community in which you reside. Using internet resources, assess whether your local and state health departments have made preparations for dealing with climate change's impact on infectious diseases.
4. Using internet resources assess and prepare a report of appropriate depth that describes various global programs that focus on malaria control. Include material in your report that describes the purpose and degree of success of each program.

5. Contact a local pest control company and ascertain their level of effort in containing rodent populations in the area in which you reside.
6. Research why Dr. John Snow is often championed as the father of modern epidemiology. Discuss his place in your pantheon of public health heroes.
7. Access the CDC website and locate the agency's principal programs for the prevention of environment-related infectious diseases. Capsulize each program and analyze its principal accomplishments in the prevention of these diseases.
8. Access the NIAID website and locate the agency's principal programs for the prevention of environment-related infectious diseases. Capsulize each program and analyze its principal accomplishments in the prevention of these diseases.
9. Discuss the etiology, illness symptoms, treatment, and prevalence of persons who have contracted cholera. What is the primary tool to prevent the occurrence of cholera?
10. Discuss the etiology, illness symptoms, treatment, and prevalence of persons who have contracted typhus. What is the primary tool to prevent the occurrence of typhus?
11. Assume you are a seasoned wildlife biologist. Would you be concerned about programs of mosquito and tick control? What concerns, if any, might you have? List the pros and cons of such vector control programs.
12. As presented in the chapter, several infectious diseases can have tick-borne origins. Using the internet and other resources ascertain which tick-borne diseases are prevalent in your state of residence. Cite data and evaluate any public health prevention programs.
13. Explore whether your university has any ongoing research that is directed to the study of environment-related infectious diseases. Discuss the purpose and extent of any such research and forecast the potential contribution of the identified research. If your university has no such research, explore other universities in your state or adjacent states.
14. Following a hasty, tasty review of Chapter 6, together with this chapter's content on climate change, discuss your primary human health concerns about whether climate change could affect your well-being and that of your family members.
15. Some Members of Congress have advocated for a reduction in the funds from the U.S. that support activities and programs of the United Nations. Do you agree with this funding policy? If so, why? If not, why? In your opinion should any particular UN programs be excluded from funding cuts by the U.S.?
16. Assume your local health department has proposed to utilize area-wide spraying of pesticides for purpose of mosquito control. Assuming that you are the leader of a local environmental group, what actions would your organization take in regard to the proposed spraying? Be specific and provide an analysis of your actions.
17. Assume that you are the director of a local health department that has proposed to utilize area-wide spraying of pesticides for purpose of mosquito

control. What actions would you take in regard to the proposed spraying? Be specific and provide a critical thinking analysis of your actions.
18. Assume that you are a pediatrician who works for PAHO. The organization's director has come to your office and asked that you undertake a review of environment-related infectious diseases that can cause birth defects in babies born to mothers who reside in the areas covered by PAHO. Provide a copy of the report that you prepared in response.
19. Using the internet and other resources, research the life and work of Drs. Walter Reed and Ronald Ross. In an essay of appropriate depth, discuss how their work relates to the content of this chapter. In particular, describe why the public health work of Dr. Ross could be classified as heroic. Were you familiar with these two medical doctors' public health contributions?

CHAPTER 2: SOLID AND HAZARDOUS WASTE

1. Contact your county or municipal waste management department to obtain the amount of household waste collected per week. Does your county or municipality operate one or more incinerators for disposal of waste? How many landfills are present in your survey? What is the average cost to households of waste removal and disposal?
2. Plastics waste is discussed in this chapter as a major global hazard to the ocean and other water-based ecosystems. In your review of this chapter and other information resources, are there any policies that should be adopted as U.S. federal policies for purpose of controlling the amount of plastics waste left in the environment?
3. E-waste is an increasingly important burden for waste managers. Survey your personal possession of devices that could contribute in time to e-waste. Discuss each device's potential for recycling or reuse. Be specific. Ascertain if the locale in which you reside has capacity for management of e-wastes.
4. The RCR Act, as amended, covers the management of municipal solid waste and other wastes. Discuss five actions that you personally can take to reduce the volume of your household waste. Estimate in pounds the annual volume of waste reduction that you can achieve.
5. Using internet resources, ascertain your state's programs and policies that pertain to the management of hazardous waste. Discuss what you consider to be the single most effective policy.
6. The CERCL Act (also called the Superfund Act), as amended, requires EPA to rank uncontrolled hazardous waste sites and place the most hazardous sites on the National Priorities List (NPL). Using EPA resources, ascertain the number of NPL sites in your state. Discuss the status of any site's remediation. If there are no NPL sites in your state, select an adjacent state that contains one or more NPL sites.
7. Many states have state-based CERCLA programs. Ascertain if your state has such a program. If so, how many "state-lead" uncontrolled hazardous

waste sites are being remediated in your state. Select the site nearest to your residence and describe its remediation status.
8. In a public health context, in your opinion, do the successes of the federal CERCLA statute, as amended, outweigh its criticisms? Why? Be specific.
9. Discuss the essential public health differences between the RCR Act and the CERCL Act. Be specific.
10. What U.S. Public Health Service agency was created by the CERCL Act? Why?
11. Love Canal, New York, was found to be an uncontrolled hazardous waste, which became the impetus for Congress enacting the Superfund Act. Using internet resources, discuss the actions taken by the federal government in mediating the Love Canal site. Discuss the current condition of the site.
12. As a person concerned about environmental health, keep a log for one week of the items and their estimated weight that you discard. For each item, discuss the potential for its recyclability. Discuss any methods that you use to reduce your personal amount of household waste.
13. Contact the county or municipal waste management department and ascertain if and how they use waste recycling for municipal waste reduction. Given the department's reply, discuss, in your opinion, any gaps in the department's waste reduction programs.
14. You have been asked by a local community group to advise them on the efficacy of doubling the size of the county's solid waste landfill. The community group is especially concerned about public health implications. Discuss your response to the group.
15. Children in many developing countries are exposed to hazardous substances dumped in landfills. Discuss the ways that children in these countries can come into contact with landfill waste. Describe five substances in municipal landfills that could be a health hazard to children.
16. Assume that your county or province or parish or tribe is considering an ordinance that would require household waste to be separated into categories of paper/paper products, glass, metals, hazardous chemicals, and miscellaneous. Your neighborhood's residents have organized into two disparate camps: one group supports the ordinance; the other group opposes. Which group will you join and why? Does your answer change, assuming you are a senior public health specialist? Elaborate your answers.
17. Give five examples of commercial products derived from recycled waste. Are any of these products that you personally use? Assume all five products cost 10% more than products made without the use of recycled materials. Will you purchase them? If so, why? If not, why?
18. Some cities or other governed entities have banned the use of plastic bags by retailers such as grocery stores, apparel shops, and such. Assuming that you reside in such a locale, what steps will you take, if any, in compliance with the ban on plastic bags?
19. As a design engineer with a Fortune 500 company, you are head of a group that has been asked to update a product that historically ends its life wholly

deposited in a landfill. Led by the engineer genius that you claim to be, your group has invented a product update that produces 50% less waste, but will cost 15% more than the current version of the product. What arguments will you make to "the suits" in support of your group's proposed design? Will environmental ethics be an element of your argument?

CHAPTER 3: DRUGS AND ALCOHOL

1. Are illegal drugs a problem in your community of residence? If so, what illegal drugs are the most prevalent? How can you personally help combat the drug problem in your community?
2. Bad news has been brought to your attention. A friend tells you that your younger brother has begun experimenting with an addictive illegal drug. Discuss in a two-page essay what your reaction will be. Be specific.
3. Assume that you are a senior official in the local public health department. The area's school superintendent has become concerned about the prevalence of illegal drugs being experimented with by students. She asks for your help. Discuss your response to the superintendent.
4. Your college roommate has become discouraged with school events and has turned to pot smoking for alleged relief. Do you approve or disapprove of the roommate's reliance on pot? Defend your position. In a two-page essay, discuss your moral and legal responsibilities, if any, in regard to dissuading the roommate's behavior.
5. Sadly, you were raised in a family in which one parent abused alcohol, often exhibiting signs of alcoholism. Discuss how this familial experience might shape your own attitudes toward alcoholic beverages.
6. Several U.S. states and some cities have decided to approve recreational use of marijuana. Has your state taken or considered a similar action? As a public health specialist, opine in a two-page essay your position regarding legalizing marijuana.
7. Discuss in a two-page essay the consequences, if any, of the 2020 COVID-19 pandemic on illegal drug and alcohol use.
8. You have become aware that your college is considering selling alcoholic beverages at sporting events, as is the case with professional sports. In a two-page essay, discuss the pros and cons of the school's proposal.
9. Well done! You have completed your junior year in college as a pre-med major. You and your best friend have chosen to take a well-deserved miniature vacation trip to Mexico. Your friend's uncle has asked him to bring back a small container of high-grade narcotic, promising to donate a large sum of money to a children's charity foundation. What do you do? In a one-page reply, discuss the implications to you of a decision to bring illegal contraband into your country of residence.
10. Assume you are the director of the local department of public health. Budget cuts have forced your department to discontinue a program of free vaccinations for children of indigent parents. Learning of this program's termination, a local beer and spirits distributor has offered to fund the vaccinations

program, but with the proviso that they be cited as the program's sponsor. Do you accept the offer? If so, why? If not, why? Be specific and give details of the public health principles you used in making your decision.
11. Using internet resources, research the pattern of alcohol-related arrests in your community of residence. Provide details on the number of arrests, nature of those arrests, and demographics of offenders.
12. Imagine yourself as a young person residing in a major U.S. city during the Prohibition period. Further image that you desire an active social life, but you pride yourself on being a law-abiding person. Discuss how you would handle yourself during social outings such as club dances and weekend parties.
13. Using contemporary knowledge about the adverse health effects of alcoholism, assume that you were a U.S. senator in 1921, serving on a Congressional committee that will decide on whether to repeal the 18th Amendment to the Constitution. Further, consider that your vote will decide the action that Congress takes. Discuss how you would vote and provide details of your decision.
14. Assume you have completed your academic program of public health. Because of a family member's loss of life due to a heroin overdose you desire to work for a federal government agency that combats illegal drugs. Which agency would you select and state why?
15. In your opinion should advertisements of alcoholic beverages be banned from television and other media that are known to influence children? Document the reasons behind your decision.
16. There was considerable attention given by members of Congress in 2019 to the matter of alleged high costs of prescription drugs. In your opinion, should government be a party to setting drug costs? State and defend the rationale for your decision.
17. Using the material in this chapter, augmented with material you researched on the internet, compare the punishment meted to illegal drug traffickers between the U.S. federal government and the government of China. Which system do you prefer and why?
18. Research the amount of excuse taxes your state of residence derives from alcohol production and sales. Further research the cost to your state of alcohol abuses such as DUI arrests and health care for indigents who abused alcohol. Based on your analysis, should your state increase or decrease its excise tax on alcoholic beverages.
19. In 2019 a major opioid epidemic became a major public health concern to the U.S. public, as was discussed in this chapter. One pharmaceutical company was found duplicitous in causing the epidemic. In your opinion, what action should ensue in regard to the company? Be specific.

CHAPTER 4: FIREARMS VIOLENCE

1. Let us develop a thought exercise. Assume gunpowder had never been discovered. In a two-page essay, opine how global history might have developed differently. Be specific and justify your opinions.

2. Using the material in this chapter and internet resources, discuss your state's statutory policies on control of firearms. Discuss the specifics of the policies and opine whether the policies are sufficiently protective of public health.
3. This chapter asserts that firearms are a matter of public health. As a student of public health, do you agree with this assertion? If so, why? If not, why? Be specific in your two-page essay reply.
4. Using your own understanding and personal (if any) experiences with firearms, list five reasons why firearms are beneficial to you. And in the same reply, list five reasons why firearms are detrimental to you. Be specific and use data where available.
5. Sadly, two young children playing with an unsecured pistol unintentionally discharged the weapon, fatally wounding a teenage babysitter. You, as a senior member of the local public health agency, have been asked by a county commissioner to appear before the commission and present a draft program of firearms safety in the home. In a two-page essay, describe the key elements of your presentation.
6. Referring to question #5, a local gun rights club has become aware of your proposed presentation to the county commission. The group has requested to meet with you for purpose of helping develop your presentation. As a senior public health officer, do you accept the offer of assistance? If so, why? If not, why? Explain your reasons for acceptance or declination.
7. Assume that you were a member of the team that wrote the Second Amendment to the U.S. Constitution. Using internet resources forms an opinion of the politics at play during the consideration of the Second Amendment. And further assume that you have been designated as the author of the subject amendment. Provide your version of the Second Amendment and compare your draft with the actual Second Amendment. Contrast your version with the actual version. Cite how your draft improves the clarity of the Second Amendment.
8. Regrettably, mass shootings have occurred in schools in the U.S. and in a few other countries. A local TV station has asked you as a senior member of county health department to debate a local parent who advocates that school administrators and teachers be permitted to carry firearms in defense against school terrorists. Do you accept the invitation to debate? If so, what will be your debating position? If not, why not? Explain your reasons for acceptance or declination.
9. Sadly, a high school classmate was fatally wounded during an altercation at a local restaurant. This loss of a friend has led to your determination to emigrate to a new country of residence. Which country do you choose? And why? In your two-page essay, cite firearms data that influenced your decision of which country to choose.
10. The local school district has invited you as the director of the county's health department to present a firearms seminar to first-year public high school students. You have been advised in advance of your seminar that some parents who support gun rights will be in attendance. Describe in a two-page essay the essential points of your seminar presentation.

11. Your sister, a pediatrician, has sent an announcement of her forthcoming visit for the December holidays. This is good news, since your sister has not previously visited your new residence. Your family of two young children is looking forward to greeting their aunt. Shortly after her arrival, she observed the presence of several firearms in your residence. In your opinion, does the pediatrician have a moral obligation to take any action? If so, what? If not, why? Use public health principles on which to base your answer.
12. In most U.S. localities, convicted sex offenders must be registered with local and state authorities, comprising a data base that is accessible to the public. As a public health professional, in your opinion, should possession of firearms be a similar public registry, accessible to the public? Explain your response in a two-page essay.
13. Using internet resources and your state of residence's records, identify the number, if any, of mass casualty shootings that have occurred in your state during the previous 5 years. If your state has had no mass shootings, select a neighboring state that has suffered one or more shootings. Select one of the mass shootings and describe the weapon(s) used by the shooter and the extent of firearm violence. Describe how in your opinion the shooting could have been prevented.
14. As noted in the chapter, persons with alleged mental disorders have sometimes been shooters in mass casualty shootings. In a two-page essay, discuss the services provided by your state of residence in regard to mental health services. If no such services are provided, use your essay to opine on whether such services should be provided as a resource by the state's health department.
15. Your best friend in high school has shared with you that his home life has become intolerable. He further confides that he plans to bring a gun to school and confront two teachers who have given him low grades. What do you do?
16. Your new college roommate has recently arrived from India. She has observed that local TV stations present a rather steady stream of reports of gun violence. She has asked how the U.S. population has access to so many firearms, observing that her country keeps a tight control over weapons possessed by its citizens. Provide your answer to her via a two-page essay. Be sure to contrast gun control in India versus that in the U.S.
17. As noted in this chapter, the state of California has proposed developing regulations that would control the sources of firearm ammunition. In your opinion, what are the pros and cons of this approach toward reducing gun violence?
18. Your brother is a newly graduated pediatrician and has accepted a position with a local children's health practice. You are a senior manager in the local health department. Gun violence is a major concern of your department, owing to an escalating pattern of gun deaths of young children. Your brother's health practice eschews mentioning gun control to parents and custodians of young patients. How do you persuade your brother to incorporate gun control into his interactions with patients?

Environmental Policy and Public Health Workbook

19. Well, time has passed and you have received a visit from your four grandchildren. They are each rather curious about "the old days" when you were their current age. One of the grandchildren, a somewhat precocious granddaughter, wants to know why so many people had guns in those earlier days. In a two-page essay, provide your response to your granddaughter.

CHAPTER 5: NOISE AND LIGHT POLLUTION

1. The material in this chapter relates both noise pollution and light pollution as relatively new hazards to human and ecological health. Do you agree with this characterization? If so, provide details of your agreement or disagreement. Be specific.
2. Using the material in this chapter and internet resources, discuss your state's statutory policies on the control of noise pollution. Discuss the specifics of the policies and opine whether the policies are sufficiently protective of public health.
3. Does the locale in which you reside have either a noise or light pollution policy? If so, summarize the details of the policy. If no policy exists, opine on whether such a policy or policies should be developed and promulgated. Provide details of your proposed policy.
4. Assume that you are the Chairperson of SHUSH (Students Helping Urge Silence and Harmony). The group wants the DeKalb County Commissioners to enact a county regulation to control environmental outdoor noise levels. Using the PACM model, discuss how you would achieve your goal. Be specific.
5. Do you reside near a major thoroughfare? If so, discuss the noise patterns of vehicle traffic on the thoroughfare. Present the impacts of the vehicle traffic noise on your well-being. What adjustments have you made in order to cope with the vehicle noise? If you do not reside near a major thoroughfare, interview a person who does and interview him/her as to the impact of vehicle noise.
6. As discussed in this chapter, the federal Noise Control Act of 1972 and the Quiet Communities Act of 1978 remain on the legislative books of Congress, but both are unfunded, making them dormant. In a two-page essay, argue whether one or both of these acts should be reinvigorated by Congress. Give specifics for your argument, citing health data in your reply.
7. Develop a list of the noise sources that exist in your residential environment. Provide the specifics of each source, estimating the frequency and intensity of each noise source.
8. Develop a list of the light sources that exist in your residential environment. Provide the specifics of each source, estimating the frequency and duration of each source. Estimate how each source of light can be better controlled.
9. As noted in this chapter, the European Union has developed a directive on control of noise pollution. Recalling that EU directives are mandates to member countries for their adoption and implementation of a directive's

specifics, use internet resources to access an EU member country's implementation of the EU directive.
10. Using the crystal ball that you inherited from your fortune teller great aunt, describe in a two-page essay describing how your life would be changed if there were no sources of noise and light pollution.
11. The light-emitting diode (LED) has become a popular choice as a source for both indoor and outdoor light. In a two-page essay, identify the presence of LEDs in your daily life and discuss the potential economic and health implications of LEDs as a light source.
12. Discuss the consequences to one's personal health and well-being of hearing loss. Include in your discussion the causes of hearing loss and methods to prevent the loss.
13. Humans and other animals require sleep. Using internet resources, prepare a short analysis of the purpose of sleep and the consequences of sleep deprivation. Include in your discussion the amount and quality of sleep that you consider normal.
14. The occupational safety and health administration (OSHA) has developed and promulgated a standard for the control of workplace noise. Using internet resources, select another country that has developed a similar occupational standard for noise and compare their standard with that of the OSHA.
15. Assume that you are a senior member of your local health department. Discuss the methods and resources that you would use for purpose of educating local residents about the adverse health consequences of excessive noise and light pollution.
16. Good news! You just received notice that your graduate committee has given approval for you to prepare a thesis in support of your M.P.H. degree, but with the proviso that your thesis research must be in the area of light pollution. The good news is that you can proceed with your graduate work. The bad news is that you know nothing about light pollution.
17. The Noise Control Act of 1972 authorized the EPA to develop noise standards for commercial products such as vacuum cleaners. The authority was made dormant when Congress transferred noise control to the U.S. states. In your opinion, should Congress resurrect EPA's authority to regulate noise generated by commercial products? Be specific and reference health data to the extent possible in your analysis.
18. Good news! You were admitted to your college of choice. Oh, happy days! Regrettably, your happiness began to wane after spending the first month in the dormitory assigned to you. Seems your dormitory room is adjacent to a construction site – there is noise and light pollution almost 24/7. You resort to wearing earplugs and an eye mask at night. Discuss the pros and cons of your decisions.
19. Both noise and light pollution have the potential to disrupt circadian rhythms. In a two-page essay, discuss the nature of circadian rhythms and the consequences of disruptions to these rhythms. Have you personally ever experienced any disruptions? If so, describe the experience. If not, describe any measure you took to avoid such disruptions.

Chapter 6: The Built Environment

1. Describe the room in which you are sitting and list the potential problems that could contribute to environmental and public health issues.
2. As countries continue to develop around the globe, how might the built environment problems they face be similar to the problems faced by the rapidly industrialized U.S.? How might they be different?
3. What are the features of a healthy community? What are the tools available to make your community healthier?
4. What are the barriers to implement programs aimed at addressing a built environment? Why do those barriers exist?
5. How do the principles of mobility and accessibility change how built environments are created?
6. Urban sprawl was discussed as a problem for the built environment. Using internet and other sources, research how your state of residence is assisting in controlling urban sprawl. Should government have a presence in what is a local policy issue?
7. Using the material in Chapter 6, together with other resources, prepare an essay of appropriate depth that examines the projected impacts of climate change on the global built environment. Be specific and cite data in support of your thesis.
8. The U.S. Department of Housing and Urban Development (HUD) is the principal federal agency involved with urban built environments. Research to what extend your community of residence is receiving built environment assistance from HUD. If your community is not receiving assistance, select a city that does.
9. Sweden was cited as a country that has implemented holistic built environment policies. Prepare an essay that lists the most significant policies that you believe should be implemented in your country of residence.
10. Does your community of residence support biking lanes on the major thoroughfares? If so, detail the policies that permit bikers sharing local traffic resources. If not, opine whether such policies should be adopted in your community.
11. Assume that your county commissioners have announced their potential consideration of a property developer's proposal to remove 10 acres of old forest and thereon to build a complex of apartment and condo building. Further, assume that you chair a local environmental quality advocacy group. Using the PACM model discussed in Chapter 2, discuss the actions your group would take.
12. The LEED program also operates a LEED Neighborhood focus. Research whether your community of residence is working with a LEED project and provide details on the project. If your community is not involved with LEED, select a community that does.
13. Prepare a comparative analysis that contrasts China's and India's building policies. Does either set of policies provide better environmental protection than the other?

14. Make contact with your local planning commission and ascertain its composition and evaluate to what extent the group has public health representation. If the commission has no public health member, examine the reasons why.
15. Describe the social climate in the community in which you reside, using the material in this chapter that indicates a positive social climate has beneficial effects. Opine in an essay how you perceive your social climate's beneficial effects on yourself.
16. Develop an abbreviated Complete Streets policy for your community of residence. Discuss in brief the elements of your policy and the surmised benefits to your community.
17. Describe in your opinion the two most important changes that could be made to your community of residence in order to improve its benefits to residents. Provide rationales for each of your proposed changes.
18. Some cities in Europe have banned motor vehicle traffic from entering city center areas. Select two such cities and summarize the outcomes of this kind of policy. Additionally, opine on whether your city of residence should adopt a similar policy.
19. Ascertain if your university or other school of higher learning offers a program in urban planning. If so, summarize the purpose and resources associated with the program. If your school does not offer such a program, select a university that does.

Index

airbags 275
aircraft noise 200
alcohol 123
American Prosperity initiative 228
antibiotics 38
AR-15 rifles 183
asbestos 229
automobile 255
autonomous vehicles 278

bacterial infection 41
Benz, Karl 256
Bill of Rights 174
Brady, James 178
brownfields 98
built environment 219

carbon footprint 245
chemicals age 55
chikungunya virus 24
child safety seat 275
cholera 2
cholera epidemic 4
cigarette butts 72
circular economy 111
city farming 243
city planning 221
coal 83
coal ash 84
coast guard 136
complete streets 240
controlled substances 134
corona viruses 9
COVID-19 11
crash avoidance 278

Daimler, Gottlieb 256
dengue 21
Department of Transportation Act 263
diarrheal illnesses 42
domesticated horse 254
drinking water 2
drug abuse 125
Drug Enforcement Administration 135
drug overdose 128
drugs 123
drunk driving 269

earthquake 3
Ebola 13

electric vehicles 294
e-waste 70

federal aviation administration 283
federal highway administration 281
federal trade commission 133
firearm injuries 156
firearms 155
fleas 5
folk remedies 124
Food and Drug Act 132
food wastage 68
Ford, Henry 256
Ford Model T 256
forest service 230
French Pasteur Institute 2

Garden City movement 224
gasoline engine 257
gin 141
glare 208
greenroof policies 241
greenspace 230
Guinea worm disease 32
gun powder 156
gun storage 159
gun violence deaths 161

Haiti 3
handguns 157
hands free law 272
hazardous waste 78
health impact assessment 234
helminth infections 32
highway safety act 285

impaired hearing 194
industrialization 223
infectious diseases 1
influenza 7
insecticides 27
interstate highway system 262
Islamic Republics 150

kissing bugs 34

land-use policy 220
lead 229
Legionnaires disease 42
Leonardo da Vinci 255
light pollution 191, 206

321

London Protocol 106
low-risk drinking 151

malaria 18
marijuana 126
mass casualty shootings 164
measles 15
MERS 9
microbeads 60
Minimum Drinking Age Act 264
mites 4
mosquitoes 1
motor vehicle fatalities 258

National Firearms Act 175
National Priorities List 95
Noise Control Act 200
noise in workplaces 196
noise pollution 191, 193
nuclear waste 90
nuisance laws 222

ocean debris 104
oil spills 100
opium 125

permitted landfills 73
plastic bags 66
plastic bottle 63
plastic pollution 62
plastics 58
plastics waste 57
President Franklin Roosevelt 133
President Ronald Reagan 178
prohibition 148
psychotropic substances 139

quarantine 17
Quiet Communities Act 202

Ransome Eli Olds 256
rats 30
RCR Act 75
recycling 109

red flag gun seizure 182
rifles 157
Right-to-Know Act 96
road traffic deaths 258

SARS 9
scabies 5
school shootings 166
seatbelts 273
Second Amendment 174
Small Arms Survey 158
social justice 233
speed limits 268
superbug 34
Superfund 91
sustainable development 237

text messaging 271
ticks 5
transportation 253
Triangle of Disease 2
tropical diseases 44
typhus 4

uplight 208
urban agriculture 243
urban forest 231
urban planning 237
urban traffic control 297

vector-borne diseases 18
vehicle crash designs 277

waste generation 55
westnile virus 23
wine 140
worms 31

yellow fever 20
Yemen 4

Zika virus 24
Zoning Enabling Act 225
zoonosis 6